T0230829

Topics on
Continua

Topics on
Continua

Sergio Macías

Instituto de Matemáticas
Universidad Nacional Autónoma de México
Ciudad Universitaria, México

CRC Press
Taylor & Francis Group
Boca Raton London New York

CRC Press is an imprint of the
Taylor & Francis Group, an **informa** business

A CHAPMAN & HALL BOOK

First published in 2005 by Chapman & Hall

Published in 2019 by CRC Press
Taylor & Francis Group
6000 Broken Sound Parkway NW, Suite 300
Boca Raton, FL 33487-2742

© 2005 by Taylor & Francis Group, LLC
CRC Press is an imprint of Taylor & Francis Group, an Informa business

First issued in paperback 2019

No claim to original U.S. Government works

ISBN-13: 978-0-367-45412-8 (pbk)
ISBN-13: 978-0-8493-3738-3 (hbk)

This book contains information obtained from authentic and highly regarded sources. Reprinted material is quoted with permission, and sources are indicated. A wide variety of references are listed. Reasonable efforts have been made to publish reliable data and information, but the author and the publisher cannot assume responsibility for the validity of all materials or for the consequences of their use.

No part of this book may be reprinted, reproduced, transmitted, or utilized in any form by any electronic, mechanical, or other means, now known or hereafter invented, including photocopying, microfilming, and recording, or in any information storage or retrieval system, without written permission from the publishers.

For permission to photocopy or use material electronically from this work, please access www.copyright.com (http://www.copyright.com/) or contact the Copyright Clearance Center, Inc. (CCC) 222 Rosewood Drive, Danvers, MA 01923, 978-750-8400. CCC is a not-for-profit organization that provides licenses and registration for a variety of users. For organizations that have been granted a photocopy license by the CCC, a separate system of payment has been arranged.

Trademark Notice: Product or corporate names may be trademarks or registered trademarks, and are used only for identification and explanation without intent to infringe.

**Visit the Taylor & Francis Web site at
http://www.taylorandfrancis.com**

**and the CRC Press Web site at
http://www.crcpress.com**

Library of Congress Cataloging-in-Publication Data

Catalog record is available from the Library of Congress

To Elsa

*León Felipe escribió un
tributo, no al héroe de la
historia, sino a su fiel
caballo Rocinante,
quien lo llevó en su lomo
por las tierras de España.*

*El héroe es, por supuesto,
Don Quijote de la Mancha:*

"El Caballero de la Triste Figura"

*¡Yo quería ese nombre!
pero me lo ganaron,
llegué a este mundo casi
trescientos cincuenta
años tarde...*

Ya sólo me queda ser:

"El Caballero de la Triste Locura..."

S. M.

Preface

My aim is to present four of my favorite topics in continuum theory: inverse limits, Professor Jones's set function $\mathcal{T}$, homogeneous continua and n–fold hyperspaces.

Most topics treated in this book are not covered in Professor Sam B. Nadler Jr.'s book: *Continuum Theory: An Introduction*, Monographs and Textbooks in Pure and Applied Math., Vol. 158, Marcel Dekker, New York, Basel, Hong Kong, 1992.

The reader is assumed to have taken a one year course on general topology.

The book has seven chapters. In Chapter 1, we include the basic background to be used in the rest of the book. The experienced readers may prefer to skip this chapter and jump right to the study of their favorite subject. This can be done without any problem. The topics of Chapter 1 are essentially independent of one another and can be read at any time.

Chapter 2 is for most part about inverse limits of continua. We present the basic results on inverse limits. Some theorems are stated without proof in Professor W. Tom Ingram's book: *Inverse Limits*, Aportaciones Matemáticas, Textos # 18, Sociedad Matemática Mexicana, 2000. We show that the operation of taking inverse limits commutes with the operations of taking finite products, cones and hyperspaces. We also include some applications of inverse limits.

In Chapter 3 we discuss Professor F. Burton Jones's set function $\mathcal{T}$. After giving the basic properties of this function, we present properties of continua in terms of $\mathcal{T}$, such as connectedness im kleinen, local connectedness and semi–local connectedness. We also study continua for which the set function $\mathcal{T}$ is continuous. In the last section we present some applications of $\mathcal{T}$.

In Chapter 4 we start our study of homogeneous continua. We present a topological proof of a Theorem of E. G. Effros given by

F. D. Ancel. We include a brief introduction to topological groups and group actions.

Chapter 5 contains our main study of homogeneous continua. We present two Decomposition Theorems of such continua, whose proofs are applications of Jones's set function $\mathcal{T}$ and Effros's Theorem. These theorems have narrowed the study of homogeneous continua in such a way that they may hopefully be eventually classified. We also give examples of nontrivial homogeneous continua and their covering spaces.

In Chapter 6 we present most of what is known about n–fold hyperspaces. This chapter is slightly different from the other chapters because the proofs of many of the theorems are based on results in the literature that we do not prove; however, we give references to the appropriate places where proofs can be found. This Chapter is a complement of the two existing books–Sam B. Nadler, Jr., *Hyperspaces of Sets: A Text with Research Questions*, Monographs and Textbooks in Pure and Applied Math., Vol. 49, Marcel Dekker, New York, Basel, 1978 and Alejandro Illanes and Sam B. Nadler, Jr., *Hyperspaces: Fundamentals and Recent Advances*, Monographs and Textbooks in Pure and Applied Math., Vol. 216, Marcel Dekker, New York, Basel, 1999, in which a thorough study of hyperspaces is done.

In Chapter 6, we also prove general properties of n–fold hyperspaces. In particular, we show that n–fold hyperspaces are unicoherent and finitely aposyndetic. We study the arcwise accessibility of points of the n–fold symmetric products from their complement in n–fold hyperspaces. We give a treatment of the points that arcwise disconnect n–fold hyperspaces of indecomposable continua. Then we study continua for which the operation of taking n–fold hyperspaces is continuous ($\mathcal{C}_n^*$–smoothness). We also investigate continua for which there are retractions between their various hyperspaces. Next, we present some results about the n–fold hyperspaces of graphs. We end Chapter 6 by studying the relation between n–fold hyperspaces and cones over continua.

We end the book with a chapter containing open questions on each of the subjects presented in the book (Chapter 7).

We include figures to illustrate definitions and aspects of proofs.

The book originates from two sources – class notes I took from the course on continuum theory given by Professor James T. Rogers, Jr. at Tulane University, in the Fall Semester of 1988 and the one–year courses on continuum theory I have taught in the graduate program of mathematics at the Facultad de Ciencias of the Universidad Nacional Autónoma de México, since the Spring of 1993. I thank all the students who have taken such courses.

I thank María Antonieta Molina and Juan Carlos Macías for letting me include part of their thesis in the book. Ms. Molina's thesis was based on two talks on the set function $\mathcal{T}$ given by Professor David P. Bellamy in the *IV Research Workshop on Topology*, celebrated in Oaxaca City, Oaxaca, Mexico, November 14 through 16, 1996.

I thank Professors Sam B. Nadler, Jr. and James T. Rogers, Jr. for reading parts of the manuscript and making valuable suggestions. I also thank Ms. Gabriela Sanginés and Mr. Leonardo Espinosa for answering my questions about LaTeX, while I was typing this book.

I thank Professor Charles Hagopian and Marvi Hagopian for letting me use their living room to work on the book during my visit to California State University, Sacramento.

I thank the Instituto de Matemáticas of the Universidad Nacional Autónoma de México and the Mathematics Department of West Virginia University, for the use of resources during the preparation of the book.

Finally, I thank the people at Marcel Dekker, Inc., especially Ms. Maria Allegra and Mr. Kevin Sequeira, who were always patient and helpful.

Sergio Macías

Contents

Chapter 1

PRELIMINARIES

We gather some of the results of topology of metric spaces which will be useful for the rest of the book. We assume the reader is familiar with the notion of metric space and its elementary properties. We present the proofs of most of the results; we give an appropriate reference otherwise.

The topics reviewed in this chapter are: product topology, continuous decompositions, homotopy, fundamental group, geometric complexes, polyhedra, complete metric spaces, compacta, continua and hyperspaces.

1.1 Product Topology

The symbols $\mathbb{N}$, $\mathbb{Z}$, $\mathbb{Q}$, $\mathbb{R}$ and $\mathbb{C}$ denote the positive integers, integers, rational numbers, real numbers and complex numbers, respectively.

The word *map* means a continuous function. A *compactum* is a compact metric space.

1.1.1. Definition. Given a sequence, $\{X_n\}_{n=1}^{\infty}$, of nonempty sets, we define its *Cartesian product*, denoted by $\prod_{n=1}^{\infty} X_n$, as the set:

$$\prod_{n=1}^{\infty} X_n = \{(x_n)_{n=1}^{\infty} \mid x_n \in X_n \text{ for each } n \in \mathbb{N}\}.$$

For each $m \in \mathbb{N}$, there is a function

$$\pi_m \colon \prod_{n=1}^{\infty} X_n \twoheadrightarrow X_m$$

defined by $\pi_m((x_n)_{n=1}^{\infty}) = x_m$. This function π_m is called the *mth–projection map*.

1.1.2. Remark. Given a metric space (X, d'), there is a metric, d, which generates the same topology as d', with the property that $d(x, x') \leq 1$ for each pair of points x and x' of X. This metric d is called *bounded metric*. An example of such metric is given by $d(x, x') = \min\{1, d'(x, x')\}$.

1.1.3. Notation. Given a metric space (X, d) and a subset A of X, $Cl_X(A)$, $Int_X(A)$ and $Bd_X(A)$ denote the closure, interior and boundary of A, respectively. We omit the subindex if there is no confusion. If ε is a positive real number, then the symbol $\mathcal{V}_{\varepsilon}^d(A)$ denotes the *open ball of radius ε about A*. If $A = \{x\}$, for some $x \in X$, we write $\mathcal{V}_{\varepsilon}^d(x)$ instead of $\mathcal{V}_{\varepsilon}^d(\{x\})$.

1.1.4. Definition. If $\{(X_n, d_n)\}_{n=1}^{\infty}$ is a sequence of metric spaces, with bounded metrics, we define a metric ρ, for its Cartesian product as follows:

$$\rho((x_n)_{n=1}^{\infty}, (x_n')_{n=1}^{\infty}) = \sum_{n=1}^{\infty} \frac{1}{2^n} d_n(x_n, x_n').$$

1.1.5. Remark. Since the metrics, d_n, in Definition 1.1.4 are bounded, ρ is well defined.

1.1.6. Lemma. *If $\{(X_n, d_n)\}_{n=1}^{\infty}$ is a sequence of metric spaces, with bounded metrics, then ρ (Definition 1.1.4) is a metric and for each $m \in \mathbb{N}$, π_m is a continuous function.*

Proof. The proof of the fact that ρ is, in fact, a metric is left to the reader.

Let $m \in \mathbb{N}$ be given. We show that π_m is continuous. Let $\varepsilon > 0$ and let $\delta = \dfrac{1}{2^m}\varepsilon$. If $(x_n)_{n=1}^{\infty}$ and $(x'_n)_{n=1}^{\infty}$ are two points of $\prod\limits_{n=1}^{\infty} X_n$ such that $\rho((x_n)_{n=1}^{\infty}, (x'_n)_{n=1}^{\infty}) < \delta$, then, since $\dfrac{1}{2^m}d_m(x_m, x'_m) \le \sum\limits_{n=1}^{\infty} \dfrac{1}{2^n} d_n(x_n, x'_n)$, we have that $\dfrac{1}{2^m}d_m(x_m, x'_m) < \delta$. Hence,

$$d_m(x_m, x'_m) < 2^m \delta = \varepsilon.$$

Therefore, π_m is continuous.

Q.E.D.

1.1.7. Lemma. *If $\{(X_n, d_n)\}_{n=1}^{\infty}$ is a sequence of metric spaces, with bounded metrics, then given $\varepsilon > 0$ and a point $(x_n)_{n=1}^{\infty} \in \prod\limits_{n=1}^{\infty} X_n$, there exist $N \in \mathbb{N}$ and N positive real numbers, $\varepsilon_1, \dots, \varepsilon_N$, such that $\bigcap\limits_{j=1}^{N} \pi_j^{-1}(\mathcal{V}_{\varepsilon_j}^{d_j}(x_j)) \subset \mathcal{V}_{\varepsilon}^{\rho}((x_n)_{n=1}^{\infty})$.*

Proof. Let $N \in \mathbb{N}$ be such that $\sum\limits_{n=N+1}^{\infty} \dfrac{1}{2^n} < \dfrac{\varepsilon}{2}$. For each $j \in \{1, \dots, N\}$, let $\varepsilon_j = \dfrac{\varepsilon}{2^N}$. We assert that $\bigcap\limits_{j=1}^{N} \pi_j^{-1}\left(\mathcal{V}_{\varepsilon_j}^{d_j}(x_j)\right) \subset \mathcal{V}_{\varepsilon}^{\rho}((x_n)_{n=1}^{\infty})$. To see this, let $(y_n)_{n=1}^{\infty} \in \bigcap\limits_{j=1}^{N} \pi_j^{-1}\left(\mathcal{V}_{\varepsilon_j}^{d_j}(x_j)\right)$. We want to see that $\rho((x_n)_{n=1}^\infty, (y_n)_{n=1}^\infty) < \varepsilon$. Note that

$$
\begin{aligned}
\rho((x_n)_{n=1}^\infty, (y_n)_{n=1}^\infty) &= \sum_{n=1}^\infty \frac{1}{2^n} d_n(x_n, y_n) = \\
\sum_{n=1}^N \frac{1}{2^n} d_n(x_n, y_n) &+ \sum_{n=N+1}^\infty \frac{1}{2^n} d_n(x_n, y_n) < \\
\sum_{n=1}^N \frac{1}{2^n}\frac{1}{2^N}\varepsilon + \frac{1}{2}\varepsilon &= \left(1 - \frac{1}{2^N}\right)\frac{1}{2^N}\varepsilon + \frac{1}{2}\varepsilon \le \frac{1}{2}\varepsilon + \frac{1}{2}\varepsilon = \varepsilon.
\end{aligned}
$$

$$\textbf{Q.E.D.}$$

1.1.8. Lemma. *If $\{(X_n, d_n)\}_{n=1}^\infty$ is a sequence of metric spaces, with bounded metrics, then given a finite number of positive real numbers $\varepsilon_1, \ldots, \varepsilon_k$ and a point $(x_n)_{n=1}^\infty \in \prod_{n=1}^\infty X_n$, there exists $\varepsilon > 0$ such that $\mathcal{V}_\varepsilon^\rho((x_n)_{n=1}^\infty) \subset \bigcap_{j=1}^k \pi_j^{-1}(\mathcal{V}_{\varepsilon_j}^{d_j}(x_j))$.*

Proof. Let $(x_n)_{n=1}^\infty \in \prod_{n=1}^\infty X_n$, and let $U = \bigcap_{j=1}^k \pi_j^{-1}\left(\mathcal{V}_{\varepsilon_j}^{d_j}(x_j)\right)$. Take

$$
\varepsilon = \min\left\{\frac{1}{2}\varepsilon_1, \ldots, \frac{1}{2^k}\varepsilon_k\right\}.
$$

We show $\mathcal{V}_\varepsilon^\rho((x_n)_{n=1}^\infty) \subset U$. Let $(y_n)_{n=1}^\infty \in \mathcal{V}_\varepsilon^\rho((x_n)_{n=1}^\infty)$. Then

$$
\rho((x_n)_{n=1}^\infty, (y_n)_{n=1}^\infty) < \varepsilon, \text{ i.e., } \sum_{n=1}^\infty \frac{1}{2^n} d_n(x_n, y_n) < \varepsilon.
$$

Hence, $\frac{1}{2^j} d_j(x_j, y_j) < \varepsilon \le \frac{1}{2^j}\varepsilon_j$ for each $j \in \{1, \ldots, k\}$. Thus, if $j \in \{1, \ldots, k\}$, then $d_j(x_j, y_j) < \varepsilon_j$. Therefore, $\mathcal{V}_\varepsilon^\rho((x_n)_{n=1}^\infty) \subset U$.

$$\textbf{Q.E.D.}$$

1.1.9. Theorem. *Let Z be a metric space. If $\{(X_n, d_n)\}_{n=1}^{\infty}$ is a sequence of metric spaces, then a function $f \colon Z \to \prod_{n=1}^{\infty} X_n$ is continuous if and only if $\pi_n \circ f$ is continuous for each $n \in \mathbb{N}$.*

Proof. Clearly, if f is continuous, then $\pi_n \circ f$ is continuous for each $n \in \mathbb{N}$.

Suppose $\pi_n \circ f$ is continuous for each $n \in \mathbb{N}$. Let $\bigcap_{j=1}^{k} \pi_j^{-1}(U_j)$ be a basic open subset of $\prod_{n=1}^{\infty} X_n$. Since

$$f^{-1}\left(\bigcap_{j=1}^{k} \pi_j^{-1}(U_j)\right) = \bigcap_{j=1}^{k} f^{-1}(\pi_j^{-1}(U_j))$$
$$= \bigcap_{j=1}^{k} (\pi_j \circ f)^{-1}(U_j),$$

we have that $f^{-1}\left(\bigcap_{j=1}^{k} \pi_j^{-1}(U_j)\right)$ is open in Z. Hence, f is continuous.

$$\text{Q.E.D.}$$

1.1.10. Theorem. *Let $\{X_n\}_{n=1}^{\infty}$ and $\{Y_n\}_{n=1}^{\infty}$ be two countable collections of metric spaces. Suppose that for each $n \in \mathbb{N}$, there exists a map $f_n \colon X_n \to Y_n$. Then the function $\prod_{n=1}^{\infty} f_n \colon \prod_{n=1}^{\infty} X_n \to \prod_{n=1}^{\infty} Y_n$ given by $\prod_{n=1}^{\infty} f_n((x_n)_{n=1}^{\infty}) = (f_n(x_n))_{n=1}^{\infty}$ is continuous.*

Proof. For each $m \in \mathbb{N}$, let $\pi_m \colon \prod_{n=1}^{\infty} X_n \twoheadrightarrow X_m$ and $\pi'_m \colon \prod_{n=1}^{\infty} Y_n \twoheadrightarrow Y_m$ be the projection maps.

Let $(x_n)_{n=1}^\infty \in \prod_{n=1}^\infty X_n$, and let $m \in \mathbb{N}$. Then $\pi'_m \circ \prod_{n=1}^\infty f_n((x_n)_{n=1}^\infty)$
$= \pi'_m\left((f_n(x_n))_{n=1}^\infty\right) = f_m(x_m) = f_m \circ \pi_m((x_n)_{n=1}^\infty)$. Hence, by Theorem 1.1.9, $\prod_{n=1}^\infty f_n$ is continuous.

Q.E.D.

The following result is a particular case of *Tychonoff's Theorem*, which says that the Cartesian product of any family of compact topological spaces is compact. The proof of this theorem uses the *Axiom of Choice*. However, the case we show only uses the fact that compactness and sequential compactness are equivalent in metric spaces (Remark 3 (p. 3) of [13]).

1.1.11. Theorem. *If $\{(X_n, d_n)\}_{n=1}^\infty$ is a sequence of compacta, then $\prod_{n=1}^\infty X_n$ is compact.*

Proof. By Lemma 1.1.6, $\prod_{n=1}^\infty X_n$ is a metric space. We show that any sequence of points of $\prod_{n=1}^\infty X_n$ has a convergent subsequence.

Let $\{p^k\}_{k=1}^\infty$ be a sequence of points of $\prod_{n=1}^\infty X_n$, where $p^k = (p_n^k)_{n=1}^\infty$ for each $k \in \mathbb{N}$ (in this way, if we keep n fixed, $\{p_n^k\}_{k=1}^\infty$ is a sequence of points of X_n). Since (X_1, d_1) is sequentially compact, $\{p_1^k\}_{k=1}^\infty$ has a convergent subsequence $\{p_1^{k_j}\}_{j=1}^\infty$ converging to a point q_1 of X_1. Let us note that, implicitly, we have defined a subsequence $\{p^{k_j}\}_{j=1}^\infty$ of $\{p^k\}_{k=1}^\infty$.

Now, suppose, inductively, that for some $m \in \mathbb{N}$, we have defined a subsequence $\{p^{k_i}\}_{i=1}^\infty$ of $\{p^k\}_{k=1}^\infty$ such that $\{p_m^{k_i}\}_{i=1}^\infty$ converges to a point q_m of X_m. Since (X_{m+1}, d_{m+1}) is sequentially compact, $\{p_{m+1}^{k_i}\}_{i=1}^\infty$ has a convergent subsequence $\{p_{m+1}^{k_{i_j}}\}_{j=1}^\infty$ such that it converges to a point q_{m+1} of X_{m+1}. Hence, by the Induction Principle, we have defined a sequence of subsequences of $\{p^k\}_{k=1}^\infty$ in such a way that each subsequence is a subsequence of the preceding one.

Now, let $\Sigma = \{p^1, p^{k_2}, p^{k_{j_3}}, p^{k_{j_{i_4}}}, \ldots\}$. Clearly, Σ is a subsequence of $\{p^k\}_{k=1}^{\infty}$ which converges to the point $(q_n)_{n=1}^{\infty}$. Therefore, $\prod_{n=1}^{\infty} X_n$ is compact.

$$\textbf{Q.E.D.}$$

1.1.12. Definition. Let $\mathcal{Q} = \prod_{n=1}^{\infty} [0,1]_n$, where $[0,1]_n = [0,1]$, for each $n \in \mathbb{N}$. Then $\mathcal{Q}$ is called the *Hilbert cube*.

1.1.13. Theorem. *The Hilbert cube is a connected compactum.*

Proof. By Lemma 1.1.6, $\mathcal{Q}$ is a metric space. By Theorem 1.1.11, $\mathcal{Q}$ is compact. By Theorem 11 (p. 137) of [13], $\mathcal{Q}$ is connected.

$$\textbf{Q.E.D.}$$

1.1.14. Definition. Let $f\colon X \to Y$ be a map between metric spaces. We say that f is an *embedding* if f is a homeomorphism onto $f(X)$.

1.1.15. Definition. A map $f\colon X \to Y$ between metric spaces is said to be *closed* provided that for each closed subset K of X, $f(K)$ is closed in Y.

The next Theorem says that there is a "copy" of every compactum inside the Hilbert cube.

1.1.16. Theorem. *If X is a compactum, then X can be embedded in the Hilbert cube $\mathcal{Q}$.*

Proof. Let d be the metric of X. Without loss of generality, we assume that $\operatorname{diam}(X) \leq 1$. Since X is a compactum, it contains a countable dense subset, $\{x_n\}_{n=1}^{\infty}$. Let $h\colon X \to \mathcal{Q}$ be given by $h(x) = (d(x, x_n))_{n=1}^{\infty}$. By Theorem 1.1.9, h is continuous. Clearly, h is one–to–one. Since X is compact and $\mathcal{Q}$ is metric, h is a closed map. Therefore, h is an embedding.

$$\textbf{Q.E.D.}$$

1.2   Continuous Decompositions

We present a method to construct "new" spaces from "old" ones by "shrinking" certain subsets to points.

1.2.1. Definition. A *decomposition* of a set X is a collection of nonempty, pairwise disjoint sets whose union is X. The decomposition is said to be *closed* if each of its element is a closed subset of X.

1.2.2. Definition. Let $\mathcal{G}$ be a decomposition of a metric space X. We define $X/\mathcal{G}$ as the set whose elements are the elements of the decomposition $\mathcal{G}$. $X/\mathcal{G}$ is called the *quotient space*. The function $q\colon X \twoheadrightarrow X/\mathcal{G}$, which sends each point x of X to the unique element G of $\mathcal{G}$ such that $x \in G$, is called the *quotient map*.

1.2.3. Remark. Given a decomposition of a metric space X, note that $q(x) = q(y)$ if and only if x and y belong to the same element of $\mathcal{G}$. We give a topology to $X/\mathcal{G}$ in such a way that the function q is continuous and it is the biggest with this property.

1.2.4. Definition. Let X be a metric space, let $\mathcal{G}$ be a decomposition of X and let $q\colon X \twoheadrightarrow X/\mathcal{G}$ be the quotient map. Then the topology
$$\mathcal{U} = \{U \subset X/\mathcal{G} \mid q^{-1}(U) \text{ is open in } X\}$$
is called the *quotient topology for $X/\mathcal{G}$*.

1.2.5. Remark. Let $\mathcal{G}$ be a decomposition of a metric space X, and let $q\colon X \twoheadrightarrow X/\mathcal{G}$ be the quotient map. Then a subset U of $X/\mathcal{G}$ is open (closed, respectively) if and only if $q^{-1}(U)$ is an open (closed, respectively) subset of X.

1.2.6. Definition. Let $f\colon X \twoheadrightarrow Y$ be a surjective map between metric spaces. Since f is a function, $\mathcal{G}_f = \{f^{-1}(y) \mid y \in Y\}$ is a decomposition of X. The function $\varphi_f\colon X/\mathcal{G}_f \to Y$ given by $\varphi_f(q(x)) = f(x)$ is of special interest. Note that φ_f is well defined; in fact, it is a bijection and the following diagram:

$$
\begin{array}{ccc}
X & \xrightarrow{\ f\ } & Y \\
 & \searrow_{\ q} \quad \nearrow_{\varphi_f} & \\
 & X/\mathcal{G}_f &
\end{array}
$$

is commutative.

The next Lemma is a special case of the Transgression Lemma (3.22 of [23]).

1.2.7. Lemma. *Let $f\colon X \twoheadrightarrow Y$ be a surjective map between metric spaces. If $X/\mathcal{G}_f$ has the quotient topology, then the function φ_f is continuous.*

Proof. If U is an open subset of Y, then $\varphi_f^{-1}(U) = qf^{-1}(U)$. Since $q^{-1}\varphi_f^{-1}(U) = q^{-1}qf^{-1}(U) = f^{-1}(U)$ and $f^{-1}(U)$ is an open subset of X, we have, by the definition of quotient topology, that $\varphi_f^{-1}(U)$ is an open subset of $X/\mathcal{G}_f$. Therefore, φ_f is continuous.

$\qquad\qquad$ **Q.E.D.**

1.2.8. Example. Let $X = [0, 2\pi)$ and let $f\colon X \twoheadrightarrow \mathcal{S}^1$, where $\mathcal{S}^1$ is the unit circle, be given by $f(t) = \exp(t) = e^{it}$. Then f is a continuous bijection. Since $\mathcal{G}_f$ is, "essentially," X, it follows that $X/\mathcal{G}_f$ is homeomorphic to X. On the other hand, X is not homeomorphic to $\mathcal{S}^1$, since X is not compact and $\mathcal{S}^1$ is. Therefore, φ_f is not a homeomorphism.

1.2.9. Definition. A map $f\colon X \to Y$ between metric spaces is said to be *open* provided that for each open subset K of X, $f(K)$ is open in Y.

The following Theorem gives sufficient conditions to ensure that φ_f is a homeomorphism:

1.2.10. Theorem. *Let $f\colon X \twoheadrightarrow Y$ be a surjective map between metric spaces. If f is open or closed, then $\varphi_f\colon X/\mathcal{G}_f \twoheadrightarrow Y$ is a homeomorphism.*

Proof. Suppose f is an open map. Since φ is a bijective map, it is enough to show that φ is open. Let A be an open subset of $X/\mathcal{G}$. Since $\varphi_f(A) = fq^{-1}(A)$, $\varphi_f(A)$ is an open subset of Y. Therefore, φ_f is an open map.

The proof of the case when f is closed is similar.

$\hspace{11cm}$ **Q.E.D.**

Decompositions are also used to construct the cone and suspension over a given space.

1.2.11. Definition. Let X be a metric space and let $\mathcal{G} = \{\{(x, t)\} \mid x \in X \text{ and } t \in [0, 1)\} \cup \{(X \times \{1\})\}$. Then $\mathcal{G}$ is a decomposition of $X \times [0, 1]$. The *cone over* X, denoted by $K(X)$, is the quotient space $(X \times [0, 1])/\mathcal{G}$. The element $\{X \times \{1\}\}$ of $(X \times [0, 1])/\mathcal{G}$ is called the *vertex* of the cone and it is denoted by ν_X.

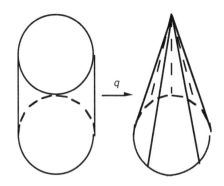

A proof of the following Proposition may be found in 5.2 (p. 127) of [9].

1.2.12. Proposition. *Let $f\colon X \to Y$ be a map between metric spaces. Then f induces a map $K(f)\colon K(X) \to K(Y)$ by*

$$K(f)(\omega) = \begin{cases} \nu_Y & \text{if } \omega = \nu_X \in K(X); \\ (f(x), t) & \text{if } \omega = (x, t) \in K(X) \setminus \{\nu_X\}. \end{cases}$$

1.2.13. Definition. Let X be a metric space and let $\mathcal{G} = \{\{(x, t)\} \mid x \in X \text{ and } t \in (0, 1)\} \cup \{(X \times \{0\}), (X \times \{1\})\}$. Then $\mathcal{G}$ is a decomposition of $X \times [0, 1]$. The *suspension over X*, denoted by $\Sigma(X)$, is the quotient space $(X \times [0, 1])/\mathcal{G}$.

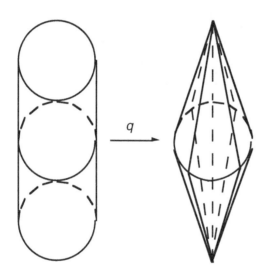

1.2.14. Definition. Let X be a metric space and let $\mathcal{G}$ be a decomposition of X. We say that $\mathcal{G}$ is *upper semicontinuous* if for each $G \in \mathcal{G}$ and each open subset U of X such that $G \subset U$, there exists an open subset V of X such that $G \subset V$ and such that if $G' \in \mathcal{G}$ and $G' \cap V \neq \emptyset$, then $G' \subset U$. We say that $\mathcal{G}$ is *lower semicontinuous* provided that for each $G \in \mathcal{G}$ any two points x and y of G and each open set U of X such that $x \in U$, there exists an open set V of X such that $y \in V$ and such that if $G' \in \mathcal{G}$ and $G' \cap V \neq \emptyset$, then $G \cap U \neq \emptyset$. Finally, we say that $\mathcal{G}$ is *continuous* if $\mathcal{G}$ is both upper and lower semicontinuous.

1.2.15. Example. Let $X = ([-1,1] \times [0,1]) \cup (\{0\} \times [0,2])$. For each $t \in [-1,1] \setminus \{0\}$, let $G_t = \{t\} \times [0,1]$, and for $t = 0$, let $G_0 = \{0\} \times [0,2]$. Let $\mathcal{G} = \{G_t \mid t \in [0,1]\}$. Then $\mathcal{G}$ is an upper semicontinuous decomposition of X.

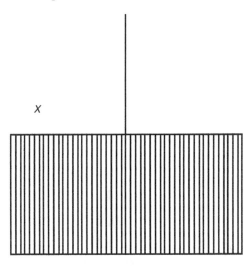

Upper semicontinuous decompostion

1.2.16. Example. Let

$$X = ([0,1] \times [0,1)) \cup \left(\{1\} \times \left[0, \frac{1}{3}\right] \cup \{1\} \times \left[\frac{2}{3}, 1\right] \right).$$

For each $t \in [0,1)$, let $G_t = \{t\} \times [0,1]$, let $G_1 = \{1\} \times \left[0, \frac{1}{3}\right]$ and let $G_1' = \{1\} \times \left[\frac{2}{3}, 1\right]$. Let $\mathcal{G} = \{G_t \mid t \in [0,1]\} \cup \{G_1'\}$. Then $\mathcal{G}$ is a lower semicontinuous decomposition of X.

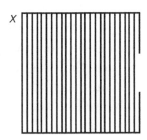

Lower semicontinuous decomposition

The following Theorem gives us a way to obtain upper semicontinuous decompositions of compacta.

1.2.17. Theorem. *Let* $f\colon X \twoheadrightarrow Y$ *be a surjective map between compacta. If* $\mathcal{G}_f = \{f^{-1}(y) \mid y \in Y\}$, *then* $\mathcal{G}_f$ *is an upper semicontinuous decomposition of* X.

Proof. Let U be an open subset of X such that $f^{-1}(y) \subset U$. Note that $X \setminus U$ is a closed subset, hence compact, of X. Then $f(X \setminus U)$ is a compact subset, hence closed, of Y such that $y \notin f(X \setminus U)$. Thus, $Y \setminus f(X \setminus U)$ is an open subset of Y containing y.

If $V = f^{-1}(Y \setminus f(X \setminus U))$, then V is an open subset of X such that $f^{-1}(y) \subset V \subset U$. Since $V = \bigcup \{f^{-1}(y) \mid y \in Y \setminus f(X \setminus U)\}$, clearly V satisfies the required property of the definition of upper semicontinuous decomposition.

Q.E.D.

1.2.18. Remark. Let us note that Theorem 1.2.17 is not true without the compactness of X. Let X be the Euclidean plane $\mathbb{R}^2$ and let $\pi\colon \mathbb{R}^2 \twoheadrightarrow \mathbb{R}$ be given by $\pi((x, y)) = x$. Then $\mathcal{G}_\pi$ is a decomposition of X which is not upper semicontinuous. To see this, let $U = \left\{ (x, y) \in X \mid x \neq 0 \text{ and } y < \dfrac{1}{x} \right\} \cup \{0\} \times \mathbb{R}$. Then U is an open set of X such that $\pi^{-1}(0) \subset U$, whose boundary is asymptotic to $\pi^{-1}(0)$. Hence, for each $t \in \mathbb{R} \setminus \{0\}$, $\pi^{-1}(t) \cap (X \setminus U) \neq \emptyset$.

The next Theorem gives three other ways to think about upper semicontinuous decompositions.

1.2.19. Theorem. *If X is a metric space and $\mathcal{G}$ is a decomposition of X, then the following conditions are equivalent:*

(a) $\mathcal{G}$ is an upper semicontinuous decomposition;

(b) the quotient map $q\colon X \twoheadrightarrow X/\mathcal{G}$ is closed;

(c) if U is an open subset of X, then $W_U = \bigcup \{G \in \mathcal{G} \mid G \subset U\}$ is an open subset of X;

(d) if D is a closed subset of X, then $K_D = \bigcup\{G \in \mathcal{G} \mid G \cap D \neq \emptyset\}$ is a closed subset of X.

Proof. Suppose $\mathcal{G}$ is an upper semicontinuous decomposition. Let D be a closed subset of X. By Remark 1.2.5, we have that $q(D)$ is closed in $X/\mathcal{G}$ if and only if $q^{-1}(q(D))$ is closed in X. We show that $X \setminus q^{-1}(q(D))$ is open in X. Let $x \in X \setminus q^{-1}(q(D))$. Then $q(x) \in X/\mathcal{G} \setminus q(D)$ and hence $q^{-1}(q(x)) \subset X \setminus D$. Therefore, since $X \setminus D$ is open, by Definition 1.2.14, there exists an open set V of X such that $q^{-1}(q(x)) \subset V$ and for each $y \in V$, $q^{-1}(q(y)) \subset X \setminus D$. Clearly, $x \in V$ and $q(V) \subset X/\mathcal{G} \setminus q(D)$. Thus, $V \subset X \setminus q^{-1}(q(D))$. Therefore, $X \setminus q^{-1}(q(D))$ is open, since $x \in V \subset X \setminus q^{-1}(q(D))$.

Now, suppose q is a closed map. Let U be an open subset of X. Since q is a closed map, we have that $q^{-1}(X/\mathcal{G} \setminus q(X \setminus U))$ is an open subset of X such that $q^{-1}(X/\mathcal{G} \setminus q(X \setminus U)) = W_U$. (If $x \in q^{-1}(X/\mathcal{G} \setminus q(X \setminus U))$, then $q(x) \in X/\mathcal{G} \setminus q(X \setminus U)$. Hence, $q^{-1}(q(x)) \subset X \setminus q^{-1}(q(X \setminus U)) \subset X \setminus (X \setminus U) = U$. Thus, $x \in W_U$. The other inclusion is obvious.)

Next, suppose W_U is open for each open subset U of X. Let D be a closed subset of X. Then $X \setminus D$ is open in X. Hence, $W_{X \setminus D}$ is open in X. Since, clearly, $K_D = X \setminus W_{X \setminus D}$, we have that K_D is closed.

Finally, suppose K_D is closed for each closed subset D of X. To see $\mathcal{G}$ is upper semicontinuous, let $G \in \mathcal{G}$ and let U be an open subset of X such that $G \subset U$. Note that $X \setminus U$ is a closed subset of X. Hence, $K_{X \setminus U}$ is a closed subset of X. Let $V = X \setminus K_{X \setminus U}$. Then V is open, $G \subset V \subset U$ and if $G' \in \mathcal{G}$ and $G' \cap V \neq \emptyset$, then $G' \subset V$. Therefore, $\mathcal{G}$ is upper semicontinuous.

Q.E.D.

1.2.20. Corollary. *Let X be a metric space. If $\mathcal{G}$ is an upper semicontinuous decomposition of X, then the elements of $\mathcal{G}$ are closed.*

Proof. Let $G \in \mathcal{G}$. Take $x \in G$ and let $q \colon X \twoheadrightarrow X/\mathcal{G}$ be the quotient map. Since X is a metric space, $\{x\}$ is closed in X. By Theorem 1.2.19, $q(\{x\})$ is closed in $X/\mathcal{G}$. Since q is continuous and $q^{-1}(q(\{x\})) = G$, G is a closed subset of X.

Q.E.D.

1.2.21. Theorem. *If X is a compactum and $\mathcal{G}$ is an upper semi-continuous decomposition of X, then $X/\mathcal{G}$ has a countable basis.*

Proof. Let $q\colon X \twoheadrightarrow X/\mathcal{G}$ be the quotient map. Since X is a compactum, it has a countable basis $\mathcal{U}$. Let

$$\mathcal{B} = \left\{ \bigcup_{j=1}^{n} U_j \,\middle|\, U_1,\dots,U_n \in \mathcal{U} \text{ and } n \in \mathbb{N} \right\}.$$

Note that $\mathcal{B}$ is a countable family of open subsets of X.
Let

$$\boldsymbol{\mathcal{B}} = \{X/\mathcal{G} \setminus q(X \setminus U) \mid U \in \mathcal{B}\}.$$

We see that $\boldsymbol{\mathcal{B}}$ is a countable basis for $X/\mathcal{G}$. Clearly, $\boldsymbol{\mathcal{B}}$ is a countable family of open subsets of $X/\mathcal{G}$. Let $\boldsymbol{U}$ be an open subset of $X/\mathcal{G}$ and let $\boldsymbol{x} \in \boldsymbol{U}$. Then $q^{-1}(\boldsymbol{U})$ is an open subset of X and $q^{-1}(\boldsymbol{x}) \subset q^{-1}(\boldsymbol{U})$. Since $q^{-1}(\boldsymbol{x})$ is compact, there exist $U_1,\dots,U_k \in \mathcal{U}$ such that $q^{-1}(\boldsymbol{x}) \subset \bigcup_{j=1}^{k} U_j \subset q^{-1}(\boldsymbol{U})$. Let $U = \bigcup_{j=1}^{k} U_j$. Then $U \in \mathcal{B}$. Hence, $X/\mathcal{G} \setminus q(X \setminus U) \in \boldsymbol{\mathcal{B}}$. Also, $\boldsymbol{x} \in X/\mathcal{G} \setminus q(X \setminus U) \subset \boldsymbol{U}$. Therefore, $\boldsymbol{\mathcal{B}}$ is a countable basis for $X/\mathcal{G}$.

$$\textbf{Q.E.D.}$$

1.2.22. Corollary. *If X is a compactum and $\mathcal{G}$ is an upper semi-continuous decomposition of X, then $X/\mathcal{G}$ is metrizable.*

Proof. By Theorem 1.2.21, we have that $X/\mathcal{G}$ has a countable basis. By Theorem 1 (p. 241) of [12], it suffices to show that $X/\mathcal{G}$ is a Hausdorff space. Let $\boldsymbol{x}$ and $\boldsymbol{y}$ be two distinct points of $X/\mathcal{G}$. Then $q^{-1}(\boldsymbol{x})$ and $q^{-1}(\boldsymbol{y})$ are two disjoint closed subsets of X. Since X is a metric space, there exist two disjoint open subsets, U_1 and U_2, of X such that $q^{-1}(\boldsymbol{x}) \subset U_1$ and $q^{-1}(\boldsymbol{y}) \subset U_2$. Note that, by Theorem 1.2.19 (c), W_{U_1} and W_{U_2} are open subsets of X such that $q^{-1}(\boldsymbol{x}) \subset W_{U_1} \subset U_1$, $q^{-1}(\boldsymbol{y}) \subset W_{U_2} \subset U_2$, and $q(W_{U_1})$ and $q(W_{U_2})$ are open subsets of $X/\mathcal{G}$. Since $U_1 \cap U_2 = \emptyset$, $q(W_{U_1}) \cap q(W_{U_2}) = \emptyset$. Therefore, $X/\mathcal{G}$ is a Hausdorff space.

$$\textbf{Q.E.D.}$$

The next Theorem gives a characterization of lower semicontinuous decompositions.

1.2.23. Theorem. *Let X be a metric space and let $\mathcal{G}$ be a decomposition of X. Then $\mathcal{G}$ is lower semicontinuous if and only if the quotient map $q\colon X \twoheadrightarrow X/\mathcal{G}$ is open.*

Proof. Suppose $\mathcal{G}$ is lower semicontinuous. Let U be an open subset of X. We show $q(U)$ is an open subset of $X/\mathcal{G}$. To this end, by Remark 1.2.5, we only need to show that $q^{-1}(q(U))$ is an open subset of X.

Let $y \in q^{-1}(q(U))$. Then $q(y) \in q(U)$, and there exists a point x in U such that $q(x) = q(y)$. Since $\mathcal{G}$ is a lower semicontinuous decomposition, there exists an open subset V of X containing y such that if $G \in \mathcal{G}$ y $G \cap V \neq \emptyset$, then $G \cap U \neq \emptyset$. Hence, $V \subset q^{-1}(q(U))$. Therefore, q is open.

Now, suppose q is open. Let $G \in \mathcal{G}$. Take $x, y \in G$ and let U be an open subset of X such that $x \in U$. Since q is open, $V = q^{-1}(q(U))$ is an open subset of X such that $G \subset V$. In particular, $y \in V$. Let $G' \in \mathcal{G}$ such that $G' \cap V \neq \emptyset$. Then $G' \subset V$. Thus, $q(G') \in q(U)$. Hence, there exists $u \in U$ such that $q(u) = q(G')$. Since $q^{-1}(q(G')) = G'$, $u \in G'$. Thus, $G' \cap U \neq \emptyset$. Therefore, $\mathcal{G}$ is lower semicontinuous.

$$\textbf{Q.E.D.}$$

The following Corollary is a consequence of Theorems 1.2.19 and 1.2.23:

1.2.24. Corollary. *Let X be a metric space and let $\mathcal{G}$ be a decomposition of X. Then $\mathcal{G}$ is continuous if and only if the quotient map is both open and closed.*

The following Theorem gives us a necessary and sufficient condition on a map $f\colon X \twoheadrightarrow Y$ between compacta, to have that $\mathcal{G}_f = \{f^{-1}(y) \mid y \in Y\}$ is a continuous decomposition.

1.2.25. Theorem. *Let X and Y be compacta and let $f\colon X \twoheadrightarrow Y$ be a surjective map. Then $\mathcal{G}_f = \{f^{-1}(y) \mid y \in Y\}$ is continuous if and only if f is open.*

Proof. If $\mathcal{G}_f$ is a continuous decomposition of X, by Theorem 1.2.23, the quotient map $q\colon X \twoheadrightarrow X/\mathcal{G}_f$ is open. By Theorem 1.2.10,

$\varphi_f \colon X/\mathcal{G}_f \twoheadrightarrow Y$ is a homeomorphism. Hence, $f = \varphi_f \circ q$ is an open map.

Now, suppose f is open. By Theorem 1.2.17, $\mathcal{G}_f$ is upper semicontinuous. Since $q = \varphi_f^{-1} \circ f$ and f is open, q is open. By Theorem 1.2.23, $\mathcal{G}_f$ is a lower semicontinuous decomposition. Therefore, $\mathcal{G}_f$ is continuous.

Q.E.D.

In the following Definition a notion of convergence of sets is introduced.

1.2.26. Definition. Let $\{X_n\}_{n=1}^{\infty}$ be a sequence of subsets of the metric space X. Then:

(1) the *limit inferior* of the sequence $\{X_n\}_{n=1}^{\infty}$ is defined as follows:

$$\liminf X_n = \{x \in X \mid \text{for each open subset } U \text{ of } X \text{ such that}$$

$x \in U, U \cap X_n \neq \emptyset$ for each $n \in \mathbb{N}$, save, possibly, finitely many$\}$.

(2) the *limit superior* of the sequence $\{X_n\}_{n=1}^{\infty}$ is defined as follows:

$$\limsup X_n = \{x \in X \mid \text{for each open subset } U \text{ of } X \text{ such that}$$

$x \in U, U \cap X_n \neq \emptyset$ for infinitely many indices $n \in \mathbb{N}\}$.

Clearly, $\liminf X_n \subset \limsup X_n$. If $\liminf X_n = \limsup X_n = L$, then we say that the sequence $\{X_n\}_{n=1}^{\infty}$ is a *convergent sequence* with limit $L = \lim\limits_{n \to \infty} X_n$.

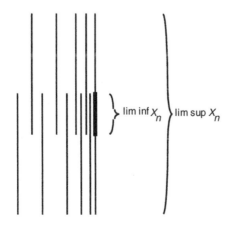

1.2.27. Lemma. *Let $\{X_n\}_{n=1}^{\infty}$ be a sequence of subsets of the metric space X. Then $\liminf X_n$ and $\limsup X_n$ are both closed subsets of X.*

Proof. Let $x \in Cl(\liminf X_n)$. Let U be an open subset of X such that $x \in U$. Since $x \in Cl(\liminf X_n) \cap U$, we have that $\liminf X_n \cap U \neq \emptyset$. Hence, $U \cap X_n \neq \emptyset$ for each $n \in \mathbb{N}$, save, possibly, finitely many. Therefore, $x \in \liminf X_n$. The proof for $\limsup$ is similar.

Q.E.D.

The next Theorem tells us that separable metric spaces behave like sequentially compact spaces using the notion of convergence just introduced.

1.2.28. Theorem. *Each sequence $\{X_n\}_{n=1}^{\infty}$ of closed subsets of a separable metric space X has a convergent subsequence.*

Proof. Let $\{U_m\}_{m=1}^{\infty}$ be a countable basis for X. Let $\{X_n^1\}_{n=1}^{\infty} = \{X_n\}_{n=1}^{\infty}$. Suppose, inductively, that we have defined the sequence $\{X_n^m\}_{n=1}^{\infty}$. We define the sequence $\{X_n^{m+1}\}_{n=1}^{\infty}$ as follows:

(1) If $\{X_n^m\}_{n=1}^{\infty}$ has a subsequence $\{X_{n_k}^m\}_{k=1}^{\infty}$ such that $\limsup X_{n_k}^m \cap U_m = \emptyset$, then let $\{X_n^{m+1}\}_{n=1}^{\infty}$ be such subsequence of $\{X_n^m\}_{n=1}^{\infty}$.

(2) If for each subsequence $\{X_{n_k}^m\}_{k=1}^{\infty}$ of $\{X_n^m\}_{n=1}^{\infty}$, we have that $\limsup X_{n_k}^m \cap U_m \neq \emptyset$, we define $\{X_n^{m+1}\}_{n=1}^{\infty}$ as $\{X_n^m\}_{n=1}^{\infty}$.

Since we have the subsequences $\{X_n^m\}_{n=1}^{\infty}$, let us consider the "diagonal subsequence" $\{X_m^m\}_{m=1}^{\infty}$. By construction, $\{X_m^m\}_{m=1}^{\infty}$ is a subsequence of $\{X_n\}_{n=1}^{\infty}$. We see that $\{X_m^m\}_{m=1}^{\infty}$ converges.

Let us assume that $\{X_m^m\}_{m=1}^{\infty}$ does not converge. Hence, there exists $p \in \limsup X_m^m \setminus \liminf X_m^m$. Let U_k be a basic open set such that $p \in U_k$ and $U_k \cap X_{m_\ell}^{m_\ell} = \emptyset$ for some subsequence $\{X_{m_\ell}^{m_\ell}\}_{\ell=1}^{\infty}$ of $\{X_m^m\}_{m=1}^{\infty}$ ($\liminf X_n$ is a closed subset of X by Lemma 1.2.27). Clearly, $\{X_{m_\ell}^{m_\ell}\}_{\ell=1}^{\infty}$ is a subsequence of $\{X_n^k\}_{n=1}^{\infty}$. Thus, $\{X_n^k\}_{n=1}^{\infty}$ satisfies condition (1), with k in place of m. Hence, $\limsup X_n^{k+1} \cap U_k = \emptyset$. Since $\{X_m^m\}_{m=1}^{\infty}$ is a subsequence of $\{X_n^{k+1}\}_{n=1}^{\infty}$ and $\limsup X_m^m \subset \limsup X_n^{k+1}$, it follows that $\limsup X_m^m \cap U_k = \emptyset$. Now, recall that $p \in \limsup X_m^m \cap U_k$. Thus, we obtain a contradiction. Therefore, $\{X_m^m\}_{m=1}^{\infty}$ converges.

Q.E.D.

1.2.29. Theorem. *Let X be a compactum. If $\{X_n\}_{n=1}^{\infty}$ is a sequence of connected subsets of X and $\liminf X_n \neq \emptyset$, then $\limsup X_n$ is connected.*

Proof. Suppose, to the contrary, that $\limsup X_n$ is not connected. Since $\limsup X_n$ is closed, by Lemma 1.2.27, we assume, without loss of generality, that there are two disjoint closed subsets A and B of X such that $\limsup X_n = A \cup B$. Since X is a metric space, there exist two disjoint open subsets U and V of X such that $A \subset U$ and $B \subset V$. Then there exists $N' \in \mathbb{N}$ such that if $n \geq N'$, then $X_n \subset U \cup V$. To show this, suppose it is not true. Then for each $n \in \mathbb{N}$, there is $m_n > n$ such that $X_{m_n} \setminus (U \cup V) \neq \emptyset$. Let $x_{m_n} \in X_{m_n} \setminus (U \cup V)$ for each $n \in \mathbb{N}$. Since X is compact, without loss of generality, we assume that the sequence $\{x_{m_n}\}_{n=1}^{\infty}$ converges to a point x of X. Note that $x \in X \setminus (U \cup V)$ and, by construction, $x \in \limsup X_n$, a contradiction. Therefore, there exists $N' \in \mathbb{N}$ such that if $n \geq N'$, then $X_n \subset U \cup V$.

Since $\liminf X_n \neq \emptyset$ and $\liminf X_n \subset \limsup X_n$, we assume, without loss of generality, that $\liminf X_n \cap U \neq \emptyset$. Then there exists $N'' \in \mathbb{N}$ such that if $n \geq N''$, $U \cap X_n \neq \emptyset$. Let $N = \max\{N', N''\}$. Hence, if $n \geq N$, then $X_n \subset U \cup V$ and $U \cap X_n \neq \emptyset$. Since X_n is connected for every $n \in \mathbb{N}$, $X_n \cap V = \emptyset$ for each $n \geq N$, a contradiction. Therefore, $\limsup X_n$ is connected.

Q.E.D.

The following Theorem gives us a characterization of an upper semicontinuous decomposition of a compactum in terms of limits inferior and superior.

1.2.30. Theorem. *Let X be a compactum, with metric d. Then a decomposition $\mathcal{G}$ of X is upper semicontinuous if and only if $\mathcal{G}$ is a closed decomposition and for each sequence $\{X_n\}_{n=1}^{\infty}$ of elements of $\mathcal{G}$ and each element Y of $\mathcal{G}$ such that $\liminf X_n \cap Y \neq \emptyset$, then $\limsup X_n \subset Y$.*

Proof. Suppose $\mathcal{G}$ is an upper semicontinuous decomposition of X. By Corollary 1.2.20, $\mathcal{G}$ is a closed decomposition. Let $\{X_n\}_{n=1}^{\infty}$ be a sequence of elements of $\mathcal{G}$ and let Y be an element of $\mathcal{G}$ such that $\liminf X_n \cap Y \neq \emptyset$.

Suppose there exists $p \in \limsup X_n \setminus Y$. Since Y is closed and p is not an element of Y, there exists an open set W of X such that $p \in W$ and $Cl(W) \cap Y = \emptyset$. Let $U = X \setminus Cl(W)$. Since $\mathcal{G}$ is upper semicontinuous, there is an open set V of X such that $Y \subset V$ and if $G \in \mathcal{G}$ such that $G \cap V \neq \emptyset$, $G \subset U$.

Let $q \in \liminf X_n \cap Y$. Then $q \in \liminf X_n \cap V$. Hence, $V \cap X_n \neq \emptyset$ for each $n \in \mathbb{N}$, save, possibly, finitely many. Thus, $W \cap X_n = \emptyset$ for each $n \in \mathbb{N}$, save, possibly, finitely many. This contradicts the fact that $p \in W \cap \limsup X_n$. Therefore, $\limsup X_n \subset Y$.

Now, suppose $\mathcal{G}$ is a closed decomposition and let Y be an element of $\mathcal{G}$. Suppose that if $\{X_n\}_{n=1}^{\infty}$ is a sequence of elements of $\mathcal{G}$ such that $\liminf X_n \cap Y \neq \emptyset$, then $\limsup X_n \subset Y$.

To see $\mathcal{G}$ is upper semicontinuous, let U be an open subset of X such that $Y \subset U$. For each $n \in \mathbb{N}$, let $V_n = \mathcal{V}_{\frac{1}{n}}^d(Y)$. Suppose that for each $n \in \mathbb{N}$, there is an element X_n of $\mathcal{G}$ such that $X_n \cap V_n \neq \emptyset$ and $X_n \not\subset U$. For each $n \in \mathbb{N}$, let $p_n \in X_n \cap V_n$. Since X is compact, $\{p_n\}_{n=1}^{\infty}$ has a convergent subsequence $\{p_{n_k}\}_{k=1}^{\infty}$. Let p be the point of convergence of $\{p_{n_k}\}_{k=1}^{\infty}$. Note that $p \in \liminf X_{n_k} \cap Y$. Hence, $\limsup X_{n_k} \subset Y$.

For each $k \in \mathbb{N}$, let $q_k \in X_{n_k} \setminus U$. Since X is compact, the sequence $\{q_{n_k}\}_{k=1}^{\infty}$ has a convergent subsequence $\{q_{n_{k_\ell}}\}_{\ell=1}^{\infty}$. Let q be the point of convergence of $\{q_{n_{k_\ell}}\}_{\ell=1}^{\infty}$. Note that $q \notin Y$ and $q \in \limsup X_{n_{k_\ell}} \subset \limsup X_{n_k}$, a contradiction. Therefore, $\mathcal{G}$ is upper semicontinuous.

$$\textbf{Q.E.D.}$$

The following Theorem gives us a characterization of a continuous decomposition of a compactum in terms of limits inferior and superior.

1.2.31. Theorem. *Let X be a compactum, with metric d. Then a decomposition $\mathcal{G}$ of X is continuous if and only if $\mathcal{G}$ is a closed decomposition and for each sequence $\{X_n\}_{n=1}^{\infty}$ of elements of $\mathcal{G}$ and each element Y of $\mathcal{G}$ such that $\liminf X_n \cap Y \neq \emptyset$, then $\limsup X_n = Y$.*

Proof. Suppose $\mathcal{G}$ is a continuous decomposition of X. By Corollary 1.2.20, $\mathcal{G}$ is a closed decomposition. Let $\{X_n\}_{n=1}^{\infty}$ be a sequence of elements of $\mathcal{G}$ and let Y be an element of $\mathcal{G}$ such that

$\liminf X_n \cap Y \neq \emptyset$. By Theorem 1.2.30, $\limsup X_n \subset Y$. Suppose there exists $p \in Y \setminus \limsup X_n$. Let U be an open subset of X such that $p \in U$ and $U \cap \limsup X_n = \emptyset$ (by Lemma 1.2.27 $\limsup X_n$ is closed). Let $q \in \limsup X_n \subset Y$. Since $\mathcal{G}$ is a lower semicontinuous decomposition, there is an open set V of X such that $q \in V$ and if $G \in \mathcal{G}$ and $G \cap V \neq \emptyset$, then $G \cap U \neq \emptyset$. Since $q \in \limsup X_n$ and V is an open set containing q, $V \cap X_n \neq \emptyset$ for infinitely many indices $n \in \mathbb{N}$. Hence, $U \cap X_n \neq \emptyset$ for infinitely many indices $n \in \mathbb{N}$. Then $U \cap \limsup X_n \neq \emptyset$, a contradiction. Therefore, $\limsup X_n = Y$.

Now, suppose that $\mathcal{G}$ is a closed decomposition of X such that for each sequence $\{X_n\}_{n=1}^{\infty}$ of elements of $\mathcal{G}$ and each element Y of $\mathcal{G}$ such that $\liminf X_n \cap Y \neq \emptyset$, then $\limsup X_n = Y$. By Theorem 1.2.30, $\mathcal{G}$ is upper semicontinuous.

Let Y be an element of $\mathcal{G}$. Let p and q be two points of Y, and let U be an open subset of X such that $p \in U$. For each $n \in \mathbb{N}$, let $V_n = \mathcal{V}_{\frac{1}{n}}^d(q)$. Suppose that for each $n \in \mathbb{N}$, there is an element X_n of $\mathcal{G}$ such that $X_n \cap V_n \neq \emptyset$ and $X_n \cap U = \emptyset$. Hence, $\limsup X_n \cap U = \emptyset$. For each $n \in \mathbb{N}$, let $q_n \in X_n \cap V_n$. Clearly, $\{q_n\}_{n=1}^{\infty}$ converges to q. Thus, $q \in \liminf X_n \cap Y$. By hypothesis, $Y = \limsup X_n$. Hence, $Y \cap U \neq \emptyset$, a contradiction. Therefore, $\mathcal{G}$ is a continuous decomposition.

Q.E.D.

1.3 Homotopy and Fundamental Group

We introduce the fundamental group of a metric space. We show that the fundamental group of the unit circle $\mathcal{S}^1$ is isomorphic to the group of integers $\mathbb{Z}$.

We assume that the reader is familiar with the elementary concepts of group theory. The reader may find more than enough information about groups in [25].

1.3.1. Definition. Given a metric space X and two points x and y of X, a *path* joining x and y is a map $\alpha\colon [0,1] \to X$ such that

$\alpha(0) = x$ and $\alpha(1) = y$. In this case, x and y are the *end points* of α. When α is one–to–one, α is called an *arc*. Some times, we identify a path or an arc α with its image $\alpha([0, 1])$.

1.3.2. Definition. Let X and Y be metric spaces. Let $g, f \colon X \to Y$ be two maps. We say that *f is homotopic to g* provided that there is a map $G \colon X \times [0, 1] \to Y$ such that $G((x, 0)) = f(x)$ and $G((x, 1)) = g(x)$ for each $x \in X$. The map G is called a *homotopy between f and g*. If for each $t \in [0, 1]$, $G(\cdot, t) \colon X \twoheadrightarrow Y$ is a homeomorphism, G is called an *isotopy between f and g*.

1.3.3. Example. Let $g, f \colon X \to Y$ be two constant maps between metric spaces, say $g(x) = q$ and $f(x) = p$ for each $x \in X$. Then f and g are homotopic if and only if p and q both belong to the same path component of Y. To show this, suppose first that p and q belong to the same path component of Y. Then there is a map $\alpha \colon [0, 1] \to Y$ such that $\alpha(0) = p$ and $\alpha(1) = q$. Hence, $G \colon X \times [0, 1] \to Y$ given by $G((x, t)) = \alpha(t)$ is a homotopy between f and g. Now, suppose f and g are homotopic. Then there is a map $G' \colon X \times [0, 1] \to Y$ such that $G'((x, 0)) = f(x) = p$ and $G'((x, 1)) = g(x) = q$ for each $x \in X$. Let $x_0 \in X$. Then the map $\beta \colon [0, 1] \to Y$ given by $\beta(t) = G'((x_0, t))$ is a path such that $\beta(0) = p$ and $\beta(1) = q$.

1.3.4. Example. Let X be a metric space. If $g, f \colon X \to \mathbb{R}^n$ are two maps, for some $n \in \mathbb{N}$. Then f and g are homotopic. To see this, let $G \colon X \times [0, 1] \to \mathbb{R}^n$ be given by $G((x, t)) = (1 - t)f(x) + tg(x)$. Then G is a homotopy between f and g.

The next Theorem is known as *Borsuk's homotopy extension theorem*. A proof of this result may be found in Theorem 4–4 of [10].

1.3.5. Theorem. *Let A be a closed subset of a separable metric space X, and let $g', f' : A \to \mathcal{S}^n$ be homotopic maps of A into the n–sphere $\mathcal{S}^n$. If there exists a map $f : M \to \mathcal{S}^n$ such that $f|_A = f'$ (i.e., f extends f'), then there also exists a map $g : M \to \mathcal{S}^n$ such that $g|_A = g'$, and f and g may be chosen to be homotopic also.*

1.3.6. Theorem. *Let Y be a closed subset of a compactum X. If $f : X \to \mathcal{S}^1$ is a map such that $f|_Y : Y \to \mathcal{S}^1$ is homotopic to a constant map, then there exists an open set U of X such that $Y \subset U$ and $f|_U : U \to \mathcal{S}^1$ is homotopic to a constant map.*

Proof. Let $f : X \to \mathcal{S}^1$ be a map such that $f|_Y : Y \to \mathcal{S}^1$ is homotopic to a constant map $g' : Y \to \mathcal{S}^1$ given by $g'(y) = b$ for every $y \in Y$.

Since f is a map which extends $f|_Y$ to X, by Theorem 1.3.5, there exists a map $g : X \to \mathcal{S}^1$ such that $g|_Y = g'$ and f and g are homotopic.

Let V be a proper open subset of $\mathcal{S}^1$ such that $b \in V$. By continuity of g, there exists an open set U of X such that $Y \subset U$ and $g(U) \subset V$. Since V is an arc, it is easy to see that $g|_U$ is homotopic to a constant map. Since f and g are homotopic, $f|_U$ is also homotopic to a constant map.

Q.E.D.

1.3.7. Theorem. *Let X be a compactum. If $f : X \twoheadrightarrow \mathcal{S}^1$ is a map not homotopic to a constant map, then there exists a closed connected subset C of X such that $f|_C : C \twoheadrightarrow \mathcal{S}^1$ is not homotopic to a constant map and for each proper closed subset C' of C, $f|_{C'} : C \to \mathcal{S}^1$ is homotopic to a constant map.*

Proof. Let $\mathcal{C}$ be the collection of all closed subsets, K, of X such that $f|_K : K \twoheadrightarrow \mathcal{S}^1$ is not homotopic to a constant map. Since $X \in \mathcal{C}$, $\mathcal{C} \neq \emptyset$. Define the following partial order on $\mathcal{C}$. If $K_1, K_2 \in \mathcal{C}$, then $K_1 > K_2$ if $K_1 \subset K_2$. Let $\mathcal{K}$ be a (set theoretic) chain of elements of $\mathcal{C}$. We show that $\mathcal{K}$ has an upper bound. Let $Y = \bigcap_{K \in \mathcal{K}} K$. We assert that $Y \in \mathcal{C}$. If Y does not belong to $\mathcal{C}$, then

$f|_Y \colon Y \to \mathcal{S}^1$ is homotopic to a constant map. Hence, by Theorem 1.3.6 there exists an open subset U of X such that $f|_U \colon U \to \mathcal{S}^1$ is homotopic to a constant map. Note that $\{X \setminus K \mid K \in \mathcal{K}\}$ is an open cover of $X \setminus U$. Then there exist $K_1, \ldots, K_m \in \mathcal{K}$ such that $X \setminus U \subset \bigcup_{j=1}^{m} X \setminus K_j$. Hence, $K = \bigcap_{j=1}^{m} K_j \subset U$. Since $\mathcal{K}$ is a chain, $K \in \mathcal{K}$. Since $K \subset U$, $f|_K \colon K \to \mathcal{S}^1$ is homotopic to a constant map, a contradiction to the fact that $K \in \mathcal{C}$. Therefore, $Y \in \mathcal{C}$ and Y is an upper bound for $\mathcal{K}$. Thus, by Kuratowski–Zorn Lemma, there exists an element $C \in \mathcal{C}$ such that $f|_C \colon C \twoheadrightarrow \mathcal{S}^1$ is not homotopic to a constant map and for each proper closed subset C' of C, $f|_{C'} \colon C \to \mathcal{S}^1$ is homotopic to a constant map. It remains to show that C is connected.

Suppose C is not connected. Thus, there exist two closed subsets C_1 and C_2 of X such that $C = C_1 \cup C_2$. Since C is a maximal element of $\mathcal{C}$, $f|_{C_j} \colon C_j \to \mathcal{S}^1$ is homotopic to a constant map $g_j \colon C_j \to \mathcal{S}^1$ by a homotopy H_j such that $H_j((c,0)) = (f|_{C_j})(c)$ and $H_j((c,1)) = g_j(c)$ for each $c \in C_j$ and $j \in \{1,2\}$. Since $\mathcal{S}^1$ is arcwise connected, we assume, without loss of generality, that g_1 and g_2 have the same image $\{b\} \subset \mathcal{S}^1$. Let $g \colon C \to \mathcal{S}^1$ be given by $g(c) = b$ for each $c \in C$. Let $H \colon C \times [0,1] \to \mathcal{S}^1$ be given by $H((c,t)) = H_j((c,t))$ if $c \in C_j$, $j \in \{1,2\}$. Then H is a homotopy such that $H((c,0)) = (f|_{C_j})(c) = (f|_C)(c)$ and $H((c,1)) = g_j(c) = b$ for every $c \in C$. Thus, $f|_C$ is homotopic to a constant map, a contradiction. Therefore, C is connected.

Q.E.D.

1.3.8. Theorem. *Let X and Y be metric spaces. The relation of homotopy is an equivalence relation in the set of maps between X and Y. The equivalence classes of this equivalence relation are called homotopy classes.*

Proof. Let $f \colon X \to Y$ be a map. Then $G \colon X \times [0,1] \to Y$, given by $G((x,t)) = f(x)$, is a homotopy between f and f. Hence, the relation is reflexive.

Now, let $g, f \colon X \to Y$ be two maps and suppose f is homotopic to g. Then there is a homotopy $H \colon X \times [0,1] \to Y$ such that $H((x,0)) = f(x)$ and $H((x,1)) = g(x)$ for each $x \in X$. Hence, the map $K\colon X \times [0,1] \to Y$ given by $K((x,t)) = H((x, 1-t))$ is a homotopy between g and f, since $K((x,0)) = g(x)$ and $K((x,1)) = f(x)$ for each $x \in X$. Thus, the relation is symmetric.

Finally, let $h, g, f\colon X \to Y$ be three maps and suppose that f is homotopic to g and g is homotopic to h. Then there exist two homotopies $J, L\colon X \times [0,1] \to Y$ such that $J((x,0)) = f(x)$, $J((x,1)) = g(x)$, $L((x,0)) = g(x)$ and $L((x,1)) = h(x)$ for each $x \in X$. Thus, the map $R\colon X \times [0,1] \to Y$ given by

$$R((x,t)) = \begin{cases} J((x, 2t)) & \text{if } t \in \left[0, \dfrac{1}{2}\right]; \\[2ex] L((x, 2t-1)) & \text{if } t \in \left[\dfrac{1}{2}, 1\right) \end{cases}$$

is a homotopy between f and h, since for each $x \in X$, $R((x,0)) = f(x)$ and $R((x,1)) = h(x)$. Hence, the relation is transitive.

Q.E.D.

1.3.9. Notation. If $f\colon X \to Y$ is a map, then the homotopy class to which f belongs is denoted by $[f]$.

1.3.10. Definition. A metric space X is said to be *contractible* provided that the identity map, 1_X, of X is homotopic to a constant map g. We say that X is *locally contractible at p* if for each neighborhood U of p in X, there exist a neighborhood V of p in X and a homotopy $G\colon V \times [0,1] \to U$ such that $G((x,0)) = x$ and $G((x,1)) = x_0$ for each $x \in V$ and some $x_0 \in U$. The metric space X is *locally contractible* if it is locally contractible at each of its points.

The following two Theorems present some consequences of the contractibility of a space.

1.3.11. Theorem. *If X is a contractible metric space, then X is path connected.*

Proof. Since X is contractible, there exists a map $G\colon X \times [0,1] \to X$ such that $G((x,0)) = x$ and $G((x,1)) = p$ for each $x \in X$ and some point p of X. Let x and y be two points of X. Then the map $\alpha\colon [0,1] \to X$ given by

$$\alpha(t) = \begin{cases} G((x,2t)) & \text{if } t \in \left[0, \dfrac{1}{2}\right]; \\[2ex] G((y, 2-2t)) & \text{if } t \in \left[\dfrac{1}{2}, 1\right] \end{cases}$$

is a path such that $\alpha(0) = x$ and $\alpha(1) = y$. Therefore, X is path connected.

Q.E.D.

1.3.12. Theorem. *Let X and Y be metric spaces, where Y is arcwise connected. If either X or Y is contractible and if $g, f\colon X \to Y$ are two maps, then f and g are homotopic.*

Proof. Suppose Y is contractible. Then there exists a map $G\colon Y \times [0,1] \to Y$ such that $G((y,0)) = y$ and $G((y,1)) = q$ for each $y \in Y$ and some point q of Y. Then the map $K\colon X \times [0,1] \to Y$ given by

$$K((x,t)) = \begin{cases} G((f(x),2t)) & \text{if } t \in \left[0, \dfrac{1}{2}\right]; \\[2ex] G((g(x), 2-2t)) & \text{if } t \in \left[\dfrac{1}{2}, 1\right] \end{cases}$$

is a homotopy between f and g.

The proof when X is contractible is similar.

Q.E.D.

Now, we consider a particular case of homotopy; namely, we study the homotopies between paths.

1.3.13. Definition. Let X be a metric space. We say that two paths $\beta, \alpha\colon [0,1] \to X$ are *homotopic relative to* $\{0,1\}$ provided that there is a homotopy $G\colon [0,1] \times [0,1] \to X$ such that $G((s,0)) = \alpha(s)$, $G((s,1)) = \beta(s)$, $G((0,t)) = \alpha(0) = \beta(0)$ and $G((1,t)) = \alpha(1) = \beta(1)$ for each $s, t \in [0,1]$.

In order to define the fundamental group we need the following definitions:

1.3.14. Definition. Let X be a metric space and let x_0 be a point in X. The pair (X, x_0) is called *pointed space.*

1.3.15. Definition. Let X be a metric space. We say that a path $\alpha \colon [0, 1] \to X$ is *closed* provided that $\alpha(0) = \alpha(1)$.

1.3.16. Notation. If (X, x_0) is a pointed space and $\alpha \colon [0, 1] \to X$ is a closed path, we assume that $\alpha(0) = \alpha(1) = x_0$. The point x_0 is called the *base of the closed path.*

1.3.17. Definition. Let X be a metric space. Given two closed paths $\beta, \alpha \colon [0, 1] \to X$ such that $\alpha(0) = \beta(0)$, we define their *product*, denoted by $\alpha * \beta$, as the closed path given by:

$$(\alpha * \beta)(t) = \begin{cases} \alpha(2t) & \text{if } t \in \left[0, \dfrac{1}{2}\right]; \\ \beta(2t - 1) & \text{if } t \in \left[\dfrac{1}{2}, 1\right]. \end{cases}$$

1.3.18. Definition. Let X be a metric space. Given a closed path $\alpha \colon [0, 1] \to X$, we define its *inverse*, denoted by α^{-1}, as the closed path $\alpha^{-1} \colon [0, 1] \to X$ given by $\alpha^{-1}(t) = \alpha(1 - t)$.

1.3.19. Theorem. *Let X be a metric space. If $\beta', \beta, \alpha', \alpha \colon [0, 1] \to X$ are closed paths such that $\alpha(0) = \alpha'(0) = \beta(0) = \beta'(0)$ and such that α is homotopic to α' relative to $\{0, 1\}$ and β is homotopic to β' relative to $\{0, 1\}$, then $\alpha * \beta$ is homotopic to $\alpha' * \beta'$ relative to $\{0, 1\}$ and α^{-1} is homotopic to $(\alpha')^{-1}$ relative to $\{0, 1\}$.*

Proof. Since α is homotopic to α' relative to $\{0, 1\}$ and β is homotopic to β' relative to $\{0, 1\}$, there exist two homotopies $K, G \colon [0, 1] \times [0, 1] \to X$ such that $G((s, 0)) = \alpha(s)$, $G((s, 1)) = \alpha'(s)$, $G((0, t)) = G((1, t)) = \alpha(0) = \alpha'(0)$, $K((s, 0)) = \beta(s)$, $K((s, 1)) = \beta'(s)$, and $K((0, t)) = K((1, t)) = \beta(0) = \beta'(0)$. Let $L \colon [0, 1] \times [0, 1] \to X$ be given by

$$
L((s, t)) = \begin{cases} G((2s, t)) & \text{if } s \in \left[0, \dfrac{1}{2}\right]; \\[2ex] K((2s - 1, t)) & \text{if } s \in \left[\dfrac{1}{2}, 1\right]. \end{cases}
$$

Then L is the required homotopy between $\alpha * \beta$ and $\alpha' * \beta'$, relative to $\{0, 1\}$.

Next, let $R \colon [0, 1] \times [0, 1] \to X$ be given by $R((s, t)) = G((1 - s, t))$. Then R is the required homotopy between α^{-1} and $(\alpha')^{-1}$, relative to $\{0, 1\}$.

Q.E.D.

We are ready to define the fundamental group of a metric space.

1.3.20. Definition. Let (X, x_0) be a pointed space. The *fundamental group of* (X, x_0), denoted by $\pi_1(X, x_0)$, is the family of all homotopy classes of closed paths whose base is x_0. The group operation is given by $[\alpha] * [\beta] = [\alpha * \beta]$.

1.3.21. Remark. If (X, x_0) is a pointed space, then, by Theorem 1.3.19, the operation defined on $\pi_1(X, x_0)$ is well defined. Clearly, the identity element of $\pi_1(X, x_0)$ is the homotopy class of the constant path "x_0."

1.3.22. Notation. Let (X, x_0) and (Y, y_0) be two pointed spaces. By a *map, $f \colon (X, x_0) \to (Y, y_0)$, between the pointed spaces (X, x_0) and (Y, y_0)*, we mean a map $f \colon X \to Y$ such that $f(x_0) = y_0$.

1.3.23. Lemma. *Let $f\colon (X, x_0) \to (Y, y_0)$ be a map between pointed spaces. If α and β are two closed paths whose base is x_0 which are homotopic relative to $\{0, 1\}$, then $f \circ \alpha$ and $f \circ \beta$ are two closed paths whose base is y_0 which are homotopic relative to $\{0, 1\}$.*

Proof. Clearly, $f \circ \alpha$ and $f \circ \beta$ are two closed paths whose base is y_0.

Since α and β are homotopic relative to $\{0, 1\}$, there exists a homotopy $G\colon [0, 1] \times [0, 1] \to X$ such that $G((s, 0)) = \alpha(s)$, $G((s, 1)) = \beta(s)$ and $G((0, t)) = G((1, t)) = x_0$ for each $s, t \in [0, 1]$. Then the map $K\colon [0, 1] \times [0, 1] \to Y$ given by $K((s, t)) = f(G((s, t)))$ is a homotopy between $f \circ \alpha$ and $f \circ \beta$ relative to $\{0, 1\}$.

$\hfill$ **Q.E.D.**

1.3.24. Definition. If $f\colon (X, x_0) \to (Y, y_0)$ is a map between pointed spaces, then f *induces a homomorphism* $\pi_1(f)\colon \pi_1(X, x_0) \to \pi_1(Y, y_0)$ given by $\pi_1(f)([\alpha]) = [f \circ \alpha]$.

1.3.25. Remark. Let $f\colon (X, x_0) \to (Y, y_0)$ be a map between pointed spaces. If α and β are two closed paths whose base is x_0, then, clearly, $f \circ (\alpha * \beta) = (f \circ \alpha) * (f \circ \beta)$. Hence, the induced map defined in Definition 1.3.24 is a well defined group homomorphism.

The next Lemma says that the induced map of a composition is the composition of the induced maps.

1.3.26. Lemma. *If $f\colon (X, x_0) \to (Y, y_0)$ and $g\colon (Y, y_0) \to (Z, z_0)$ are maps between pointed spaces, then $\pi_1(g \circ f) = \pi_1(g) \circ \pi_1(f)$.*

Proof. Let $[\alpha] \in \pi_1(X, x_0)$. Then $\pi_1(g \circ f)([\alpha]) = [(g \circ f) \circ \alpha] = [g \circ (f \circ \alpha)] = \pi_1(g)([f \circ \alpha]) = \pi_1(g)(\pi_1(f)([\alpha])) = (\pi_1(g) \circ \pi_1(f))([\alpha])$.

$\hfill$ **Q.E.D.**

In order to show that the fundamental group of the unit circle $\mathcal{S}^1$ is isomorphic to $\mathbb{Z}$, we associate to each closed path α in $\mathcal{S}^1$ a number $\eta(\alpha)$, which is called the *degree of α*, in such a way that two closed paths are homotopic if and only if they have the same degree. We use the exponential map $\exp\colon \mathbb{R} \twoheadrightarrow \mathcal{S}^1$ given by $\exp(t) = e^{it}$.

1.3.27. Remark. Recall that given $z \in \mathcal{S}^1$, $\exp^{-1}(z) = \{t + 2\pi n \mid n \in \mathbb{Z}\}$, where t is any real number such that $\exp(t) = z$.

1.3.28. Notation. Let A be a nonempty subset of $\mathbb{R}$ and let $t \in \mathbb{R}$. Then $A + t = \{a + t \mid a \in A\}$.

1.3.29. Lemma. *The exponential map is open.*

Proof. Let U be an open subset of $\mathbb{R}$, and let $F = \mathcal{S}^1 \setminus \exp(U)$. We show that F is closed in $\mathcal{S}^1$.

Note that $\exp^{-1}(\exp(U)) = \bigcup\{U + 2\pi n \mid n \in \mathbb{Z}\}$, which is an open subset of $\mathbb{R}$. Hence, its complement, $\exp^{-1}(F)$, is closed in $\mathbb{R}$. Since for each $t \in \exp^{-1}(F)$, there exists $t' \in [0, 2\pi]$ such that $\exp(t') = \exp(t)$, $F = \exp(\exp^{-1}(F)) = \exp(\exp^{-1}(F) \cap [0, 2\pi])$. Since $\exp^{-1}(F) \cap [0, 2\pi]$ is compact, F is compact. Thus, F is closed in $\mathcal{S}^1$.

Q.E.D.

1.3.30. Corollary. *If $t \in \mathbb{R}$, then the restriction,*

$$\exp|_{(t, t+2\pi)} \colon (t, t + 2\pi) \twoheadrightarrow \mathcal{S}^1 \setminus \{\exp(t)\},$$

of the exponential map to $(t, t + 2\pi)$ is a homeomorphism onto $\mathcal{S}^1 \setminus \{\exp(t)\}$.

1.3.31. Theorem. *Let $\alpha \colon [0, 1] \to \mathcal{S}^1$ be a closed path whose base is $(1, 0)$. Then there exists a unique map $\alpha^\star \colon [0, 1] \to \mathbb{R}$ such that $\alpha^\star(0) = 0$ and $\alpha(t) = \exp(\alpha^\star(t))$. The map $\alpha^\star$ is called the lifting of α beginning at 0.*

Proof. First, suppose that $\alpha([0, 1]) \neq \mathcal{S}^1$. Let A be the component of $\exp^{-1}(\alpha([0, 1]))$ containing 0. Then $\exp|_A \colon A \twoheadrightarrow \alpha([0, 1])$ is a homeomorphism (Corollary 1.3.30). Hence, the map $\alpha^\star \colon [0, 1] \to \mathbb{R}$ given by $\alpha^\star(t) = (\exp|_A)^{-1}(\alpha(t))$ is the required map.

Next, suppose $\alpha([0, 1]) = \mathcal{S}^1$. Let $t_0 = 0 < t_1 < \cdots < t_{n-1} < t_n = 1$ be a subdivision of $[0, 1]$ such that $\alpha([t_{j-1}, t_j]) \neq \mathcal{S}^1$ for each $j \in \{1, \ldots, n\}$.

Let A_0 be the component of $\exp^{-1}(\alpha([0, t_1]))$ containing 0. Then, as before, $\exp|_{A_0}\colon A_0 \twoheadrightarrow \alpha([0, t_1])$ is a homeomorphism. Let $\alpha_0^\star\colon [0, t_1] \to \mathbb{R}$ be given by $\alpha_0^\star(t) = (\exp|_{A_0})^{-1}(\alpha(t))$. Let A_1 be the component of $\exp^{-1}(\alpha([t_1, t_2]))$ containing $\alpha_0^\star(t_1)$, and let $\alpha_1^\star\colon [t_1, t_2] \to \mathbb{R}$ be given by $\alpha_1^\star(t) = (\exp|_{A_1})^{-1}(\alpha(t))$. Repeating this process, for each $j \in \{0, \dots, n-1\}$, we define maps $\alpha_j^\star\colon [t_j, t_{j+1}] \to \mathbb{R}$ given by $\alpha_j^\star(t) = (\exp|_{A_j})^{-1}(\alpha(t))$.

Let $\alpha^\star\colon [0, 1] \to \mathbb{R}$ be given by $\alpha^\star(t) = \alpha_j^\star(t)$ if $t \in [t_j, t_{j+1}]$. Then $\alpha^\star(0) = 0$ and $\alpha(t) = \exp(\alpha^\star(t))$.

To see $\alpha^\star$ is unique, suppose $\beta\colon [0, 1] \to \mathbb{R}$ is another map such that $\beta(0) = 0$ and $\alpha(t) = \exp(\beta(t))$. Hence, $\exp(\alpha^\star(t)) = \exp(\beta(t))$ for each $t \in [0, 1]$. Consider the map $\gamma\colon [0, 1] \to \mathbb{R}$ given by $\gamma(t) = \dfrac{\alpha^\star(t) - \beta(t)}{2\pi}$. Then $\gamma([0, 1]) \subset \mathbb{Z}$. Since $\alpha^\star(0) = \beta(0)$, $\gamma([0, 1]) = \{0\}$. Therefore, $\alpha^\star(t) = \beta(t)$ for each $t \in [0, 1]$.

$$\textbf{Q.E.D.}$$

1.3.32. Remark. If in Theorem 1.3.31 we do not require that the map $\alpha^\star$ satisfies that $\alpha^\star(0) = 0$, we may have many liftings of the map α. However, any other lifting α' of α satisfies that $\alpha'(t) = \alpha^\star(t) + 2\pi k$ for some $k \in \mathbb{Z}$ and each $t \in [0, 1]$.

1.3.33. Definition. Let $\alpha\colon [0, 1] \to \mathcal{S}^1$ be a closed path, and let α' be a lifting of α. Then

$$\eta(\alpha) = \frac{\alpha'(1) - \alpha'(0)}{2\pi}$$

is an integer, and it is called the *degree of* α.

1.3.34. Remark. Observe that for each closed path α, the definition of $\eta(\alpha)$ does not depend on the lifting of α. Since for any two liftings α' and α'' of α, we have that $\alpha'(1) - \alpha'(0) = \alpha''(1) - \alpha''(0)$ by Remark 1.3.32. Intuitively, the degree of a closed path "counts" the number of times that the closed path wraps $[0, 1]$ around $\mathcal{S}^1$.

1.3.35. Theorem. *Let $\beta, \alpha \colon [0,1] \to \mathcal{S}^1$ be two closed paths whose base is $(1,0)$. Then:*

(1) $\eta(\alpha * \beta) = \eta(\alpha) + \eta(\beta)$;

(2) If α and β are homotopic relative to $\{0,1\}$ if and only if $\eta(\alpha) = \eta(\beta)$;

(3) Given $k \in \mathbb{Z}$, there exists a closed path γ, whose base is $(1,0)$, such that $\eta(\gamma) = k$.

Proof. Let $\alpha^\star \colon [0,1] \to \mathbb{R}$ be a lifting of α such that $\alpha^\star(0) = 0$ (Theorem 1.3.31), and let $\beta^\star \colon [0,1] \to \mathbb{R}$ be a lifting of β such that $\beta^\star(0) = \alpha^\star(1)$. Define $\alpha^\star * \beta^\star \colon [0,1] \to \mathbb{R}$ by

$$(\alpha^\star * \beta^\star)(s) = \begin{cases} \alpha^\star(2s) & \text{if } s \in \left[0, \dfrac{1}{2}\right]; \\[2ex] \beta^\star(2s-1) & \text{if } s \in \left[\dfrac{1}{2}, 1\right]. \end{cases}$$

Then it is easy to see that $\alpha^\star * \beta^\star$ is a lifting of $\alpha * \beta$. Since $2\pi\eta(\alpha * \beta) = (\alpha^\star * \beta^\star)(1) - (\alpha^\star * \beta^\star)(0) = \beta^\star(1) - \alpha^\star(0) = (\beta^\star(1) - \beta^\star(0)) + (\alpha^\star(1) - \alpha^\star(0)) = 2\pi(\eta(\beta) + \eta(\alpha))$, we have that $\eta(\alpha * \beta) = \eta(\alpha) + \eta(\beta)$.

Suppose $\eta(\alpha) = \eta(\beta)$. Let α' and β' be liftings of α and β, respectively, such that $\alpha'(0) = \beta'(0) = 0$. Since $\eta(\alpha) = \eta(\beta)$, $\alpha'(1) - \alpha'(0) = \beta'(1) - \beta'(0)$. In particular, $\alpha'(1) = \beta'(1)$. By Example 1.3.4, the map $G \colon [0,1] \times [0,1] \to \mathbb{R}$ given by $G((s,t)) = (1-t)\alpha'(s) + t\beta'(s)$ is a homotopy between α' and β'. Note that for each $t \in [0,1]$, $G((1,t)) - G((0,t)) = (1-t)\left[\alpha'(1) - \alpha'(0)\right] + t\left[\beta'(1) - \beta'(0)\right] = (1-t)2\pi\eta(\alpha) + t2\pi\eta(\beta) = 2\pi\eta(\alpha)$. Hence, the map $K = \exp \circ G$ is a homotopy between α and β such that $K((0,t)) = K((1,t)) = (1,0)$. Thus, K is a homotopy between α and β relative to $\{0,1\}$.

Now, assume α and β are homotopic relative to $\{0,1\}$. First, suppose that $\|\alpha(s) - \beta(s)\| < 2$ for each $s \in [0,1]$, i.e., $\alpha(s)$ and $\beta(s)$ are never antipodal points. Let α'' and β'' be liftings of α and β, respectively, such that $\alpha''(0) = \beta''(0) = 0$. Since $\|\alpha(s) - \beta(s)\| < 2$, $|\alpha''(s) - \beta''(s)| < \pi$ for each $s \in [0,1]$. Hence, $2\pi|\eta(\alpha) - \eta(\beta)| = |\alpha''(1) - \alpha''(0) - \beta''(1) + \beta''(0)| \leq |\alpha''(1) - \beta''(1)| + |\alpha''(0) - \beta''(0)| < \pi + \pi = 2\pi$. Thus, $|\eta(\alpha) - \eta(\beta)| = 0$, i.e., $\eta(\alpha) = \eta(\beta)$. Next, assume there is an $s \in [0,1]$ such that $\|\alpha(s) - \beta(s)\| = 2$. Let

$H\colon [0,1] \times [0,1] \to \mathcal{S}^1$ be a homotopy between α and β relative to $\{0,1\}$. Since H is uniformly continuous, there exists $\delta > 0$ such that if $|t - t'| < \delta$, then $\|H((s,t)) - H((s,t'))\| < 2$ for each $s \in [0,1]$. Let $t_0 = 0 < t_1 < \cdots < t_{k-1} < t_k = 1$ be a subdivision of $[0,1]$ such that $t_j - t_{j-1} < \delta$ for each $j \in \{1, \dots, k\}$. Let $\alpha_j\colon [0,1] \to \mathcal{S}^1$ be given by $\alpha_j(s) = H((s,t_j))$ for each $j \in \{0, \dots, k\}$. Note that $\alpha_0 = \alpha$ and $\alpha_k = \beta$. By construction, $\|\alpha_{j-1}(s) - \alpha_j(s)\| < 2$ for each $j \in \{1, \dots, k\}$. Then, applying the above argument, we obtain that $\eta(\alpha) = \eta(\alpha_0) = \eta(\alpha_1) = \dots = \eta(\alpha_k) = \eta(\beta)$.

Finally, let $k \in \mathbb{Z}$. Define $\gamma\colon [0,1] \to \mathcal{S}^1$ by $\gamma(s) = \exp(2\pi k s)$. Then γ is a closed path whose base is $(1,0)$. Note that the map $\gamma^\star\colon [0,1] \to \mathbb{R}$ given by $\gamma^\star(s) = 2\pi k s$ is a lifting of γ such that $\gamma^\star(0) = 0$. Hence, $\eta(\gamma) = \dfrac{\gamma^\star(1) - \gamma^\star(0)}{2\pi} = k$.

Q.E.D.

Now, we are ready to show that the fundamental group of the unit circle $\mathcal{S}^1$ is isomorphic to $\mathbb{Z}$.

1.3.36. Theorem. *The fundamental group of the unit circle $\mathcal{S}^1$ is isomorphic to $\mathbb{Z}$.*

Proof. Let $\Sigma\colon \pi_1(\mathcal{S}^1) \twoheadrightarrow \mathbb{Z}$ be given by $\Sigma([\alpha]) = \eta(\alpha)$. Note that, by Theorem 1.3.35 (2), Σ is well defined and by (3) of the same Theorem, Σ is, indeed, a surjection. By Theorem 1.3.35 (1), Σ is a homomorphism. Finally, by Theorem 1.3.35 (2), Σ is one–to–one. Therefore, Σ is an isomorphism.

Q.E.D.

Next, we define the degree of a map between simple closed curves.

1.3.37. Definition. Let $f\colon \mathcal{S}^1 \to \mathcal{S}^1$ be a map. The *degree of* f, denoted by $\deg(f)$, is defined as follows. Consider the induced map $\pi_1(f)\colon \pi_1(\mathcal{S}^1) \to \pi_1(\mathcal{S}^1)$. Then $\deg(f) = (\Sigma \circ \pi_1(f))\,(\Sigma^{-1}(1))$, where Σ is defined in Theorem 1.3.36.

1.3.38. Remark. Let $f\colon \mathcal{S}^1 \to \mathcal{S}^1$ be a map. If $\deg(f) = 0$, then f is homotopic to a constant map (see Theorem 7.4 (p. 352) of [9]).

1.3.39. Lemma. *If $g, f \colon \mathcal{S}^1 \to \mathcal{S}^1$ are two maps, then $\deg(g \circ f) = \deg(g) \cdot \deg(f)$.*

Proof. Recall that, by Lemma 1.3.26, $\pi_1(g \circ f) = \pi_1(g) \circ \pi_1(f)$.
By definition,

$$\deg(g \circ f) = \left[\Sigma \circ \pi_1(g \circ f)\right]\left(\Sigma^{-1}(1)\right) =$$

$$\left[\Sigma \circ (\pi_1(g) \circ \pi_1(f))\right]\left(\Sigma^{-1}(1)\right) = \left[\Sigma \circ \pi_1(g)\right]\left(\pi_1(f)(\Sigma^{-1}(1))\right) =$$

$$\left[\Sigma \circ \pi_1(g)\right]\left(\pi_1(f)(\Sigma^{-1}(1)) * \Sigma^{-1}(1)\right) =$$

$$\left[\Sigma \circ \pi_1(f)(\Sigma^{-1}(1))\right] * \left[\Sigma \circ \pi_1(g)(\Sigma^{-1}(1))\right] = \deg(f) \cdot \deg(g).$$

Q.E.D.

We end this section with the following Theorem.

1.3.40. Theorem. *A map $f \colon \mathcal{S}^1 \to \mathcal{S}^1$ is homotopic to a constant map if and only if there exists a map $\tilde{f} \colon \mathcal{S}^1 \to \mathbb{R}$ such that $f = \exp \circ \tilde{f}$.*

Proof. If $\tilde{f} \colon \mathcal{S}^1 \to \mathbb{R}$ exists such that $f = \exp \circ \tilde{f}$, then $\tilde{f}$ is homotopic to a constant map since $\mathbb{R}$ is contractible (Theorem 1.3.12). Hence, $f = \exp \circ \tilde{f}$ is also homotopic to a constant map.

Now, suppose f is homotopic to a constant map. Without loss of generality, we assume that $f((1, 0)) = (1, 0)$.

Since f is homotopic to a constant map, $\deg(f) = 0$. Let $g \colon [0, 1] \twoheadrightarrow \mathcal{S}^1$ be given by $g(t) = \exp(2\pi t)$. Since f is homotopic to a constant map, $f \circ g$ is homotopic to a constant map. By Theorem 1.3.35 (2), $\eta(f \circ g) = 0$. Let $\xi \colon [0, 1] \to \mathbb{R}$ be a map such that $\exp \circ \xi = f \circ g$, i.e., ξ is a lifting of $f \circ g$ (Theorem 1.3.31). Since $\eta(f \circ g) = 0$, $\xi(0) = \xi(1)$. Hence, the function $\tilde{f} \colon \mathcal{S}^1 \to \mathbb{R}$ given by $\tilde{f}(z) = \xi(g^{-1}(z))$ is well defined, continuous, and satisfies that $(\exp \circ \tilde{f})(z) = (\exp \circ \xi)(g^{-1}(z)) = (f \circ g)(g^{-1}(z)) = f(z)$ for each $z \in \mathcal{S}^1$.

Q.E.D.

1.4 Geometric Complexes and Polyhedra

This is a very small section. We present the definitions of polyhedra and the nerve of a finite collection of sets, which are used in Chapter 2.

1.4.1. Definition. We say that $r + 1$ points, $\{x_0, \dots, x_r\}$, of $\mathbb{R}^n$ are *affinely independent* provided that the set $\{x_1 - x_0, \dots, x_r - x_0\}$ is linearly independent.

The following Theorem gives us an alternative way, in terms of linear algebra, to see affinely independent subsets of $\mathbb{R}^n$.

1.4.2. Theorem. *The set of points $\{x_0, \dots, x_r\}$ of $\mathbb{R}^n$ is affinely independent if and only if each time*

$$\sum_{j=0}^{r} \xi_j x_j = 0 \text{ and } \sum_{j=0}^{r} \xi_j = 0,$$

where $\xi_j \in \mathbb{R}^n$, we have that $\xi_j = 0$ for each $j \in \{0, \dots, r\}$.

Proof. Suppose $\{x_0, \dots, x_r\}$ is affinely independent. Let $\xi_0, \dots, \xi_r$ be real numbers such that $\sum_{j=0}^{r} \xi_j x_j = 0$ and $\sum_{j=0}^{r} \xi_j = 0$.

Since $\sum_{j=0}^{r} \xi_j = 0$, $\xi_0 = -\sum_{j=1}^{r} \xi_j$. Then: $0 = \sum_{j=0}^{r} \xi_j x_j = \xi_0 x_0 + \sum_{j=1}^{r} \xi_j x_j = \left(-\sum_{j=1}^{r} \xi_j\right) x_0 + \sum_{j=1}^{r} \xi_j x_j = \sum_{j=1}^{r} \xi_j (x_j - x_0)$. Since $\{x_1 - x_0, \dots, x_r - x_0\}$ is linearly independent, $\xi_1 = \dots = \xi_r = 0$. Since $\xi_0 = -\sum_{j=1}^{r} \xi_j$, $\xi_0 = 0$.

Next, suppose $\{x_0, \dots, x_r\}$ satisfies the hypothesis of the Theorem. Let $\xi_1, \dots, \xi_r$ be real numbers such that $\sum_{j=1}^{r} \xi_j (x_j - x_0) = 0$.

Note that $\displaystyle\sum_{j=1}^{r}\xi_j(x_j - x_0) = \sum_{j=1}^{r}\xi_j x_j - \left(\sum_{j=1}^{r}\xi_j\right)x_0.$ Let $\xi_0 =$
$-\left(\displaystyle\sum_{j=1}^{r}\xi_j\right)$. Then $\displaystyle\sum_{j=1}^{r}\xi_j x_j-\left(\sum_{j=1}^{r}\xi_j\right)x_0 = \sum_{j=0}^{r}\xi_j x_j$ and $\displaystyle\sum_{j=0}^{r}\xi_j = 0.$
Then, by hypothesis, $\xi_0 = \ldots = \xi_r = 0$. In particular, $\xi_1 = \ldots = \xi_r = 0$. Thus, $\{x_1 - x_0, \ldots, x_r - x_0\}$ is linearly independent. Therefore, $\{x_0, \ldots, x_r\}$ is affinely independent.

Q.E.D.

1.4.3. Definition. Let $\{x_0, \ldots, x_r\}$ be an affinely independent subset of $\mathbb{R}^n$. We define the *(geometric) r–simplex generated by* $\{x_0, \ldots, x_r\}$, denoted by $[x_0, \ldots, x_r]$, as the following subset of $\mathbb{R}^n$:

$$[x_0, \ldots, x_r] = \left\{ \sum_{j=0}^{r}\xi_j x_j \ \middle|\ \text{for each } j \in \{0, \ldots, r\},\right.$$

$$\left.\xi_j \in [0,1] \text{ and } \sum_{j=0}^{r}\xi_j = 1\right\}.$$

The set $\{x_0, \ldots, x_r\}$ is called the *set of vertexes* of the r–simplex. The point $x = \displaystyle\sum_{j=0}^{r}\frac{1}{r+1}x_j$ is called the *barycenter* of the r–simplex. Each s–simplex generated by $s+1$ points taken from $\{x_0, \ldots, x_r\}$ is called an *s–face* of $[x_0, \ldots, x_r]$.

1.4.4. Remark. It is easy to see that a 0–simplex is a point, a 1–simplex is a line segment, a 2–simplex is a triangle and a 3–simplex is a tetrahedron.

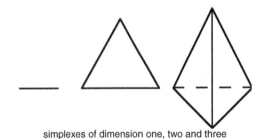

simplexes of dimension one, two and three

1.4.5. Definition. A *(geometric) complex*, $\mathcal{K}$, is a finite collection of simplexes in $\mathbb{R}^n$ such that:

(1) each face of a simplex in $\mathcal{K}$ also belongs to $\mathcal{K}$ and

(2) the intersection of any two simplexes is either empty or it is a face of both simplexes.

not a complex   complex

1.4.6. Definition. A *polyhedron* in $\mathbb{R}^n$ is the union of the simplexes of a geometric complex.

1.4.7. Definition. The *barycentric subdivision of an r–simplex $\mathcal{R}$* is a (geometric) complex $\mathcal{K}$ obtained as follows:

(a) If $r = 0$, then $\mathcal{K}$ is just $\mathcal{R}$.

(b) Suppose $r > 0$. If $\mathcal{S}_0, \ldots, \mathcal{S}_r$ are the $(r-1)$–faces of $\mathcal{R}$ and if x_b is the barycenter of $\mathcal{R}$, then $\mathcal{K}$ consists of all r–simplexes generated by x_b and the vertexes of all $(r-1)$–simplexes of the barycentric subdivision of $\mathcal{S}_j$ for each $j \in \{0, \ldots, r\}$. The *barycentric subdivision of a complex $\mathcal{K}_0$* is a (geometric) complex $\mathcal{K}_1$ consisting of all the simplexes obtained by the barycentric subdivision of each simplex of $\mathcal{K}_0$.

1.4.8. Remark. It is well known that a complex $\mathcal{K}_1$ obtained from a complex $\mathcal{K}_0$ by barycentric subdivision is, in fact, a complex (15.2 of [20]). Some times the barycentric subdivision is performed several times, i.e., one can apply the barycentric subdivision to $\mathcal{K}_1$ to obtain a complex $\mathcal{K}_2$, and then apply the barycentric subdivision to $\mathcal{K}_2$ to obtain a complex $\mathcal{K}_3$, etc. It is also possible to show that the diameters of the simplexes of the complex $\mathcal{K}_m$, obtained after m barycentric subdivisions, tend to zero as m tends to infinity (15.4 of [20]).

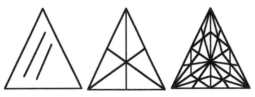
Barycentric subdivisions

1.4.9. Definition. Let X be a compactum and let $\mathcal{U} = \{U_1, \ldots, U_n\}$ be a finite collection of subsets of X. Then the *nerve* of $\mathcal{U}$, denoted by $\mathcal{N}(\mathcal{U})$, is the complex defined as follows. To each $U_j \in \mathcal{U}$ associate the point e_j of $\mathbb{R}^n$, where $e_j = (0, \ldots, 0, 1, 0 \ldots, 0)$ (the number "1" appears in the jth coordinate). Hence, the vertexes of the complex are $\{e_1, \ldots, e_n\}$. For each subfamily $\{U_{j_1}, \ldots, U_{j_k}\}$ of $\mathcal{U}$ such that $\bigcap_{\ell=1}^{k} U_{j_\ell} \neq \emptyset$, we consider the simplex whose vertexes are the points $\{e_{j_1}, \ldots, e_{j_k}\}$. Then $\mathcal{N}(\mathcal{U})$ consists of all such possible simplexes. We denote by $\mathcal{N}^\star(\mathcal{U})$ the polyhedron associated to $\mathcal{N}(\mathcal{U})$.

1.5 Complete Metric Spaces

We present results about complete metric spaces.

1.5.1. Definition. Let X be a metric space, with metric d. A sequence, $\{x_n\}_{n=1}^{\infty}$, of elements of X is said to be a *Cauchy sequence* provided that for each $\varepsilon > 0$, there exists $N \in \mathbb{N}$ such that if $n, m \geq N$, then $d(x_n, x_m) < \varepsilon$.

1.5.2. Definition. A metric space X, with metric d, is said to be *complete* provided that every Cauchy sequence of elements of X converges to a point of X. We say that X is *topologically complete* if there exists an equivalent metric d_1 for X such that (X, d_1) is a complete space.

The following results give some basic properties of complete metric spaces.

1.5.3. Lemma. *Let X be a complete metric space. If F is a closed subset of X, then F is complete.*

Proof. Let $\{x_n\}_{n=1}^{\infty}$ be a Cauchy sequence of elements of F. Since X is complete, there exists a point $x \in X$ such that $\{x_n\}_{n=1}^{\infty}$ converges to x. Since F is a closed subset of X, $x \in F$. Therefore, F is complete.

Q.E.D.

1.5.4. Proposition. *Let X be a metric space. If Y is a subset of X and Y is complete, then Y is closed in X.*

Proof. Let $x \in Cl(Y)$. Then there exists a sequence $\{y_n\}_{n=1}^{\infty}$ of elements of Y converging to x. Since a convergent sequence is a Cauchy sequence, and Y is complete, we have that $x \in Y$. Therefore, Y is closed in X.

Q.E.D.

1.5.5. Proposition. *Let X be a complete metric space, with metric d. If $\{F_n\}_{n=1}^{\infty}$ is a sequence of closed subsets of X such that $F_{n+1} \subset F_n$ and $\lim_{n\to\infty} \text{diam}(F_n) = 0$, then there exists a point $x \in X$ such that $\bigcap_{n=1}^{\infty} F_n = \{x\}$.*

Proof. For each $n \in \mathbb{N}$, let $x_n \in F_n$. Then $\{x_n\}_{n=1}^{\infty}$ is a sequence of points of X such that for each $N \in \mathbb{N}$, if $n, m \geq N$, then $x_n, x_m \in F_N$. Thus, since $\lim_{n\to\infty} \text{diam}(F_n) = 0$, $\{x_n\}_{n=1}^{\infty}$ is a Cauchy sequence. Hence, there exists $x \in X$ such that $\lim_{n\to\infty} x_n = x$. Note that, since for every $N \in \mathbb{N}$, $\{x_n\}_{n=N}^{\infty} \subset F_N$, $x \in \bigcap_{n=1}^{\infty} F_n$. Now, suppose there exists $y \in \bigcap_{n=1}^{\infty} F_n \setminus \{x\}$. Then for each $n \in \mathbb{N}$, $0 < d(x,y) \leq$

$\mathrm{diam}\left(\bigcap_{n=1}^{\infty} F_n\right) \leq \mathrm{diam}(F_n)$, a contradiction. Therefore, $\bigcap_{n=1}^{\infty} F_n = \{x\}$.

Q.E.D.

1.5.6. Definition. A subset of a metric space X is said to be a G_δ *set* provided that it is a countable intersection of open sets.

1.5.7. Lemma. *Let X be a metric space, with metric d. If F is a closed subset of X, then F is a G_δ subset of X.*

Proof. It follows from the fact that $F = \bigcap_{n=1}^{\infty} \mathcal{V}_{\frac{1}{n}}^{d}(F)$.

Q.E.D.

The following Theorem is due to Mazurkiewicz, a proof of which may be found in Theorem 8.3 (p. 308) of [9]:

1.5.8. Theorem. *Let Y be a complete metric space. Then a nonempty subset A of Y is topologically complete if and only if A is a G_δ subset of Y.*

1.5.9. Lemma. *If X is a complete metric space, with metric d, and $\{D_n\}_{n=1}^{\infty}$ is sequence of dense open subsets of X, then $\bigcap_{n=1}^{\infty} D_n$ is a dense subset of X.*

Proof. Let U be an open subset of X. We show that $U \cap \left(\bigcap_{n=1}^{\infty} D_n\right) \neq \emptyset$.

Since D_1 is a dense subset of X, $D_1 \cap U \neq \emptyset$. In fact, $D_1 \cap U$ is an open subset of X. Hence, there exist $x_1 \in D_1 \cap U$ and $\varepsilon_1 > 0$ such that $\varepsilon_1 < 1$ and $Cl(\mathcal{V}_{\varepsilon_1}^{d}(x_1)) \subset D_1 \cap U$. Since D_2 is a dense open subset of X and $\mathcal{V}_{\varepsilon_1}^{d}(x_1)$ is open, $D_2 \cap \mathcal{V}_{\varepsilon_1}^{d}(x_1) \neq \emptyset$. So, there exist $x_2 \in D_2 \cap \mathcal{V}_{\varepsilon_1}^{d}(x_1)$ and $\varepsilon_2 > 0$ such that $\varepsilon_2 < \dfrac{1}{2}$ and

$Cl(\mathcal{V}^d_{\varepsilon_2}(x_2)) \subset D_2 \cap \mathcal{V}^d_{\varepsilon_1}(x_1)$. If we continue with this process, we construct a decreasing family of closed subset of X, whose diameters tend to zero. Hence, by Proposition 1.5.5, there exists a point $x \in X$ such that $\{x\} = \bigcap\limits_{n=1}^{\infty} Cl(\mathcal{V}^d_{\varepsilon_n}(x_n)) \subset \left(\bigcap\limits_{n=1}^{\infty} D_n \right) \cap U.$

Since $\bigcap\limits_{n=1}^{\infty} Cl(\mathcal{V}^d_{\varepsilon_n}(x_n)) \subset \left(\bigcap\limits_{n=1}^{\infty} D_n \right) \cap U$, we obtain that $U \cap$
$\left(\bigcap\limits_{n=1}^{\infty} D_n \right) \neq \emptyset.$

Q.E.D.

1.5.10. Definition. Let X be a metric space. A subset A of X is said to be *nowhere dense* provided that $Int(Cl(A)) = \emptyset$.

1.5.11. Definition. Let X be a metric space. A subset A of X is said to be of the *first category* if it is the countable union of nowhere dense subsets of X. A subset of X that is not of the first category is said to be of the *second category*.

The following result is known as the Baire Category Theorem:

1.5.12. Theorem. *If X is a complete metric space, then X is of the second category.*

Proof. Suppose X is of the first category. Then there exists a sequence $\{A_n\}_{n=1}^{\infty}$ of nowhere dense subsets of X such that $X = \bigcup\limits_{n=1}^{\infty} A_n$. Since, for each $n \in \mathbb{N}$, A_n is nowhere dense, each set $X \setminus Cl(A_n)$ is an open dense subset of X. By Lemma 1.5.9, $\bigcap\limits_{n=1}^{\infty}(X \setminus Cl(A_n)) \neq \emptyset$.

Since $X = \bigcup\limits_{n=1}^{\infty} A_n$, we have that

$$\bigcap_{n=1}^{\infty}(X \setminus Cl(A_n)) = X \setminus \bigcup_{n=1}^{\infty} Cl(A_n) \subset X \setminus \bigcup_{n=1}^{\infty} A_n = X \setminus X = \emptyset,$$

a contradiction. Therefore, X is of the second category.

Q.E.D.

The following Theorem is due to Hausdorff, a proof of which may be found in the Appendix of [2]:

1.5.13. Theorem. *Let X and Y be metric spaces. If X is complete and $f\colon X \twoheadrightarrow Y$ is a surjective open map, then Y is topologically complete.*

1.6   Compacta

We give basic properties of compacta. We construct the Cantor set and present some of its properties.

A proof of the following Theorem may be found in Corolário 2 (p. 212) of [14].

1.6.1. Theorem. *If X is a compactum, then X is complete.*

1.6.2. Lemma. *Let X be a compactum, and let A be a closed subset of X with a finite number of components. If $x \in Int(A)$ and C is the component of A such that $x \in C$, then $x \in Int(C)$.*

Proof. Let $C, C_1, \ldots, C_n$ be the components of A. Since $x \in Int(A)$, there exists an open subset U of X such that $x \in U \subset A$. Since A is closed in X, each of $C, C_1, \ldots, C_n$ is closed in X. Let $V = U \cap \left(X \setminus \bigcup_{j=1}^{n} C_j \right)$. Then V is an open subset of X such that $x \in V \subset A$ and $V \cap \left(\bigcup_{j=1}^{n} C_j \right) = \emptyset$. Thus, $V \subset C$. Therefore, $x \in Int(C)$.

Q.E.D.

1.6.3. Definition. Let X be a metric space. A subset Y of X is said to be *perfect* if Y is closed and every point of Y is a limit point of Y.

1.6.4. Example. We construct the Cantor set and prove some of its properties. Let $C_0 = [0,1]$. Remove $\left(\dfrac{1}{3}, \dfrac{2}{3}\right)$, and let $C_1 = \left[0, \dfrac{1}{3}\right] \cup \left[\dfrac{2}{3}, 1\right]$. Remove the middle thirds of these intervals, and let $C_2 = \left[0, \dfrac{1}{9}\right] \cup \left[\dfrac{2}{9}, \dfrac{1}{3}\right] \cup \left[\dfrac{2}{3}, \dfrac{7}{9}\right] \cup \left[\dfrac{8}{9}, 1\right]$. Continuing in this way, we obtain a sequence, $\{C_n\}_{n=0}^{\infty}$, of compact sets such that for each $n \in \mathbb{N}$, $C_{n+1} \subset C_n$ and C_n is the union of 2^n intervals, $I_{n,0}, \ldots, I_{n,2^n-1}$, of length 3^{-n}. The set

$$\mathcal{C} = \bigcap_{n=0}^{\infty} C_n$$

is called the *Cantor set*. $\mathcal{C}$ is clearly compact and nonempty.

No interval of the form

$$(*) \qquad \left(\frac{3k+1}{3^m}, \frac{3k+2}{3^m}\right),$$

where $k, m \in \mathbb{N}$, has a point in common with $\mathcal{C}$. Since every interval (x,y) contains an interval of the form $(*)$, if $3^{-m} < \dfrac{y-x}{6}$, $\mathcal{C}$ does not contain a nondegenerate interval.

We show that $\mathcal{C}$ is perfect. Let $x \in \mathcal{C}$, and let A be any interval containing x. Let I_n be the interval in C_n that contains x. Let n large enough, so that $I_n \subset A$. Let x_n be an end point of I_n such that $x_n \neq x$. It follows from the construction that $x_n \in \mathcal{C}$. Hence, x is a limit point of $\mathcal{C}$. Therefore, $\mathcal{C}$ is perfect.

It follows, from the construction and the fact that $\mathcal{C}$ does not contain any nondegenerate interval, that $\mathcal{C}$ is totally disconnected. In Chapter 2, we present a characterization of the Cantor set as the only totally disconnected and perfect compactum.

1.6.5. Definition. Let X be a metric space, and let $\mathcal{U}$ be an open cover of X. We say that a number $\lambda > 0$ is a *Lebesgue number for the open cover* $\mathcal{U}$ provided that if A is a nonempty subset of X such that $\operatorname{diam}(A) < \lambda$, then there exists $U \in \mathcal{U}$ such that $A \subset U$.

1.6.6. Theorem. *If X is a compactum, with metric d, and $\mathcal{U}$ is an open cover of X, then there exists a Lebesgue number for $\mathcal{U}$.*

Proof. Suppose that no such number exists. Then for each $n \in \mathbb{N}$, there exists a nonempty subset A_n of X such that $\operatorname{diam}(A_n) < \dfrac{1}{n}$ and A_n is not contained in any element of $\mathcal{U}$. Let $x_n \in A_n$. Since X is compact, without loss of generality, we assume that the sequence $\{x_n\}_{n=1}^{\infty}$ converges to a point $x \in X$.

Since $\mathcal{U}$ covers X, there exists $U \in \mathcal{U}$ such that $x \in U$. Hence, there exists $\varepsilon > 0$ such that $\mathcal{V}_\varepsilon^d(x) \subset U$ (U is open in X). Let $n \in \mathbb{N}$ such that $\dfrac{1}{n} < \dfrac{\varepsilon}{2}$ and $d(x_n, x) < \dfrac{\varepsilon}{2}$. Then for each $y \in A_n$,

$$d(y, x) \leq d(y, x_n) + d(x_n, x) < \frac{1}{n} + \frac{\varepsilon}{2} < \varepsilon.$$

Hence, $A_n \subset \mathcal{V}_\varepsilon^d(x) \subset U$, a contradiction. Therefore, there exists a Lebesgue number for $\mathcal{U}$.

$$\textbf{Q.E.D.}$$

The following Lemma is very useful in continuum theory.

1.6.7. Lemma. *Let Z be a compactum and let $\{X_n\}_{n=1}^{\infty}$ be a sequence of closed subsets of Z such that $X_{n+1} \subset X_n$ for each $n \in \mathbb{N}$. If U is an open subset of Z such that $\bigcap\limits_{n=1}^{\infty} X_n \subset U$, then there exists $N \in \mathbb{N}$ such that $X_n \subset U$ for each $n \geq N$.*

Proof. Since U is an open subset of Z, $Z \setminus U$ is closed in Z. Note that, $Z \setminus U \subset Z \setminus \bigcap\limits_{n=1}^{\infty} X_n = \bigcup\limits_{n=1}^{\infty}(Z \setminus X_n)$. Hence $\{Z \setminus X_n\}_{n=1}^{\infty}$ is an open cover of $Z \setminus U$. Since $Z \setminus U$ is compact, there exist

$n_1, \ldots, n_k \in \mathbb{N}$ such that $Z \setminus U \subset \bigcup_{j=1}^{k} (Z \setminus X_{n_j})$. Thus, $\bigcap_{j=1}^{k} X_{n_j} \subset U$.

Let $N = \max\{n_1, \ldots, n_k\}$. Then $X_N = \bigcap_{j=1}^{k} X_{n_j}$, and $X_N \subset U$.

Therefore, $X_n \subset U$ for each $n \geq N$.

Q.E.D.

The following result is known as *The Cut Wire Theorem*; it is very useful in continuum theory. A proof of this result may be found in 5.2 of [23].

1.6.8. Theorem. *Let X be a compactum and let A and B be closed subsets of X. If no connected subset of X intersects both A and B, then there exist two disjoint closed subsets X_1 and X_2 of X such that $A \subset X_1$, $B \subset X_2$ and $X = X_1 \cup X_2$.*

1.7 Continua

We define the type of spaces we are more interested in, namely, continua, and give some of its main properties.

1.7.1. Definition. *A continuum* is a connected compactum. *A subcontinuum* is a continuum contained in some metric space.

The following Theorem provides a method to construct continua.

1.7.2. Theorem. *Let Z be a compactum and let $\{X_n\}_{n=1}^{\infty}$ be a sequence of subcontinua of Z such that $X_{n+1} \subset X_n$ for each $n \in \mathbb{N}$. If $X = \bigcap_{n=1}^{\infty} X_n$, then X is a subcontinuum of Z.*

Proof. Clearly, X is a closed, hence compact, subset of Z. Suppose X is not connected. Then there exist two disjoint closed subsets A and B of Z such that $X = A \cup B$. Since Z is a metric space, there exist two disjoint open subsets U and V of Z such that $A \subset U$ and $B \subset V$. Hence, $X \subset U \cup V$. By Lemma 1.6.7, there exists $N \in \mathbb{N}$, such that $X_N \subset U \cup V$. Since X_N is connected, either $X_N \subset U$ or $X_N \subset V$. Assume, without loss of generality, that $X_N \subset U$. Since $X \subset X_N \subset U$ and $X = A \cup B$, we have that $B \subset U$, a contradiction. Therefore, X is connected. Hence, X is a subcontinuum of Z.

$$\textbf{Q.E.D.}$$

1.7.3. Theorem. *If X is a continuum and $\mathcal{G}$ is an upper semicontinuous decomposition of X, then $X/\mathcal{G}$ is a continuum.*

Proof. By Corollary 1.2.22, $X/\mathcal{G}$ is a metric space. Since continuous images of compact and connected spaces are compact and connected, X is a continuum.

$$\textbf{Q.E.D.}$$

The following definition is due to Davis and Doyle and they used it in their study of invertible continua [8]:

1.7.4. Definition. Let X be a continuum, and let $x \in X$. We say that X *is almost connected im kleinen at x* provided that for each open subset U of X containing x, there exists a subcontinuum W of X such that $Int(W) \neq \emptyset$ and $W \subset U$. We say that X is *almost connected im kleinen* if it is connected im kleinen at each of its points.

1.7.5. Example. Let X be the cone over the closure of the harmonic sequence $\{0\} \cup \left\{ \dfrac{1}{n} \right\}_{n=1}^{\infty}$. Then X is called *harmonic fan*. It is easy to see that X is almost connected im kleinen at each of its points.

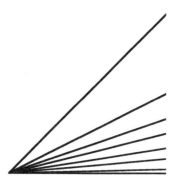

Almost connected im kleinen

1.7.6. Definition. Let X be a continuum, and let $x \in X$. We say that X *is connected im kleinen at* x if for each closed subset F of X such that $F \subset X \setminus \{x\}$, there exists a subcontinuum W of X such that $x \in Int(W) \subset W \subset X \setminus F$.

1.7.7. Example. Let X be a sequence of harmonic fans converging to a point p; see picture below. Then X is connected im kleinen at p but X is not locally connected at that point.

Connected im kleinen at p

1.7.8. Remark. Note that we do not define connectedness im kleinen globally because, by Theorem 1.7.12, a continuum is connected im kleinen globally if and only if it is locally connected (Definition 1.7.10).

The way we defined connectedness im kleinen in Definition 1.7.6 is not the usual one. This definition provides the author the answer to the fact that aposyndesis (Definition 1.7.15) is a generalization of connectedness im kleinen. The following Theorem presents the usual definition of this concept:

1.7.9. Theorem. *If X is a continuum and $x \in X$, then the following are equivalent:*

(a) X is connected im kleinen at x.

(b) For each open subset U of X such that $x \in U$, there exists an open subset V of X such that $x \in V \subset U$ and with the property that for each $y \in V$, there exists a connected subset C_y of X such that $\{x, y\} \subset C_y \subset U$.

(c) For each open subset U of X such that $x \in U$, there exists a subcontinuum W of X such that $x \in Int(W) \subset W \subset U$.

Proof.   Suppose X is connected im kleinen at x. We show (b). Let U be an open subset of X such that $x \in U$. Then $X \setminus U$ is a closed subset of X not containing x. By hypothesis, there exists a subcontinuum W of X such that $x \in Int(W) \subset W \subset X \setminus (X \setminus U) = U$. Thus, $V = Int(W)$ satisfies the required properties.

Next, suppose (b). We show (c). Let U be an open subset of X such that $x \in U$. Let U' be an open subset of X such that $x \in U' \subset Cl(U') \subset U$. By hypothesis, there exists an open subset V of X such that $x \in V \subset U'$ with the property that for each $y \in V$, there exists a connected subset C_y of X such that $\{x, y\} \subset C_y \subset U'$. Let $W = Cl\left(\bigcup_{y \in V} C_y\right)$. Then W is a subcontinuum of X such that $x \in V \subset W \subset Cl(U') \subset U$.

Finally, suppose (c). We show X is connected im kleinen at x. Let F be a closed subset of X such that $F \subset X \setminus \{x\}$. Then $X \setminus F$ is an open subset of X containing x. By hypothesis, there exists a subcontinuum W of X such that $x \in Int(W) \subset W \subset X \setminus F$. Therefore, X is connected im kleinen at x.

Q.E.D.

1.7.10. Definition. Let X be a continuum, and let $x \in X$. We say that *X is locally connected at x* provided that for each open subset U of X such that $x \in U$, there exists a connected open subset V of X such that $x \in V \subset U$. We say *X is locally connected* if it is locally connected at each of its points.

1.7.11. Lemma. *A continuum X is locally connected if and only if the components of the open subsets of X are open.*

Proof. Suppose X is locally connected. Let U be an open subset of X, and let C be a component of U. For each point $x \in C$, there exists an open connected set V_x such that $x \in V_x \subset U$. Then $C \cup V_x$ is a connected subset of U. Hence, $V_x \subset C$. Therefore, each point of C is an interior point of C. Thus, C is open.

Next, suppose the components of open sets are open. Let $x \in X$, and let U be an open subset of X such that $x \in U$. By hypothesis, the component, C, of U containing x is open. Then C is an open connected set such that $x \in C \subset U$. Therefore, X is locally connected.

Q.E.D.

1.7.12. Theorem. *A continuum X is connected im kleinen at each of its points if and only if X is locally connected.*

Proof. Clearly, if X is locally connected, then X is connected im kleinen at each of its points.

Suppose X is connected im kleinen at each of its points. Let U be an open subset of X, and let C be a component of U. If $x \in C$, then there exists a subcontinuum W of X such that $x \in Int(W) \subset W \subset U$, by Theorem 1.7.9. Since W is a connected subset of U and $W \cap C \neq \emptyset$, $W \subset C$. Then x is an interior point of C. Thus, C is open. Therefore, by Lemma 1.7.11, X is locally connected.

Q.E.D.

The notions of semi–aposyndesis and aposyndesis resembles those of T_0 and T_1 topological spaces.

1.7.13. Definition. Let X be a continuum, and let $p, q \in X$. We say that X *is semi–aposyndetic at p and q* provided that there exists a subcontinuum W of X such that $\{p, q\} \cap Int(W) \neq \emptyset$ and $\{p, q\} \setminus W \neq \emptyset$. X is *semi–aposyndetic* if it is semi–aposyndetic at each pair of its points.

1.7.14. Example. Let X be the harmonic fan (Example 1.7.5). Then X is semi–aposyndetic at each pair of its points.

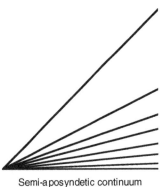

Semi-aposyndetic continuum

1.7.15. Definition. Let X be a continuum, and let $p, q \in X$. We say that *X is aposyndetic at p with respect to q* provided that there exists a subcontinuum W of X such that $p \in Int(W) \subset W \subset X \setminus \{q\}$. Now, *$X$ is aposyndetic at p* if X is aposyndetic at p with respect to each point of $X \setminus \{p\}$. We say that *X is aposyndetic* provided that X is aposyndetic at each of its points.

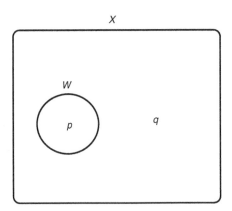

Aposyndetic at p with respect to q

1.7.16. Definition. Let X be a continuum, and let $p \in X$. We say that *X is semi–locally connected at p* provided that for each open subset U of X such that $p \in U$, there exists an open subset V of X such that $p \in V \subset U$ and $X \setminus V$ has a finite number of components.

We say X *is semi–locally connected* if X is semi–locally connected at each of its points.

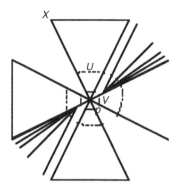

Semi–locally connected continuum

Even though aposyndesis and semi–local connectedness seem to be different concepts, it turns out that globally they are equivalent:

1.7.17. Theorem. *A continuum X is aposyndetic if and only if it is semi–locally connected.*

Proof. Suppose X is aposyndetic. Let $x \in X$. We show that X is semi–locally connected at x. Let U be an open subset of X such that $x \in U$. Since X is aposyndetic, for each $y \in X \setminus U$, there exists a subcontinuum W_y of X such that $y \in Int(W_y) \subset W_y \subset X \setminus \{x\}$. Since $X \setminus U$ is compact, there exists $y_1, \ldots, y_m \in X \setminus U$ such that $X \setminus U \subset \bigcup_{j=1}^{m} Int(W_{y_j})$. Let $V = X \setminus \left(\bigcup_{j=1}^{m} W_{y_j} \right)$. Then V is an open subset of X such that $x \in V \subset U$ and $X \setminus V$ has a finite number of components. Therefore, X is semi–locally connected, since x was an arbitrary point of X.

Now, suppose X is semi–locally connected. Let $x, y \in X$. We show that X is aposyndetic at x with respect to y. Let U be an open subset of X such that $y \in U \subset X \setminus \{x\}$. Let U' be an open subset of X such that $y \in U' \subset Cl(U') \subset U$. Since X is semi–locally connected, there exists an open subset V of X such that $y \in V \subset U'$ and $X \setminus V$ has a finite number of components. By Lemma 1.6.2, x is contained in the interior of the component of $X \setminus V$ containing x. Hence, X is aposyndetic at x with respect to y.

Therefore, X is aposyndetic.

Q.E.D.

1.7.18. Lemma. *Let X be a continuum, and let A be a subcontinuum of X such that $X \setminus A$ is not connected. If U and V are nonempty disjoint open subsets of X such that $X \setminus A = U \cup V$, then $A \cup U$ and $A \cup V$ are subcontinua of X.*

Proof. Since $X \setminus (A \cup U) = V$, $A \cup U$ is closed, hence, compact. Similarly $A \cup V$ is compact.

We show $A \cup U$ is connected. To see this, suppose $A \cup U$ is not connected. Then there exist two nonempty disjoint closed subsets K and L of X such that $A \cup U = K \cup L$. Since A is connected, without loss of generality, we assume that $A \subset K$. Note that, in this case, $L \subset U$. Hence, $L \cap Cl(V) = \emptyset$. Thus, $X = L \cup (K \cup Cl(V))$, a contradiction, since L and $K \cup Cl(V)$ are disjoint closed subsets of X. Therefore, $A \cup U$ is connected. Similarly, $A \cup V$ is connected.

Q.E.D.

1.7.19. Definition. A continuum X is *decomposable* provided that it can be written as the union of two of its proper subcontinua. We say X is *indecomposable* if it is not decomposable. We say X is *hereditarily decomposable (indecomposable)* if each nondegenerate subcontinuum of X is decomposable (indecomposable, respectively).

1.7.20. Lemma. *A continuum X is decomposable if and only if X contains a proper subcontinuum with nonempty interior.*

Proof. Suppose X is a decomposable continuum. Then there exist two proper subcontinua, A and B, of X such that $X = A \cup B$. Note that $X \setminus B$ is an open set contained in A. Therefore, $Int(A) \neq \emptyset$.

Now, assume A is a proper subcontinuum of X with nonempty interior. If $X \setminus A$ is connected, then $Cl(X \setminus A)$ is a proper subcontinuum of X and $X = A \cup Cl(X \setminus A)$. Hence, X is decomposable.

Suppose that $X \setminus A$ is not connected. Then there exist two nonempty disjoint open subsets U and V of X such that $X \setminus A =$

$U \cup V$. By Lemma 1.7.18, $A \cup U$ and $A \cup V$ are subcontinua of X, and $X = (A \cup U) \cup (A \cup V)$. Therefore, X is decomposable.

Q.E.D.

1.7.21. Corollary. *A continuum X is indecomposable if and only if each proper subcontinuum of X has empty interior.*

The following result is known as *The Boundary Bumping Theorem*. It has many applications in continuum theory and in hyperspaces. A proof of this Theorem may be found in 5.4 of [23].

1.7.22. Theorem. *Let X be a continuum and let U be a nonempty, proper, open subset of X. If K is a component of $Cl(U)$, then $K \cap Bd(U) \neq \emptyset$.*

The following Corollary is also very useful; a proof of it may be found in 5.5 of [23].

1.7.23. Corollary. *Let X be a nondegenerate continuum. If A is a proper subcontinuum of X and U is an open subset of X such that $A \subset U$, then there is a subcontinuum B of X such that $A \subsetneq B \subset U$.*

1.7.24. Lemma. *If X is a continuum such that each of its proper subcontinuum is indecomposable, then X is indecomposable. Hence, X is hereditarily indecomposable.*

Proof. Suppose X is decomposable, then there exist two proper subcontinua A and B of X such that $X = A \cup B$. Note that $X \setminus B$ is an open subset of X contained in A. On the other hand, there exists a proper subcontinuum, H, of X containing A (Corollary 1.7.23). Hence, H is an indecomposable continuum containing a proper subcontinuum with nonempty interior, which is impossible (Corollary 1.7.21). Therefore, X is indecomposable.

Q.E.D.

1.7.25. Definition. A continuum X *is irreducible between two of its points* if no proper subcontinuum of X contains both points. A continuum is *irreducible* if it is irreducible between two of its points.

The following results present some of the properties of irreducible continua.

1.7.26. Theorem. *Let X be an irreducible continuum between a and b. If C is a subcontinuum of X such that $X \setminus C$ is not connected, then $X \setminus C$ is the union of two open and connected sets, one containing a and the other containing b. Moreover, if $a \in C$, then $X \setminus C$ is connected.*

Proof. Suppose $X \setminus C$ is not connected. Then there exist two nonempty disjoint open subsets U and V of X such that $X \setminus C = U \cup V$. By Lemma 1.7.18, $A = C \cup U$ and $B = C \cup V$ are subcontinua of X such that $X = A \cup B$, $A \cap B = C$, $A \neq X$ and $B \neq X$.

Since X is irreducible between a and b, $\{a, b\} \cap C = \emptyset$. If $\{a, b\} \cap C \neq \emptyset$, then either A or B is a proper subcontinuum of X containing $\{a, b\}$, a contradiction. Therefore, $\{a, b\} \cap C = \emptyset$. We assume that $a \in U$ and $b \in V$.

Since A and B are proper subcontinua of X, neither A nor B may contain $\{a, b\}$.

Since A is a proper subcontinuum of X and $a \in A$, we assert that $V = X \setminus A$ is connected. To see this, suppose $X \setminus A$ is not connected. Then there exist two nonempty disjoint open subsets K and L such that $X \setminus A = K \cup L$. Since $b \in X \setminus A$, we may assume that $b \in K$. Then, by Lemma 1.7.18, $A \cup K$ is subcontinuum of X which is proper and satisfies that $\{a, b\} \subset A \cup K$, a contradiction. Therefore, $V = X \setminus A$ is connected. Similarly, $U = X \setminus B$ is connected.

A similar argument shows that if $a \in C$, then $X \setminus C$ is connected.
                                                        Q.E.D.

1.7.27. Lemma. *Let X be an irreducible decomposable continuum. If Y and Z are two disjoint subcontinua of X, then $X \setminus (Y \cup Z)$ has at most three components.*

Proof. Since X is irreducible, by Theorem 1.7.26, $X \setminus Y$ has at most two components, U and V, such that they are connected open subsets of X, $Z \subset U$ and V may be empty.

Similarly, we assume that $X \setminus Z = H \cup K$, where H and K are connected open subsets of X such that $Y \subset H$ and K may be empty. Let $R = (U \setminus Z) \cap (H \setminus Y)$. Note that R is an open subset of X, $Cl(R) \cap Y \neq \emptyset$ and $Cl(R) \cap Z \neq \emptyset$. We assert that R is connected. To show this, assume R is not connected. Let C be a component of R. By Theorem 1.7.22, $Cl(C) \cap (Y \cup Z) \neq \emptyset$. If $Cl(C) \cap Y \neq \emptyset$ and $Cl(C) \cap Z \neq \emptyset$, then $V \cup Y \cup Cl(C) \cup Z \cup K$ is a subcontinuum of X containing the points of irreducibility of X. Thus, $X = V \cup Y \cup Cl(C) \cup Z \cup K$. Since $R \cap (V \cup Y \cup Z \cup K) = \emptyset$, it follows that $R \subset Cl(C)$. Hence, R is connected ($C \subset R \subset Cl(C)$), a contradiction. Therefore, either $Cl(C) \cap Y = \emptyset$ or $Cl(C) \cap Z = \emptyset$. Let

$$\mathcal{A} = \{Cl(C) \mid C \text{ is a component of } R \text{ and } Cl(C) \cap Y \neq \emptyset\}$$

and

$$\mathcal{B} = \{Cl(C) \mid C \text{ is a component of } R \text{ and } Cl(C) \cap Z \neq \emptyset\}.$$

We claim that $\mathcal{A}$ and $\mathcal{B}$ are both nonempty. Suppose, to the contrary, that $\mathcal{B}$ is empty. Note that $V \cup Y \cup \left(\bigcup \mathcal{A}\right)$ is a connected set and that $\bigcup \mathcal{A}$ is not connected. Then there exist two separated subsets J and L of $V \cup Y \cup \left(\bigcup \mathcal{A}\right)$ such that $\bigcup \mathcal{A} = J \cup L$. Note that $V \cup Y \cup J$ and $V \cup Y \cup L$ are connected sets (Lemma 1.7.18). Hence, either $(V \cup Y \cup Cl(J)) \cap Z \neq \emptyset$ or $(V \cup Y \cup Cl(L)) \cap Z \neq \emptyset$. Suppose $(V \cup Y \cup Cl(J)) \cap Z \neq \emptyset$. Then $(V \cup Y \cup Cl(J)) \cup Z \cup K$ is a proper subcontinuum of X containing the points of irreducibility of X, a contradiction. Therefore, $\mathcal{B} \neq \emptyset$. Similarly, $\mathcal{A} \neq \emptyset$.

Since X is a continuum, $Cl\left(\bigcup \mathcal{A}\right) \cap Cl\left(\bigcup \mathcal{B}\right) \neq \emptyset$. Let $x \in Cl\left(\bigcup \mathcal{A}\right) \cap Cl\left(\bigcup \mathcal{B}\right)$. Since $x \in Cl\left(\bigcup \mathcal{A}\right)$, there exists a sequence $\{a_n\}_{n=1}^{\infty}$ of elements of $\bigcup \mathcal{A}$ converging to x. For each $n \in \mathbb{N}$, let $Cl(C_n) \in \mathcal{A}$ such that $a_n \in Cl(C_n)$. By Theorem 1.8.5, we assume that the sequence $\{Cl(C_n)\}_{n=1}^{\infty}$ of subcontinua of X converges (in the Hausdorff metric) to a subcontinuum T of X. Note that $T \cap Y \neq \emptyset$ and $x \in T$. Similarly, there exists a subcontinuum T' of X such that $T' \cap Z \neq \emptyset$ and $x \in T'$. Consequently, $V \cup Y \cup T \cup T' \cup Z \cup K$ is a proper subcontinuum of X containing its points of irreducibility, a contradiction. Therefore, R is connected.

Now, observe that

$$X \setminus [(V \cup Y) \cup (Z \cup K)] = (X \setminus V) \cap (X \setminus Y) \cap (X \setminus Z) \cap (X \setminus K)$$

$$= (U \cap Y) \cap (X \setminus Y) \cap (X \setminus Z) \cap (H \cap Z)$$

$$= U \cap H = U \cap (X \setminus Y) \cap (X \setminus Z) \cap H$$

$$= U \cap (X \setminus Z) \cap (X \setminus Y) \cap H$$

$$= (U \setminus Z) \cap (H \setminus Y) = R.$$

Since $V \cap (Y \cup Z \cup K) = \emptyset$ and $K \cap (Z \cup Y \cup V) = \emptyset$, $X \setminus (Y \cup Z) = V \cup R \cup K$.

$$\textbf{Q.E.D.}$$

1.7.28. Definition. A continuum X *is weakly irreducible* provided that the complement of each finite union of subcontinua of X has a finite number of components.

1.7.29. Theorem. *If X is an irreducible continuum, then X is weakly irreducible.*

Proof. Let $Z_1, \ldots, Z_n$ be a finite family of subcontinua of X such that $Z_j \cap Z_k = \emptyset$ if $j \neq k$. For each $j \in \{1, \ldots, n\}$, by Theorem 1.7.26, we assume that $X \setminus Z_j = U_j \cup V_j$, where U_j and V_j are open connected subsets of X. Without loss of generality, we suppose that $\bigcup_{j=2}^{n} Z_j \subset V_1$ and $\bigcup_{j=1}^{n-1} Z_j \subset U_n$. Hence, U_1 and V_n may be empty. We assume also that, for each $j \in \{2, \ldots, n-1\}$, $\bigcup_{k=1}^{j-1} Z_k \subset U_j$ and $\bigcup_{k=j+1}^{n} Z_k \subset V_j$. For each $j \in \{1, \ldots, n-1\}$, let $R_j = (V_j \setminus Z_{j+1}) \cap (U_{j+1} \setminus Z_j)$. By the proof of Lemma 1.7.27, R_j is a connected open subset of X. Note that $X \setminus \left(\bigcup_{j=1}^{n} Z_j \right) =$

$U_1 \cup \left(\bigcup_{j=1}^{n-1} R_j \right) \cup V_n$. Thus, $X \setminus \left(\bigcup_{j=1}^{n} Z_j \right)$ has, at most, $n+1$ components. Therefore, X is weakly irreducible.

Q.E.D.

1.7.30. Definition. A continuum X *is unicoherent* provided that for each pair A and B of subcontinua of X such that $X = A \cup B$, $A \cap B$ is connected. We say that X *is hereditarily unicoherent* if each subcontinuum of X is unicoherent.

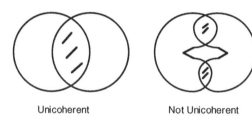

Unicoherent   Not Unicoherent

1.7.31. Lemma. *If X is a subcontinuum of $[0,1]^n$ and U is an open subset of $[0,1]^n$ such that $X \subset U$, then there exists a polyhedron P such that $X \subset Int(P) \subset P \subset U$.*

Sketch of proof. Since $[0,1]^n$ is a polyhedron, we can perform barycentric subdivisions to obtain complexes $\mathcal{K}_1, \mathcal{K}_2, \ldots$. For each $m \in \mathbb{N}$, let $\mathcal{K}_m^\star$ be the subcomplex of $\mathcal{K}_m$ consisting of all the simplexes of $\mathcal{K}_m$ which intersect X (we include all the faces of such simplexes as part of $\mathcal{K}_m^\star$). Let P_m be the polyhedron determined by $\mathcal{K}_m^\star$. Note that P_m is a continuum. (P_m is connected since it is the union of connected sets intersecting X and X itself is connected. Since P_m is a finite union of compact sets, P_m is compact.) Note that, by construction, $Int(P_{m+1}) \subset P_m$. Since the limit of the diameters of the simplexes of the complex $\mathcal{K}_m^\star$ tend to zero as m tends to infinity, $\bigcap_{m=1}^{\infty} Int(P_m) = \bigcap_{m=1}^{\infty} P_m = X$. By Lemma 1.6.7, there exists $m \in \mathbb{N}$ such that $P_m \subset U$.

Q.E.D.

1.8   Hyperspaces

We give the definition of the main hyperspaces associated to a continuum. We present some of their elementary properties.

1.8.1. Definition. Given a compactum X, we define its *hyperspaces* as the following sets:

$$2^X = \{A \subset X \mid A \text{ is closed and nonempty}\},$$

$$\mathcal{C}(X) = \{A \in 2^X \mid A \text{ is a connected}\},$$

and for each $n \in \mathbb{N}$

$$\mathcal{C}_n(X) = \{A \in 2^X \mid A \text{ has at most } n \text{ components}\},$$

$$\mathcal{F}_n(X) = \{A \in 2^X \mid A \text{ has at most } n \text{ points}\}.$$

$\mathcal{F}_n(X)$ is called *n–fold symmetric product* and $\mathcal{C}_n(X)$ is called *n–fold hyperspace*.

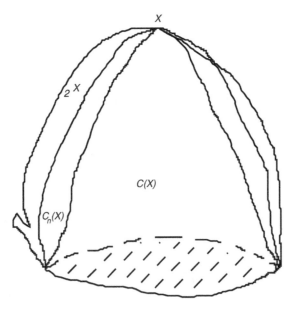

1.8.2. Remark. Given a continuum X, we agree that $\mathcal{C}(X) = \mathcal{C}_1(X)$. Let us observe that for each $n \in \mathbb{N}$,

$$\mathcal{F}_n(X) \subset \mathcal{C}_n(X),$$

$$\mathcal{C}_n(X) \subset \mathcal{C}_{n+1}(X),$$

and that

$$\mathcal{F}_n(X) \subset \mathcal{F}_{n+1}(X).$$

The n–fold symmetric products were defined by Borsuk and Ulam [3]. These hyperspaces have been studied by many people. Some recent results and references about n–fold symmetric products may be found in [16], [17], [4] and [5]. It is not clear who defined or where the n–fold hyperspaces were defined. A study of the n–fold hyperspaces is presented in Chapter 6.

1.8.3. Theorem. *Let X be a continuum with metric d. Then the map $\mathcal{H}\colon 2^X \times 2^X \to [0, \infty)$ given by:*

$$\mathcal{H}(A, B) = \inf\{\varepsilon > 0 \mid A \subset \mathcal{V}_\varepsilon^d(B) \text{ and } B \subset \mathcal{V}_\varepsilon^d(A)\}$$

is a metric for 2^X. It is called the Hausdorff metric.

Proof. We show the triangle inequality. The other two properties are obvious.

Let A, B and C be three elements of 2^X. We show that $\mathcal{H}(A, C) \leq \mathcal{H}(A, B) + \mathcal{H}(B, C)$. To this end, let α be a positive real number. Let $\beta_A = \mathcal{H}(A, B) + \dfrac{\alpha}{2}$ and $\beta_B = \mathcal{H}(B, C) + \dfrac{\alpha}{2}$. Let us observe that: $A \subset \mathcal{V}_{\beta_A}^d(B)$ and $B \subset \mathcal{V}_{\beta_B}^d(C)$. Then, given $a \in A$, there exists $b \in B$ such that $d(a, b) < \mathcal{H}(A, B) + \dfrac{\alpha}{2}$. For this b, there exists $c \in C$ such that $d(b, c) < \mathcal{H}(B, C) + \dfrac{\alpha}{2}$. Hence, $d(a, c) < \mathcal{H}(A, B) + \mathcal{H}(B, C) + \alpha$. Therefore, if $\beta = \mathcal{H}(A, B) + \mathcal{H}(B, C) + \alpha$, then $A \subset \mathcal{V}_\beta^d(C)$. A similar argument shows that $C \subset \mathcal{V}_\beta^d(A)$. Since α was an arbitrary positive number, it follows, from the definition of $\mathcal{H}$, that $\mathcal{H}(A, C) \leq \mathcal{H}(A, B) + \mathcal{H}(B, C)$.

Q.E.D.

Let X be a continuum. Note that given a sequence, $\{Y_n\}_{n=1}^{\infty}$, of elements of 2^X, we have two types of convergence of this sequence, namely, the one given in Definition 1.2.26 (the limit belongs to 2^X by Lemma 1.2.27), and the other one given by the Hausdorff metric. It is known that both types of convergence coincide, i.e., we have the following theorem, a proof of which may be found in (0.7) of [22]:

1.8.4. Theorem. *Let X be a continuum, and let $\{Y_n\}_{n=1}^{\infty}$ be a sequence of elements of 2^X. Then $\{Y_n\}_{n=1}^{\infty}$ converges to Y in the sense of Definition 1.2.26 if and only if $\{Y_n\}_{n=1}^{\infty}$ converges to Y in the Hausdorff metric.*

1.8.5. Theorem. *If X is a compactum, then 2^X and $\mathcal{C}(X)$ are compact.*

Proof. First, we show that 2^X is compact. By Theorem 1.8.4, it is enough to prove that every sequence of elements in 2^X has a convergent subsequence in the sense of Definition 1.2.26. This was already done in Theorem 1.2.28. We just need to mention that the limit of the subsequence constructed in Theorem 1.2.28 is closed by Lemma 1.2.27, and, since X is compact, such limit is nonempty.

To see that $\mathcal{C}(X)$ is compact, it suffices to show that $\mathcal{C}(X)$ is closed in 2^X. This follows from Theorem 1.2.29.

Q.E.D.

Regarding n–fold symmetric products, we have the following results:

1.8.6. Lemma. *Let X be a continuum with metric d, and let $n \in \mathbb{N}$. If D_n denotes the metric on X^n given by*

$$D_n((x_1, \ldots, x_n), (x_1', \ldots, x_n')) = \max\{d(x_1, x_1'), \ldots, d(x_n, x_n')\},$$

then the function $f_n \colon X^n \to \mathcal{F}_n(X)$ given by

$$f_n((x_1, \ldots, x_n)) = \{x_1, \ldots, x_n\}$$

is surjective and satisfies the following inequality:

$$\mathcal{H}(f_n((x_1,\dots,x_n)), f_n((x_1',\dots,x_n'))) \le$$

$$D_n((x_1,\dots,x_n),(x_1',\dots,x_n')),$$

for each $(x_1,\dots,x_n)$ and $(x_1',\dots,x_n')$ in X^n.

Proof. Clearly the map f_n is surjective. Let $(x_1,\dots,x_n)$ and $(x_1',\dots,x_n')$ be two points of X^n. Assume that

$$D_n((x_1,\dots,x_n),(x_1',\dots,x_n')) = r$$

and let $\varepsilon > 0$ be given. Hence, $D_n((x_1,\dots,x_n),(x_1',\dots,x_n')) < r + \varepsilon$. This implies that for each $j \in \{1,\dots,n\}$, $d(x_j,x_j') < r+\varepsilon$. Thus, for each $x_j \in f_n((x_1,\dots,x_n))$, we have that $x_j' \in f_n((x_1',\dots,x_n'))$ and $d(x_j,x_j') < r + \varepsilon$. This shows that

$$f_n((x_1,\dots,x_n)) \subset \mathcal{V}_{r+\varepsilon}^d(f_n((x_1',\dots,x_n'))).$$

Similarly, $f_n((x_1',\dots,x_n')) \subset \mathcal{V}_{r+\varepsilon}^d(f_n((x_1,\dots,x_n)))$. Thus,

$$\mathcal{H}(f_n((x_1,\dots,x_n)), f_n((x_1',\dots,x_n'))) \le r + \varepsilon.$$

Since the ε was arbitrary, we obtain that

$$\mathcal{H}(f_n((x_1,\dots,x_n)), f_n((x_1',\dots,x_n'))) \le r.$$

$$\textbf{Q.E.D.}$$

As an immediate consequence of Lemma 1.8.6, we have the following Corollary:

1.8.7. Corollary. *Let X be a continuum and let $n \in \mathbb{N}$. Then the function $f_n\colon X^n \twoheadrightarrow \mathcal{F}_n(X)$ given by:*

$$f_n((x_1,\dots,x_n)) = \{x_1,\dots,x_n\}$$

is continuous.

1.8.8. Corollary. *If X is a continuum, then $\mathcal{F}_n(X)$ is a continuum for each $n \in \mathbb{N}$.*

1.8.9. Corollary. *If X is a continuum, then 2^X is a continuum.*

Proof. By Corollary 1.8.8, $\mathcal{F}_n(X)$ is a continuum for each $n \in \mathbb{N}$. Then $\mathcal{F}(X) = \bigcup_{n=1}^{\infty} \mathcal{F}_n(X)$ is a connected subset of 2^X. We show that $\mathcal{F}(X)$ is dense in 2^X. To this end, let ε be a positive real number. Since X is compact, there exist $x_1, \ldots, x_m \in X$, such that $X \subset \bigcup_{n=1}^{m} \mathcal{V}_\varepsilon^d(x_n)$. Then $\{x_1, \ldots, x_m\} \in \mathcal{F}(X)$, $X \subset \mathcal{V}_\varepsilon^d(\{x_1, \ldots, x_m\})$ and, clearly, $\{x_1, \ldots, x_m\} \subset \mathcal{V}_\varepsilon^d(X)$. Hence, $\mathcal{H}(\{x_1, \ldots, x_m\}, X) < \varepsilon$. Therefore, $\mathcal{F}(X)$ is dense in 2^X. Thus, 2^X is connected. By Theorem 1.8.5, 2^X is a compactum. Therefore, 2^X is a continuum.
$\qquad$ **Q.E.D.**

The following Theorem (see (1.13) of [22]) is a better result than Corollary 1.8.9:

1.8.10. Theorem. *If X is a continuum, then 2^X and $\mathcal{C}(X)$ are both arcwise connected continua.*

Recall that given a compactum X, we defined 2^X as the family of all nonempty closed subsets of X with the Hausdorff metric $\mathcal{H}$. By Theorem 1.8.5, 2^X is a compactum. Hence, we define 2^{2^X} as the family of all nonempty closed subsets of 2^X with the Hausdorff metric $\mathcal{H}^2$. The following Lemma, a proof of which may be found in (1.48) of [22], gives a nice map from 2^{2^X} onto 2^X:

1.8.11. Lemma. *Let X be a compactum. Let $\sigma \colon 2^{2^X} \to 2^X$ be given by $\sigma(\mathcal{A}) = \bigcup \{A \mid A \in \mathcal{A}\}$. Then σ is well defined and satisfies the following inequality:*

$$\mathcal{H}(\sigma(\mathcal{A}_1), \sigma(\mathcal{A}_2)) \leq \mathcal{H}^2(\mathcal{A}_1, \mathcal{A}_2)$$

for each $\mathcal{A}_1$ and $\mathcal{A}_2$ in 2^{2^X}. In particular, σ is continuous.

1.8.12. Corollary. *Let X be a continuum. Then $\mathcal{C}_n(X)$ is an arcwise connected continuum for each $n \in \mathbb{N}$.*

Proof. Let $n \in \mathbb{N}$. Since $\mathcal{C}(X)$ is an arcwise connected continuum (Theorem 1.8.10), $\mathcal{F}_n(\mathcal{C}(X))$ is a continuum (Corollary 1.8.8). It is easy to see that, in fact, $\mathcal{F}_n(\mathcal{C}(X))$ is arcwise connected. Let σ be the union map. Then $\sigma\left(\mathcal{F}_n(\mathcal{C}(X))\right) = \mathcal{C}_n(X)$. Hence, $\mathcal{C}_n(X)$ is an arcwise connected continuum.

$\qquad\qquad\qquad\qquad\qquad\qquad\qquad\qquad\qquad$ **Q.E.D.**

1.8.13. Notation. Let X be a compactum. Given a finite collection of open sets of X, $U_1, U_2, \dots, U_m$, we define
$$\langle U_1, \dots, U_m \rangle =$$

$$\left\{ A \in 2^X \ \Big| \ A \subset \bigcup_{k=1}^m U_k \text{ and } A \cap U_k \neq \emptyset \text{ for each } k \in \{1, \dots, m\} \right\}.$$

A proof of the following Theorem may be found in (0.11) of [22]:

1.8.14. Theorem. *Let X be a compactum. If*

$$\mathcal{B} = \{ \langle U_1, \dots, U_n \rangle \mid U_1, \dots, U_n \text{ are open subsets of } X \text{ and } n \in \mathbb{N} \},$$

then $\mathcal{B}$ is a basis for a topology of 2^X.

1.8.15. Definition. The topology for 2^X given by Theorem 1.8.14 is called the *Vietoris topology.*

It is known that the topology induced by the Hausdorff metric and the Vietoris topology coincide (see (0.13) of [22]):

1.8.16. Theorem. *Let X be a compactum. Then the topology induced by the Hausdorff metric and the Vietoris topology for 2^X are the same.*

1.8.17. Definition. *Let X be a compactum. A Whitney map is a continuous function $\mu\colon 2^X \twoheadrightarrow [0,1]$ such that:*

(1) $\mu(X) = 1$,

(2) $\mu(\{x\}) = 0$ for every $x \in X$ and

(3) $\mu(A) < \mu(B)$ if $A, B \in 2^X$ and $A \subsetneq B$.

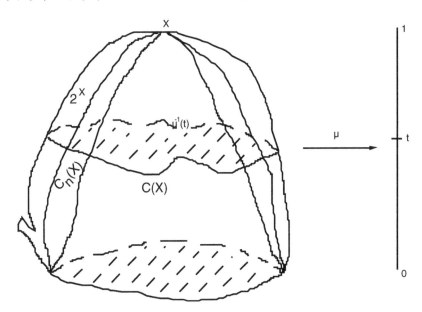

1.8.18. Remark. Whitney maps exist; constructions of them may be found in [22] and [11].

1.8.19. Definition. *Let X be a compactum, and let $A, B \in 2^X$. An order arc from A to B is a one–to–one map $\alpha\colon [0,1] \to 2^X$ such that $\alpha(0) = A$, $\alpha(1) = B$ and for each $s, t \in [0,1]$ such that $s < t$, $\alpha(s) \subsetneq \alpha(t)$.*

The following Theorem tells us when an order arc exists. A proof of it may be found in (1.8) of [22]:

1.8.20. Theorem. *Let X be a compactum, and let $A, B \in 2^X$. Then there exists an order arc from A to B if and only if $A \subset B$ and each component of B intersects A.*

1.8.21. Definition. *Let $f \colon X \to Y$ be a map between compacta. Then $2^f \colon 2^X \to 2^Y$ given by $2^f(A) = f(A)$ is called the induced map between 2^X and 2^Y. For each $n \in \mathbb{N}$, the maps $C_n(f) \colon C_n(X) \to C_n(Y)$ and $\mathcal{F}_n(f) \colon \mathcal{F}_n(X) \to \mathcal{F}_n(Y)$ given by $C_n(f) = 2^f|_{C_n(X)}$ and $\mathcal{F}_n(f) = 2^X|_{\mathcal{F}_n(X)}$ are called the induced maps between the hyperspaces $C_n(X)$ and $C_n(Y)$, and $\mathcal{F}_n(X)$ and $\mathcal{F}_n(Y)$, respectively.*

1.8.22. Theorem. *Let $f \colon X \to Y$ be a map between compacta. Then 2^f is continuous.*

Proof. Let $A \in 2^X$, and let $\varepsilon > 0$. Let $\delta > 0$ given by the uniform continuity of f. Then

$$2^f \left(\mathcal{V}_\delta^{\mathcal{H}_X}(A) \right) \subset \mathcal{V}_\varepsilon^{\mathcal{H}_Y}(2^f(A)).$$

Therefore, 2^f is continuous.

Q.E.D.

1.8.23. Corollary. *Let $f \colon X \to Y$ be a map between compacta. Then for each $n \in \mathbb{N}$, $C_n(f)$ and $\mathcal{F}_n(f)$ are continuous.*

From the proof of Theorem 20 of [7], we have the following Theorem:

1.8.24. Theorem. *Let $f \colon X \to Y$ be an open surjective map between compacta. Then the function $\Im(f) \colon 2^Y \to 2^X$ given by $(\Im(f))(B) = f^{-1}(B)$ is continuous.*

REFERENCES

[1] M. Aguilar, S. Gitler and C. Prieto, *Algebraic Topology From a Homotopical View*, Universitext, Springer–Verlag, New York, Inc., 2002.

[2] F. D. Ancel, An Alternative Proof and Applications of a Theorem of E. G. Effros, Michigan J. 34 (1987), 39–55.

[3] K. Borsuk and S. Ulam, On Symmetric Products of Topological Spaces, Bull. Amer. Math. Soc., 37 (1931), 875–882.

[4] E. Castañeda, A Unicoherent Continuum for Which its Second Symmetric Product is not Unicoherent, Topology Proceedings, 23 (1998), 61–67.

[5] E. Castañeda, *Productos Simétricos*, Tesis Doctoral, Facultad de Ciencias, U. N. A. M., 2003. (Spanish)

[6] J. J. Charatonik, On fans, Dissertationes Math. (Rozprawy Mat.), 54 (1967), 1–37.

[7] J. J. Charatonik, A. Illanes and S. Macías, Induced Mappings on the Hyperspaces $C_n(X)$ of a Continuum X, Houston J. Math., 28 (2002), 781–805.

[8] H. S. Davis, P. H. Doyle, Invertible Continua, Portugal. Math., 26 (1967), 487–491.

[9] J. Dugundji, *Topology*, Allyn and Bacon, Inc., Boston, 1966.

[10] J. G. Hocking and G. S. Young, *Topology*, Dover Publications, Inc., New York, 1988.

[11] A. Illanes and S. B. Nadler, Jr., *Hyperspaces: Fundamentals and Recent Advances,* Monographs and Textbooks in Pure and Applied Math., Vol. 216, Marcel Dekker, New York, Basel, 1999.

[12] K. Kuratowski, *Topology*, Vol. I, Academic Press, New York, N. Y., 1966.

[13] K. Kuratowski, *Topology*, Vol. II, Academic Press, New York, N. Y., 1968.

[14] E. L. Lima, *Espaços Métricos*, terceira edição, Instituto de Matemática Pura e Aplicada, CNPq, (Projeto Euclides), 1977. (Portuguese)

[15] E. L. Lima, *Grupo Fundamental e Espaços de Recobrimento*, Instituto de Matemática Pura e Aplicada, CNPq, (Projeto Euclides), 1993. (Portuguese)

[16] S. Macías, On Symmetric Products of Continua, Topology Appl., 92 (1999), 173–182.

[17] S. Macías, Aposyndetic Properties of Symmetric Products of Continua, Topology Proc., 22 (1997), 281–296.

[18] S. Macías, On the Hyperspaces $C_n(X)$ of a Continuum X, Topology Appl., 109 (2001), 237–256.

[19] S. Macías, On the Hyperspaces $C_n(X)$ of a Continuum X, II, Topology Proc., 25 (2000), 255–276.

[20] J. Munkres, *Elements of Algebraic Topology*, Addison–Wesley Publishing Company, Inc., Redwood City, California, 1984.

[21] J. Munkres, *Topology*, second edition, Prentice Hall, Upper Saddle River, NJ, 2000.

[22] S. B. Nadler, Jr., *Hyperspaces of Sets*, Monographs and Textbooks in Pure and Applied Math., Vol. 49, Marcel Dekker, New York, Basel, 1978.

[23] S. B. Nadler, Jr., *Continuum Theory: An Introduction*, Monographs and Textbooks in Pure and Applied Math., Vol. 158, Marcel Dekker, New York, Basel, Hong Kong, 1992.

[24] S. B. Nadler, Jr., *Continuum Theory*, class notes, Fall Semester of 1999, Math. 481, West Virginia University.

[25] J. J. Rotman, *An Introduction to the Theory of Groups*, fourth edition, Graduate Texts in Mathematics, Vol. 148, Springer–Verlag, New York, Inc., 1995.

[26] W. Rudin, *Principles of Mathematical Analysis*, third edition, International Series in Pure and Applied Mathematics, McGraw–Hill, Inc., New York, 1976.

Chapter 2

INVERSE LIMITS AND RELATED TOPICS

We present basic results about inverse limits and related topics. An excellent treatment of inverse limits, distinct from the one given here, was written by Professor W. Tom Ingram [10].

First, we present some basic results of inverse limits. Then a construction and a characterization of the Cantor set are given using inverse limits. With this technique, it is shown that a compactum is a continuous image of the Cantor set. Next, we show that inverse limits commute with the operation of taking finite products, cones and hyperspaces. We give some properties of chainable continua. We study circularly chainable and $\mathcal{P}$–like continua. We end the chapter presenting several properties of universal maps and AH–essential maps.

2.1   Inverse Limits

We introduce inverse limits of metric spaces. We prove basic properties of inverse limits of compacta and continua.

2.1.1. Definition. Let $\{X_n\}_{n=1}^{\infty}$ be a countable collection of metric spaces. For each $n \in \mathbb{N}$, let $f_n^{n+1} \colon X_{n+1} \to X_n$ be a map. Then the sequence $\{X_n, f_n^{n+1}\}$ of metric spaces and maps is called an *inverse sequence*. The maps f_n^{n+1} are called *bonding maps*.

$$X_1 \xleftarrow{f_1^2} X_2 \longleftarrow \cdots \longleftarrow X_{n-1} \xleftarrow{f_{n-1}^n} X_n \xleftarrow{f_n^{n+1}} X_{n+1} \longleftarrow \cdots$$

2.1.2. Notation. If $m, n \in \mathbb{N}$ and $n > m$, then $f_m^n = f_m^{m+1} \circ \cdots \circ f_{n-1}^n$ and $f_n^n = 1_{X_n}$, where 1_{X_n} denotes the identity map on X_n.

2.1.3. Definition. If $\{X_n, f_n^{n+1}\}$ be an inverse sequence of metric spaces, then *inverse limit of* $\{X_n, f_n^{n+1}\}$, denoted by $\varprojlim\{X_n, f_n^{n+1}\}$ or X_∞, is the subspace of the topological product $\prod_{n=1}^{\infty} X_n$ given by

$$\varprojlim\{X_n, f_n^{n+1}\} = \left\{ (x_n)_{n=1}^{\infty} \in \prod_{n=1}^{\infty} X_n \;\middle|\; f_n^{n+1}(x_{n+1}) = x_n \right.$$

$$\left. \text{for each } n \in \mathbb{N} \right\}.$$

2.1.4. Remark. If $\{X_n, f_n^{n+1}\}$ is an inverse sequence of metric spaces, then, by Lemma 1.1.6, its inverse limit, X_∞, is a metric space.

2.1.5. Definition. Let $\{X_n, f_n^{n+1}\}$ be an inverse sequence of metric spaces with inverse limit X_∞. Recall that there exist maps $\pi_m \colon \prod_{n=1}^{\infty} X_n \twoheadrightarrow X_m$ called projection maps. For each $m \in \mathbb{N}$, let $f_m = \pi_m|_{X_\infty}$; then f_m is a continuous function, since it is a restriction of a map (Lemma 1.1.6), and it is called a *projection map*.

2.1.6. Remark. Let $\{X_n, f_n^{n+1}\}$ be an inverse sequence of metric spaces with inverse limit X_∞. Even though the maps π_m are surjective, the projection maps f_m are not, in general, surjective. However, if all the bonding maps f_n^{n+1} are surjective, then the projection maps are surjective, and vice versa. Note that, by definition, for each $n \in \mathbb{N}$, $f_n^{n+1} \circ f_{n+1} = f_n$, i.e., the following diagram

$$
\begin{array}{ccc}
& X_\infty & \\
{}^{f_{n+1}}\swarrow & & \searrow{}^{f_n} \\
X_{n+1} & \xrightarrow[f_n^{n+1}]{} & X_n
\end{array}
$$

is commutative.

2.1.7. Definition. Let $\{X_n, f_n^{n+1}\}$ be an inverse sequence of metric spaces. For each $m \in \mathbb{N}$, let

$$
S_m = \left\{ (x_n)_{n=1}^\infty \in \prod_{n=1}^\infty X_n \;\middle|\; f_k^{k+1}(x_{k+1}) = x_k, \; 1 \le k < m \right\}.
$$

2.1.8. Proposition. *Let* $\{X_n, f_n^{n+1}\}$ *be an inverse sequence of compacta whose inverse limit is* X_∞. *Then:*

(1) For each $m \in \mathbb{N}$, S_m *is homeomorphic to* $\prod_{n=m}^\infty X_n$.

(2) $\{S_m\}_{m=1}^\infty$ *is a decreasing sequence of compacta and* $X_\infty = \bigcap_{m=1}^\infty S_m$.
In particular, $X_\infty \ne \emptyset$.
(3) If for each $n \in \mathbb{N}$, X_n *is a continuum, then* X_∞ *is a continuum.*

Proof. We show (1). Let $h \colon S_m \to \prod_{n=m}^\infty X_n$ be given by $h((x_n)_{n=1}^\infty) = (x_n)_{n=m}^\infty$. Since $\pi_n \circ h$ is continuous for each $n \ge m$, by Theorem 1.1.9, h is continuous.

Now, let $g \colon \prod_{n=m}^\infty X_n \to S_m$ be given by $g\left((x_n)_{n=m}^\infty\right) = (y_n)_{n=1}^\infty$, where $y_n = x_n$ if $n \ge m$ and $y_n = f_n^m(x_m)$ if $n < m$. Since $\pi_n \circ g$ is continuous for each $n \in \mathbb{N}$, by Theorem 1.1.9, g is continuous.

Note that $g \circ h = 1_{S_m}$ and $h \circ g = 1_{\prod_{n=m}^{\infty} X_n}$. Therefore, h is a homeomorphism.

We see (2). By definition, $S_{m+1} \subset S_m$ for each $m \in \mathbb{N}$. By (1), each S_m is a compactum (Theorem 1.1.11). Clearly, $X_\infty = \bigcap_{m=1}^{\infty} S_m$.

We prove (3). Note that (3) follows from (2) and Theorem 1.7.2.

Q.E.D.

2.1.9. Proposition. *Let $\{X_n, f_n^{n+1}\}$ be an inverse sequence of compacta with inverse limit X_∞. For each $n \in \mathbb{N}$, let*

$$\mathcal{B}_n = \left\{ f_n^{-1}(U_n) \mid U_n \text{ is an open subset of } X_n \right\}.$$

If $\mathcal{B} = \bigcup_{n=1}^{\infty} \mathcal{B}_n$, then $\mathcal{B}$ is a basis for the topology of X_∞.

Proof. Since X_∞ is a subspace of a topological product, a basic open subset of X_∞ is of the form

$$\bigcap_{j=1}^{k} f_{n_j}^{-1}(U_{n_j}),$$

where U_{n_j} is an open subset of X_{n_j}. With out loss of generality, we assume that $n_k = \max\{n_1, \ldots, n_k\}$. Let $U = \bigcap_{j=1}^{k} \left(f_{n_j}^{n_k} \right)^{-1}(U_{n_j})$. Then U is an open subset of X_{n_k} and

$$f_{n_k}^{-1}(U) = f_{n_k}^{-1}\left(\bigcap_{j=1}^{k} (f_{n_j}^{n_k})^{-1}(U_{n_j}) \right) =$$

$$\bigcap_{j=1}^{k} \left(f_{n_j}^{n_k} \circ f_{n_k} \right)^{-1}(U_{n_j}) =$$

$$\bigcap_{j=1}^{k} f_{n_j}^{-1}(U_{n_j}).$$

Q.E.D.

The next Corollary says that projection maps and bonding maps share the property of being open.

2.1.10. Corollary. *Let $\{X_n, f_n^{n+1}\}$ be an inverse sequence of compacta with surjective bonding maps, whose inverse limit is X_∞. Then the projection maps are open if and only if all the bonding maps are open.*

Proof. Suppose all the bonding maps are open. Let $m \in \mathbb{N}$. Let $\mathcal{U}$ be an open subset of X_∞, and let $(x_n)_{n=1}^\infty \in \mathcal{U}$. By Proposition 2.1.9, there exist $N \in \mathbb{N}$ and an open subset U_N of X_N such that

$$(x_n)_{n=1}^\infty \in f_N^{-1}(U_N) \subset \mathcal{U}.$$

To see that f_m is open, we consider two cases.

Suppose first that $N \geq m$. Then, by Remark 2.1.6,

$$x_m = f_m((x_n)_{n=1}^\infty) \in f_m\left(f_N^{-1}(U_N)\right) =$$

$$f_m^N f_N(f_N^{-1}(U_N)) = f_m^N(U_N) \subset f_m(\mathcal{U}).$$

Hence, since the bonding maps are open, $x_m = f_m((x_n)_{n=1}^\infty)$ is an interior point of $f_m(\mathcal{U})$. Therefore, $f_m(\mathcal{U})$ is open since $(x_n)_{n=1}^\infty$ was an arbitrary point of $\mathcal{U}$.

Now suppose that $m > N$. Then, by Remark 2.1.6,

$$x_m = f_m((x_n)_{n=1}^\infty) \in f_m\left(f_N^{-1}(U_N)\right) =$$

$$f_m f_m^{-1}(f_N^m)^{-1}(U_N) = (f_N^m)^{-1}(U_N) \subset f_m(\mathcal{U}).$$

Hence, since the bonding maps are continuous, $x_m = f_m((x_n)_{n=1}^\infty)$ is an interior point of $f_m(\mathcal{U})$. Therefore, $f_m(\mathcal{U})$ is open since $(x_n)_{n=1}^\infty$ was an arbitrary point of $\mathcal{U}$.

Next, suppose all the projection maps are open. Let $n \in \mathbb{N}$, and let U_{n+1} be an open subset of X_{n+1}. Since $f_n^{n+1}(U_{n+1}) = f_n f_{n+1}^{-1}(U_{n+1})$ (Remark 2.1.6) and the fact that the projection maps are continuous and open, $f_n^{n+1}(U_{n+1})$ is an open subset of X_n. Therefore, f_n^{n+1} is open.

Q.E.D.

2.1.11. Definition. Let X and Y be continua. A surjective map $f\colon X \twoheadrightarrow Y$ is said to be *monotone* if $f^{-1}(y)$ is connected for each $y \in Y$.

The following Lemma gives a very useful characterization of monotone maps.

2.1.12. Lemma. *Let X and Y be compacta. If $f\colon X \twoheadrightarrow Y$ is a surjective map, then f is monotone if and only if $f^{-1}(C)$ is connected for each connected subset C of Y.*

Proof. Suppose f is monotone and let C be a subset of Y such that $f^{-1}(C)$ is not connected. Then there exist two nonempty subsets A and B of X such that $f^{-1}(C) = A \cup B$, $Cl_X(A) \cap B = \emptyset$ and $A \cap Cl_X(B) = \emptyset$. Observe that if $y \in C$ and $f^{-1}(y) \cap A \neq \emptyset$, then $f^{-1}(y) \subset A$ ($f^{-1}(y)$ is connected). Let $M = \{y \in C \mid f^{-1}(y) \subset A\}$. Hence, $A = f^{-1}(M)$. Similarly, if $N = \{y \in C \mid f^{-1}(y) \subset B\}$, then $B = f^{-1}(N)$. Clearly, $C = M \cup N$. Suppose there exists $y \in Cl_Y(M) \cap N$. Since $y \in N$, $f^{-1}(y) \subset B$. Since $y \in Cl_Y(M)$, there exists a sequence $\{y_n\}_{n=1}^{\infty}$ of points of M converging to y. This implies that $f^{-1}(y_n) \subset A$ for every $n \in \mathbb{N}$. Let $x_n \in f^{-1}(y_n)$. Since X is a compactum, without loss of generality, we assume that $\{x_n\}_{n=1}^{\infty}$ converges to a point x. Note that $x \in Cl_X(A)$ and, by the continuity of f, $f(x) = y$. Thus, $x \in Cl_X(A) \cap f^{-1}(y) \subset Cl_X(A) \cap B$, a contradiction. Hence, $Cl_Y(M) \cap N = \emptyset$. Similarly, $M \cap Cl_Y(N) = \emptyset$. Therefore, C is not connected.

The other implication is obvious.

Q.E.D.

The next result says that projection maps and bonding maps share the property of being monotone.

2.1.13. Proposition. *Let $\{X_n, f_n^{n+1}\}$ be an inverse sequence of compacta with surjective bonding maps and whose inverse limit is X_∞. Then the bonding maps are monotone if and only if the projection maps are monotone.*

Proof. Suppose the bonding maps are monotone. Let $m \in \mathbb{N}$. We show f_m is monotone. To this end, let $x_m \in X_m$. For each $n \in \mathbb{N}$, let

$$W_n = \begin{cases} (f_m^n)^{-1}(x_m) & \text{if } n \geq m; \\ \{f_n^m(x_m)\} & \text{if } n < m. \end{cases}$$

Then $\{W_n, f_n^{n+1}|_{W_{n+1}}\}$ is an inverse sequence of continua. Hence, by Proposition 2.1.8 (3), $W = \varprojlim\{W_n, f_n^{n+1}|_{W_{n+1}}\}$ is a subcontinuum of X_∞. We assert that $W = f_m^{-1}(x_m)$. To see this, let $(y_n)_{n=1}^\infty \in f_m^{-1}(x_m)$. Then $y_m = x_m$, $y_n = f_n^m(x_m)$ for each $n < m$, and $y_n \in (f_m^n)^{-1}(x_m) = W_n$ for each $n > m$. Hence, $(y_n)_{n=1}^\infty \in W$. Therefore, $f_m^{-1}(x_m) \subset W$.

Since $f_m(W) = W_m = \{x_m\}$, $W \subset f_m^{-1}(x_m)$. Therefore, $W = f_m^{-1}(x_m)$ and f_m is monotone.

Now, suppose all the projection maps are monotone. Let $m \in \mathbb{N}$. We prove that f_m^{m+1} is monotone. To see this, let $x_m \in X_m$. By Remark 2.1.6, $(f_m^{m+1})^{-1}(x_m) = f_{m+1}f_m^{-1}(x_m)$. Hence, $(f_m^{m+1})^{-1}(x_m)$ is connected since the projection maps are monotone and continuous. Therefore, f_m^{m+1} is monotone.

Q.E.D.

2.1.14. Corollary. *Let $\{X_n, f_n^{n+1}\}$ be an inverse sequence of locally connected continua with surjective bonding maps and whose inverse limit is X_∞. If all the bonding maps are monotone, then X_∞ is locally connected.*

Proof. Let $(x_n)_{n=1}^\infty \in X_\infty$, and let $\mathcal{U}$ be an open subset of X_∞ such that $(x_n)_{n=1}^\infty \in \mathcal{U}$. By Proposition 2.1.9, there exist $N \in \mathbb{N}$ and an open subset U_N of X_N such that $(x_n)_{n=1}^\infty \in f_N^{-1}(U_N) \subset \mathcal{U}$. Since X_N is locally connected, there exists a connected open subset V_N of X_N such that $x_N \in V_N \subset U_N$. Hence, $(x_n)_{n=1}^\infty \in f_N^{-1}(V_N) \subset f_N^{-1}(U_N) \subset \mathcal{U}$. Since the bonding maps are monotone, by Proposition 2.1.13, f_N is monotone. Hence, $f_N^{-1}(V_N)$ is a connected (Lemma 2.1.12) open subset of X_∞. Therefore, X_∞ is locally connected.

Q.E.D.

The next Proposition says that when the projection maps are surjective, this property is no longer true for the restriction of such maps to proper closed subsets.

2.1.15. Proposition. *Let $\{X_n, f_n^{n+1}\}$ be an inverse sequence with inverse limit X_∞. If Y is a proper closed subset of X_∞, then there exists $N \in \mathbb{N}$, such that $f_m(Y) \neq X_m$ for each $m \geq N$.*

Proof. Let $(x_n)_{n=1}^{\infty} \in X_{\infty} \setminus Y$. By Proposition 2.1.9, there exist $N \in \mathbb{N}$ and an open subset U_N of X_N such that

$$(x_n)_{n=1}^{\infty} \in f_N^{-1}(U_N) \subset X_{\infty} \setminus Y.$$

Hence, $f_N(Y) \subset X_N \setminus U_N$. Also, if $m > N$, then $f_m(Y) \subset X_m \setminus (f_N^m)^{-1}(U_N)$. To see this, let $z_m \in f_m(Y)$. Then there exists $(y_n)_{n=1}^{\infty} \in Y$ such that $f_m((y_n)_{n=1}^{\infty}) = z_m$. Since $Y \subset X_{\infty}$, $f_N^m(z_m) = f_N((y_n)_{n=1}^{\infty})$. Thus, $z_m \in (f_N^m)^{-1}(f_N((y_n)_{n=1}^{\infty})) \subset (f_N^m)^{-1}(f_N(Y))$. Since $f_N(Y) \subset X_N \setminus U_N$,

$$f_m(Y) \subset (f_N^m)^{-1}(f_N(Y)) \subset (f_N^m)^{-1}(X_N \setminus U_N) \subset X_n \setminus (f_N^m)^{-1}(U_N).$$

Therefore, $f_m(Y) \neq X_m$.

Q.E.D.

The following Proposition gives us the image of the inverse limit under a projection map in terms of the factor spaces and the bonding maps.

2.1.16. Proposition. *Let $\{X_n, f_n^{n+1}\}$ be an inverse sequence of compacta whose inverse limit is X_{∞}. Then $f_n(X_{\infty}) = \bigcap_{m=n+1}^{\infty} f_n^m(X_m)$ for each $n \in \mathbb{N}$.*

Proof. Let $n \in \mathbb{N}$. Since, for each $m > n$, $f_n = f_n^m \circ f_m$, $f_n(X_{\infty}) = f_n^m \circ f_m(X_{\infty}) \subset f_n^m(X_m)$. Therefore, $f_n(X_{\infty}) \subset \bigcap_{m=n+1}^{\infty} f_n^m(X_m)$.

Now, let $x_n \in \bigcap_{m=n+1}^{\infty} f_n^m(X_m)$. We show there exists an element of X_{∞} whose nth coordinate is x_n. To this end, let $K = \prod_{k=1}^{n-1} X_k \times \{x_n\} \times \prod_{k=n+1}^{\infty} X_k$. Recall that, by Proposition 2.1.8 (2), $X_{\infty} = \bigcap_{m=1}^{\infty} S_m$ and $\{S_m\}_{m=1}^{\infty}$ is a decreasing sequence. Let $m_0 \in \mathbb{N}$. Then $K \cap S_{m_0} \neq \emptyset$. To see this, let $m_1 > \max\{n, m_0\}$. Since $x_n \in \bigcap_{m=n+1}^{\infty} f_n^m(X_m)$, $(f_n^{m_1})^{-1}(x_n) \neq \emptyset$. Let $x_{m_1} \in (f_n^{m_1})^{-1}(x_n)$, and let $x_{m_0} = f_{m_0}^{m_1}(x_{m_1})$. Now, let $(y_k)_{k=1}^{\infty} \in \prod_{k=1}^{\infty} X_k$ such that $y_{m_1} = x_{m_1}$ and $y_k = f_k^{m_1}(x_{m_1})$ for each $k \in \{1, \ldots, m_1 - 1\}$. Note that $(y_k)_{k=1}^{\infty} \in K \cap S_{m_0}$. Thus, the family $\{K\} \cup \{S_m\}_{m=1}^{\infty}$ has the finite intersection property. Therefore, $K \cap X_{\infty} \neq \emptyset$. Hence, there exists an element of X_{∞} whose nth coordinate is x_n.

Q.E.D.

2.1.17. Proposition. *Let $\{X_n, f_n^{n+1}\}$ be an inverse sequence of compacta, whose inverse limit is X_{∞}. For each $n \in \mathbb{N}$, let A_n be a closed subset of X_n, and suppose that $f_n^{n+1}(A_{n+1}) \subset A_n$. Then $\{A_n, f_n^{n+1}|_{A_{n+1}}\}$ is an inverse sequence and $\varprojlim\{A_n, f_n^{n+1}|_{A_{n+1}}\} = \bigcap_{n=1}^{\infty} f_n^{-1}(A_n)$.*

Proof. Clearly, $\{A_n, f_n^{n+1}|_{A_{n+1}}\}$ is an inverse sequence of compacta.

Observe that, since $f_n^{n+1}(A_{n+1}) \subset A_n$, $f_{n+1}^{-1}(A_{n+1}) \subset f_n^{-1}(A_n)$ for every $n \in \mathbb{N}$. Hence, $\{f_n^{-1}(A_n)\}_{n=1}^{\infty}$ is a decreasing sequence of compacta.

Let $a = (a_n)_{n=1}^{\infty} \in \varprojlim\{A_n, f_n^{n+1}|_{A_{n+1}}\}$. Since $f_n(a) = a_n \in A_n$, $a \in f_n^{-1}(A)$ for every $n \in \mathbb{N}$. Thus, $a \in \bigcap_{n=1}^{\infty} f_n^{-1}(A_n)$. Therefore,

$$\varprojlim\{A_n, f_n^{n+1}|_{A_{n+1}}\} \subset \bigcap_{n=1}^{\infty} f_n^{-1}(A_n).$$

Next, let $b = (b_n)_{n=1}^{\infty} \in \bigcap_{n=1}^{\infty} f_n^{-1}(A_n)$. Then $f_n(b) = b_n \in A_n$ for each $n \in \mathbb{N}$. Hence, $b \in \prod_{n=1}^{\infty} A_n$. Since $b \in X_{\infty}$, $f_n^{n+1}(b_{n+1}) = b_n$ for every $n \in \mathbb{N}$. This implies that $b \in \varprojlim\{A_n, f_n^{n+1}|_{A_{n+1}}\}$. Thus,

$$\bigcap_{n=1}^{\infty} f_n^{-1}(A_n) \subset \varprojlim\{A_n, f_n^{n+1}|_{A_{n+1}}\}.$$

Therefore, $\bigcap_{n=1}^{\infty} f_n^{-1}(A_n) = \varprojlim\{A_n, f_n^{n+1}|_{A_{n+1}}\}$.

Q.E.D.

2.1.18. Definition. Let $\{X_n, f_n^{n+1}\}$ be an inverse sequence of continua. Then $\{X_n, f_n^{n+1}\}$ is called an *indecomposable inverse sequence* provided that, for each $n \in \mathbb{N}$, whenever A_{n+1} and B_{n+1} are subcontinua of X_{n+1} such that $X_{n+1} = A_{n+1} \cup B_{n+1}$, we have that $f_n^{n+1}(A_{n+1}) = X_n$ or $f_n^{n+1}(B_{n+1}) = X_n$.

The motivation of the name "indecomposable inverse sequence" is given in the following result:

2.1.19. Theorem. *Let $\{X_n, f_n^{n+1}\}$ be an indecomposable inverse sequence whose inverse limit is X_∞. Then X_∞ is an indecomposable continuum.*

Proof. By Proposition 2.1.8 (3), X_∞ is a continuum. Now, suppose X_∞ is decomposable. Then there exist two proper subcontinua A and B of X_∞ such that $X_\infty = A \cup B$. By Proposition 2.1.15, there exists $n \in \mathbb{N}$ such that if $m \geq n$, then $f_m(A) \neq X_m$ and $f_m(B) \neq X_m$. Since, by definition, the bonding maps are surjective, the projection maps are surjective too (Remark 2.1.6). Hence,

$$X_{n+2} = f_{n+2}(X_\infty) = f_{n+2}(A) \cup f_{n+2}(B).$$

This implies that

$$X_{n+1} = f_{n+1}^{n+2}(X_{n+2}) = f_{n+1}^{n+2} \circ f_{n+2}(A) \cup f_{n+1}^{n+2} \circ f_{n+2}(B).$$

By hypothesis $f_{n+1}^{n+2} \circ f_{n+2}(A) = X_{n+1}$ or $f_{n+1}^{n+2} \circ f_{n+2}(B) = X_{n+1}$. Thus, $f_{n+1}(A) = X_{n+1}$ or $f_{n+1}(B) = X_{n+1}$, contradicting the election of n. Therefore, X_∞ is indecomposable.

Q.E.D.

2.1.20. Proposition. *Let $\{X_n, f_n^{n+1}\}$ be an inverse sequence of metric spaces whose inverse limit is X_∞. If A is a closed subset of X_∞, then the double sequence $\{f_n(A), f_n^{n+1}|_{f_{n+1}(A)}\}$ is an inverse sequence with surjective bonding maps and*

$$(*) \quad \varprojlim \left\{f_n(A), f_n^{n+1}|_{f_{n+1}(A)}\right\} = A = \left[\prod_{n=1}^{\infty} f_n(A)\right] \cap X_\infty.$$

Proof. By Remark 2.1.6, for each $n \in \mathbb{N}$, $f_n^{n+1} \circ f_{n+1} = f_n$. It follows that $\{f_n(A), f_n^{n+1}|_{f_{n+1}(A)}\}$ is an inverse sequence with surjective bonding maps.

Now we show ($*$). First observe that

$$\varprojlim \{f_n(A), f_n^{n+1}|_{f_{n+1}(A)}\} = \left[\prod_{n=1}^{\infty} f_n(A)\right] \cap X_{\infty}.$$

Also observe that

$$A \subset \left[\prod_{n=1}^{\infty} f_n(A)\right] \cap X_{\infty}.$$

Thus, we need to show

$$\left[\prod_{n=1}^{\infty} f_n(A)\right] \cap X_{\infty} \subset A.$$

Let $\varepsilon > 0$, and let $y = (y_n)_{n=1}^{\infty} \in \left[\prod_{n=1}^{\infty} f_n(A)\right] \cap X_{\infty}$. Now, let

$N \in \mathbb{N}$ such that $\displaystyle\sum_{n=N+1}^{\infty} \frac{1}{2^n} < \varepsilon$.

Since for each $n \in \mathbb{N}$, $y_n \in f_n(A)$, there exists $a^{(n)} = \left(a_m^{(n)}\right)_{m=1}^{\infty} \in$
A such that $f_n\left(a^{(n)}\right) = y_n$. Now observe that

$$\rho(a^{(N)}, y) = \sum_{n=1}^{\infty} \frac{1}{2^n} d_n\left(a_n^{(N)}, y_n\right) =$$

$$\sum_{n=N+1}^{\infty} \frac{1}{2^n} d_n\left(a_n^{(N)}, y_n\right) < \varepsilon.$$

Since ε was arbitrary, $y \in Cl(A) = A$.

Q.E.D.

As a consequence of Propositions 2.1.17 and 2.1.20, we have the following Corollary:

2.1.21. Corollary. *Let $\{X_n, f_n^{n+1}\}$ be an inverse sequence of compacta, whose inverse limit is X_∞. If A is a closed subset of X_∞, then*

$$A = \bigcap_{n=1}^{\infty} f_n^{-1}(f_n(A)).$$

2.1.22. Proposition. *Let $\{X_n, f_n^{n+1}\}$ be an inverse sequence of metric spaces whose inverse limit is X_∞. If A and B are two closed subsets of X_∞, $C = A \cap B$ and $C_n = f_n(A) \cap f_n(B)$ for each $n \in \mathbb{N}$, then $C = \varprojlim\{C_n, f_n^{n+1}|_{C_{n+1}}\}$.*

Proof. Let $x = (x_n)_{n=1}^{\infty} \in \varprojlim\{C_n, f_n^{n+1}|_{C_{n+1}}\}$. Then, by definition, $x_n \in C_n = f_n(A) \cap f_n(B)$ for each $n \in \mathbb{N}$. Hence, $x \in \varprojlim\{f_n(A), f_n^{n+1}|_{f_{n+1}(A)}\} = A$ and $x \in \varprojlim\{f_n(B), f_n^{n+1}|_{f_{n+1}(B)}\} = B$ (by $(*)$ of Proposition 2.1.20). Thus, $x \in C$.

Now, let $y = (y_n)_{n=1}^{\infty} \in C = A \cap B$. Then for each $n \in \mathbb{N}$, $y_n \in f_n(A)$ and $y_n \in f_n(B)$. Thus, $y_n \in C_n$ for each $n \in \mathbb{N}$. Therefore, $y \in \varprojlim\{C_n, f_n^{n+1}|_{C_{n+1}}\}$.

$$\textbf{Q.E.D.}$$

2.1.23. Definition. *Let $\{X_n, f_n^{n+1}\}$ be an inverse sequence of arcs (i.e., for each $n \in \mathbb{N}$, X_n is an arc) with surjective bonding maps. Then the inverse limit, X_∞, of $\{X_n, f_n^{n+1}\}$ is called an *arc–like continuum*.*

Propositions 2.1.20 and 2.1.22 have the following Corollaries:

2.1.24. Corollary. *If X is an arc–like continuum, then each non-degenerate subcontinuum of X is arc–like.*

Proof. Let X be an arc–like continuum, and let A be a nondegenerate subcontinuum of X. Since X is an arc–like continuum, there exists an inverse sequence $\{X_n, f_n^{n+1}\}$ of arcs such that $X = \varprojlim\{X_n, f_n^{n+1}\}$. By $(*)$ of Propositions 2.1.20, we have that $A = \varprojlim\{f_n(h(A)), f_n^{n+1}|_{f_{n+1}(h(A))}\}$. Therefore, A is an arc–like continuum.

$$\textbf{Q.E.D.}$$

2.1.25. Corollary. *If $\{X_n, f_n^{n+1}\}$ is an inverse sequence of unicoherent continua, with surjective bonding maps, whose inverse limit is X_∞, then X_∞ is a unicoherent continuum.*

Proof. Let A and B be two subcontinua of X_∞ such that $X_\infty = A \cup B$. Since the bonding maps are surjective, by Remark 2.1.6, the projection maps are surjective too. Hence, for each $n \in \mathbb{N}$, $X_n = f_n(A) \cup f_n(B)$. By the unicoherence of the spaces, $f_n(A) \cap f_n(B)$ is connected for every $n \in \mathbb{N}$. Thus, by Propositions 2.1.20 and 2.1.22, $\{f_n(A) \cap f_n(B), f_n^{n+1}|_{f_{n+1}(A) \cap f_{n+1}(B)}\}$ is an inverse sequence of continua whose inverse limit is a continuum (Proposition 2.1.8 (3)). Since $A \cap B = \varprojlim\{f_n(A) \cap f_n(B), f_n^{n+1}|_{f_{n+1}(A) \cap f_{n+1}(B)}\}$ (Proposition 2.1.22) $A \cap B$ is connected. Therefore, X_∞ is unicoherent.

Q.E.D.

2.1.26. Corollary. *Let $\{X_n, f_n^{n+1}\}$ be an inverse sequence of hereditarily unicoherent continua. If X_∞ is the inverse limit of $\{X_n, f_n^{n+1}\}$, then X_∞ is a hereditarily unicoherent continuum.*

Proof. Let A and B be two subcontinua of X_∞. If $A \cap B = \emptyset$, then $A \cap B$ is connected. Thus, suppose that $A \cap B \neq \emptyset$. Then for each $n \in \mathbb{N}$, $f_n(A)$ and $f_n(B)$ are two subcontinua of X_n such that $f_n(A) \cap f_n(B) \neq \emptyset$. Since each X_n is hereditarily unicoherent, $f_n(A) \cap f_n(B)$ is connected. Hence, by Propositions 2.1.20 and 2.1.22, $\{f_n(A) \cap f_n(B), f_n^{n+1}|_{f_{n+1}(A) \cap f_{n+1}(B)}\}$ is an inverse sequence of continua whose inverse limit is a continuum (Proposition 2.1.8 (3)). Since $A \cap B = \varprojlim\{f_n(A) \cap f_n(B), f_n^{n+1}|_{f_{n+1}(A) \cap f_{n+1}(B)}\}$ (Proposition 2.1.22) $A \cap B$ is connected. Therefore, X_∞ is hereditarily unicoherent.

Q.E.D.

2.1.27. Remark. Note that in Corollary 2.1.26 we do not require the bonding maps to be surjective.

2.1.28. Corollary. *Each arc–like continuum is hereditarily unicoherent.*

Proof. The Corollary follows from the easy fact that an arc is hereditarily unicoherent and from Corollary 2.1.26.

Q.E.D.

2.1.29. Corollary. *If X is an arc–like continuum, then X does not contain a simple closed curve.*

2.1.30. Remark. Let $\{X_n, f_n^{n+1}\}$ be an inverse sequence of arcs, with surjective bonding maps, whose inverse limit is X_∞. Note that, in this case, S_m is homeomorphic to the Hilbert cube $\mathcal{Q}$ (being a countable product of arcs). Since $X_\infty = \bigcap_{m=1}^{\infty} S_m$ (Proposition 2.1.8 (2)), X_∞ may be written as a countable intersection of Hilbert cubes.

The next Theorem tells us a way to define a map from a metric space into an inverse limit of compacta.

2.1.31. Theorem. *Let Y be a metric space. Let $\{X_n, f_n^{n+1}\}$ be a sequence of compacta. If for each $n \in \mathbb{N}$, there is a map $h_n \colon Y \to X_n$ such that $f_n^{n+1} \circ h_{n+1} = h_n$, then there exists a map $h_\infty \colon Y \to X_\infty$ such that $f_n \circ h_\infty = h_n$ for each $n \in \mathbb{N}$. The map h_∞ is called induced map, and it is denoted, also, by $\varprojlim\{h_n\}$.*

Proof. Let $h_\infty \colon Y \to X_\infty$ be given by $h_\infty(y) = (h_n(y))_{n=1}^{\infty}$. Since $f_n^{n+1} \circ h_{n+1} = h_n$ for each $n \in \mathbb{N}$, h_∞ is well defined. Clearly, $f_n \circ h_\infty = h_n$. Hence, by Theorem 1.1.9, h_∞ is continuous.

Q.E.D.

2.1.32. Theorem. *Let Y be a metric space. Let $\{X_n, f_n^{n+1}\}$ be a sequence of compacta. Suppose that for each $n \in \mathbb{N}$, there is a map $h_n \colon Y \to X_n$ such that $f_n^{n+1} \circ h_{n+1} = h_n$. If h_m is one–to–one for some $m \in \mathbb{N}$, then the induced map h_∞ is one–to–one.*

Proof. Suppose h_m is one–to–one for some $m \in \mathbb{N}$. Let $y, y' \in Y$ such that $y \neq y'$. Since h_m is one–to–one, $h_m(y) \neq h_m(y')$. Hence, $h_\infty(y) \neq h_\infty(y')$. Therefore, h_∞ is one–to–one.

Q.E.D.

2.1.33. Theorem. *Let Y be a metric space. Let $\{X_n, f_n^{n+1}\}$ be a sequence of compacta. Suppose that for each $n \in \mathbb{N}$, there is a map $h_n \colon Y \to X_n$ such that $f_n^{n+1} \circ h_{n+1} = h_n$. If each h_n is surjective, then $h_\infty(Y)$ is dense in X_∞. In particular, if Y is compact, then h_∞ is surjective.*

Proof. Let $f_n^{-1}(U_n)$ be a basic open set of X_∞, where U_n is an open subset of X_n (Proposition 2.1.9). Since h_n is surjective, there exists $y \in Y$ such that $h_n(y) \in U_n$. Hence, $h_\infty(y) \in f_n^{-1}(U_n)$. Therefore, $h_\infty(Y)$ is dense in X_∞.

Q.E.D.

2.1.34. Definition. For each $n \in \mathbb{N}$, let $X_n = \mathcal{S}^1$, and let $f_n^{n+1} \colon X_{n+1} \twoheadrightarrow X_n$ be given by $f_n^{n+1}(z) = z^2$ (complex number multiplication). Let $\Sigma_2 = \varprojlim\{X_n, f_n^{n+1}\}$. Then Σ_2 is called the *dyadic solenoid*. Note that, by Theorem 2.1.19, Σ_2 is an indecomposable continuum.

The following Example shows that an induced map may not be surjective.

2.1.35. Example. Let Σ_2 be the dyadic solenoid. For each $n \in \mathbb{N}$, let $h_n \colon \mathbb{R} \twoheadrightarrow \mathcal{S}^1$ be given by $h_n(t) = \exp\left(\dfrac{2\pi t}{2^{n-1}}\right)$. Let $n \in \mathbb{N}$. Then $f_n^{n+1} \circ h_{n+1}(t) = f_n^{n+1}\left(\exp\left(\dfrac{2\pi t}{2^n}\right)\right) = \left(\exp\left(\dfrac{2\pi t}{2^n}\right)\right)^2 = \exp\left(\dfrac{2\pi t}{2^{n-1}}\right) = h_n(t)$. Hence, $f_n^{n+1} \circ h_{n+1} = h_n$ for each $n \in \mathbb{N}$. By Theorem 2.1.31, there exists the induced map $h_\infty \colon \mathbb{R} \to \Sigma_2$. Observe that h_∞ is not surjective since $h_\infty(\mathbb{R})$ is an arcwise connected subset of Σ_2. But, since Σ_2 is indecomposable, it is not arcwise connected (This follows easily from 11.15 of [18]).

The next two Propositions present some expected results.

2.1.36. Proposition. *Let $\{X_n\}_{n=1}^{\infty}$ be a sequence of compacta such that $X_{n+1} \subset X_n$ for each $n \in \mathbb{N}$. If $f_n^{n+1}\colon X_{n+1} \to X_n$ is the inclusion map, then $\varprojlim\{X_n, f_n^{n+1}\}$ is homeomorphic to $\bigcap_{n=1}^{\infty} X_n$.*

Proof. Note that if $(x_n)_{n=1}^{\infty} \in \varprojlim\{X_n, f_n^{n+1}\}$, then $x_n = x_1$ for every $n \in \mathbb{N}$ and $x_1 \in \bigcap_{n=1}^{\infty} X_n$.

For each $n \in \mathbb{N}$, let $h_n\colon \bigcap_{n=1}^{\infty} X_n \to X_n$ be given by $h_n(x) = x$, i.e., h_n is the inclusion map. Clearly $f_n^{n+1} \circ h_{n+1} = h_n$ for every $n \in \mathbb{N}$. By Theorem 2.1.31, there exists the induced map $h_{\infty}\colon \bigcap_{n=1}^{\infty} X_n \to \varprojlim\{X_n, f_n^{n+1}\}$. Since h_1 is one–to–one, h_{∞} is one–to–one (Theorem 2.1.32). Now, if $(x_n)_{n=1}^{\infty} \in \varprojlim\{X_n, f_n^{n+1}\}$, then $h_{\infty}(x_1) = (x_n)_{n=1}^{\infty}$. Hence, h_{∞} is surjective. Since $\bigcap_{n=1}^{\infty} X_n$ is compact, h_{∞} is a homeomorphism.

Q.E.D.

2.1.37. Proposition. *Let $\{X_n, f_n^{n+1}\}$ be an inverse sequence of compacta whose inverse limit is X_{∞}. If for each $n \in \mathbb{N}$, X_n is homeomorphic to a compactum Y and all the bonding maps are homeomorphisms, then X_{∞} is homeomorphic to Y.*

Proof. Let $h_1\colon Y \to X_1$ be any homeomorphism. For $n \geq 2$, let $h_{n+1} = (f_n^{n+1})^{-1} \circ h_n$. Observe that, by definition, $f_n^{n+1} \circ h_{n+1} = h_n$ for each $n \in \mathbb{N}$. Then, by Theorem 2.1.31, there exists the induced map $h_{\infty}\colon Y \to X_{\infty}$. Since each h_n is a homeomorphism and Y is compact, by Theorems 2.1.32 and 2.1.33, h_{∞} is a homeomorphism.

Q.E.D.

The next Theorem says that certain type of subsequences of an inverse sequence converge to the same limit.

2.1.38. Theorem. *Let $\{X_n, f_n^{n+1}\}$ be an inverse sequence of compacta, with surjective bonding maps, whose inverse limit is X_∞, and let $\{m(n)\}_{n=1}^\infty$ be an increasing subsequence of $\mathbb{N}$. If $Y_\infty = \varprojlim\{Y_n, g_n^{n+1}\}$, where $Y_n = X_{m(n)}$ and $g_n^{n+1} = f_{m(n)}^{m(n+1)}$ for each $n \in \mathbb{N}$, then Y_∞ is homeomorphic to X_∞.*

Proof. For each $k \in \mathbb{N}$, let $h_k \colon X_\infty \to Y_k$ be given by $h_k((x_n)_{n=1}^\infty) = x_{m(k)}$, i.e., $h_k = f_{m(k)}$. Hence, h_k is continuous and $g_k^{k+1} \circ h_{k+1} = h_k$. Then, by Theorem 2.1.31, there exists the induced map $h_\infty \colon X_\infty \to Y_\infty$. Since the bonding maps are surjective, by Remark 2.1.6, the projections are surjective. In particular, each h_k is surjective. Thus, h_∞ is surjective (Theorem 2.1.33). To see h_∞ is one–to–one, it is enough to observe that if $(x_n)_{n=1}^\infty$ and $(x_n')_{n=1}^\infty$ are two distinct points of X_∞, then there exists $N \in \mathbb{N}$ such that $x_n \neq x_n'$ for each $n \geq N$.
Q.E.D.

2.1.39. Notation. We use the notation $\{X_n, f_n^{n+1}\}_{n=N}^\infty$ to denote the inverse subsequence obtained from $\{X_n, f_n^{n+1}\}$ by removing the first $N - 1$ factor spaces.

Next, we present some consequences of Theorem 2.1.38. First, we need the following definition:

2.1.40. Definition. A continuum X is said to be a *triod*, provided that there exists a subcontinuum M of X such that $X \setminus M = K_1 \cup K_2 \cup K_3$, where each $K_j \neq \emptyset$, $j \in \{1, 2, 3\}$, and they are mutually separated, i.e., $Cl_X(K_j) \cap K_\ell = \emptyset$, $j, \ell \in \{1, 2, 3\}$ and $j \neq \ell$.

2.1.41. Corollary. *If X is an arc–like continuum, then X does not contain a triod.*

Proof. Since X is an arc–like continuum, there exists an inverse sequence $\{X_n, f_n^{n+1}\}$ of arcs whose inverse limit is X.

Suppose Y is a triod contained in X. Then there exists a subcontinuum M of Y such that $Y \setminus M = K_1 \cup K_2 \cup K_3$, where each $K_j \neq \emptyset$, $j \in \{1, 2, 3\}$, and they are mutually separated. Note that $M \cup K_j$ is a continuum (Lemma 1.7.18).

Let $n \in \mathbb{N}$. Then $f_n(Y) = f_n(M \cup K_1 \cup K_2 \cup K_3) = f_n(M) \cup f_n(K_1) \cup f_n(K_2) \cup f_n(K_3)$. Since X_n is an arc, there exists $j_n \in \{1, 2, 3\}$ such that $f_n(M \cup K_{j_n}) \subset f_n(M \cup K_{\ell_1}) \cup f_n(M \cup K_{\ell_2})$, $\ell_1, \ell_2 \in \{1, 2, 3\} \setminus \{j_n\}$ and $\ell_1 \neq \ell_2$. Since the set of positive integers is infinite and we only have three choices, by Theorem 2.1.38, we assume, without loss of generality, that $f_n(M \cup K_3) \subset f_n(M \cup K_1) \cup f_n(M \cup K_2)$ for each $n \in \mathbb{N}$.

Note that, by $(*)$ of Proposition 2.1.20,

$$M \cup K_j = \varprojlim \{f_n(M \cup K_j), f_n^{n+1}|_{f_{n+1}(M \cup K_j)}\}.$$

Since $f_n(M \cup K_3) \subset f_n(M \cup K_1) \cup f_n(M \cup K_2)$ for each $n \in \mathbb{N}$, it follows that $M \cup K_3 \subset (M \cup K_1) \cup (M \cup K_2) = M \cup K_1 \cup K_2$, a contradiction. Therefore, X does not contain a triod.

Q.E.D.

2.1.42. Definition. Let $\{X_n, f_n^{n+1}\}$ be an inverse sequence of simple closed curves with surjective bonding maps. Then the inverse limit, X_∞, of such an inverse sequence is called a *circle–like continuum*.

2.1.43. Corollary. *If X is a circle–like continuum, then each nondegenerate proper subcontinuum of X is arc–like.*

Proof. Since X is circle–like, there exists an inverse sequence $\{X_n, f_n^{n+1}\}$ of simple closed curves whose inverse limit is X.

Let Z be a nondegenerate proper subcontinuum of X. Note that by $(*)$ of Proposition 2.1.20, $Z = \varprojlim \{f_n(Z), f_n^{n+1}|_{f_{n+1}(Z)}\}$. By Proposition 2.1.15, there exists $N \in \mathbb{N}$ such that $f_n(Z) \neq X_n$ for each $n \geq N$. Hence, $f_n(Z)$ is an arc for every $n \geq N$.

Applying Theorem 2.1.38, we have that Z is homeomorphic to $\varprojlim \{f_n(Z), f_n^{n+1}|_{f_{n+1}(Z)}\}_{n=N}^\infty$. Therefore, Z is an arc–like continuum.

Q.E.D.

2.1.44. Corollary. *If X is a circle–like continuum, then X does not contain a triod.*

2.1.45. Theorem. *Let $\{X_n, f_n^{n+1}\}$ be an inverse sequence of compacta whose inverse limit is X_∞. If for each $n \in \mathbb{N}$, X_n has at most k components, for some $k \in \mathbb{N}$, then X_∞ has at most k components.*

Proof. Suppose X_∞ has more than k components. Let $\mathcal{C}_1, \ldots, \mathcal{C}_{k+1}$ be $k + 1$ distinct components of X_∞. For each $n \in \mathbb{N}$, let us consider the set $\{f_n(\mathcal{C}_1), \ldots, f_n(\mathcal{C}_{k+1})\}$. Since X_n has at most k components, there exist $i_n, j_n \in \{1, \ldots, k+1\}$ such that $f_n(\mathcal{C}_{i_n})$ and $f_n(\mathcal{C}_{j_n})$ are contained in the same component of X_n. Since $\{1, \ldots, k+1\}$ is a finite set, there exist $i_0, j_0 \in \{1, \ldots, k+1\}$ and a subsequence $\{n_\ell\}_{\ell=1}^\infty$ of $\mathbb{N}$ such that for each $\ell \in \mathbb{N}$, $f_{n_\ell}(\mathcal{C}_{i_0})$ and $f_{n_\ell}(\mathcal{C}_{j_0})$ are contained in the same component Y_{n_ℓ} of X_{n_ℓ}. Note that, by Theorem 2.1.38 and Proposition 2.1.20, $\mathcal{C}_{i_0}$ is homeomorphic to $\mathcal{C}'_{i_0} = \varprojlim\{f_{n_\ell}(\mathcal{C}_{i_0}), f_{n_\ell}^{n_{\ell+1}}|_{f_{n_{\ell+1}}(\mathcal{C}_{i_0})}\}$ and $\mathcal{C}_{j_0}$ is homeomorphic to $\mathcal{C}'_{j_0} = \varprojlim\{f_{n_\ell}(\mathcal{C}_{j_0}), f_{n_\ell}^{n_{\ell+1}}|_{f_{n_{\ell+1}}(\mathcal{C}_{j_0})}\}$.

Let $Y_\infty = \varprojlim\{Y_{n_\ell}, f_{n_\ell}^{n_{\ell+1}}|_{Y_{n_{\ell+1}}}\}$. Then, by Proposition 2.1.8 and Theorem 2.1.38, Y_∞ is a subcontinuum of $X'_\infty = \varprojlim\{X_{n_\ell}, f_{n_\ell}^{n_{\ell+1}}\}$, which is homeomorphic to X_∞ (Theorem 2.1.38). Note that $\mathcal{C}'_{i_0} \cup \mathcal{C}'_{j_0} \subset Y_\infty$, a contradiction. Therefore, X_∞ has at most k components.

Q.E.D.

The next Theorem gives us a way to define a map between inverse limits.

2.1.46. Theorem. *Let $\{X_n, f_n^{n+1}\}$ and $\{Y_n, g_n^{n+1}\}$ be inverse sequences of compacta, whose inverse limit are X_∞ and Y_∞, respectively. If for each $n \in \mathbb{N}$, there is a map $k_n \colon X_n \to Y_n$ such that $k_n \circ f_n^{n+1} = g_n^{n+1} \circ k_{n+1}$, then there exists a map $k_\infty \colon X_\infty \to Y_\infty$ such that $g_n \circ k_\infty = k_n \circ f_n$. The map k_∞ is called* induced map, *and it is denoted, also, by $\varprojlim\{k_n\}$.*

Proof. For each $n \in \mathbb{N}$, let $h_n \colon X_\infty \to Y_n$ be given by $h_n = k_n \circ f_n$. Note that $g_n^{n+1} \circ h_{n+1} = h_n$. Hence, by Theorem 2.1.31, there exists the induced map $h_\infty \colon X_\infty \to Y_\infty$. Let $k_\infty = h_\infty$. Then $g_n \circ k_\infty = g_n \circ h_\infty = h_n = k_n \circ f_n$.

Q.E.D.

2.1.47. Theorem. *Let $\{X_n, f_n^{n+1}\}$ and $\{Y_n, g_n^{n+1}\}$ be inverse sequences of compacta, whose inverse limit are X_∞ and Y_∞, respectively. Suppose that for each $n \in \mathbb{N}$, there is a map $k_n \colon X_n \to Y_n$ such that $k_n \circ f_n^{n+1} = g_n^{n+1} \circ k_{n+1}$. If all the maps k_n are one–to–one, then the induced map k_∞ is one–to–one.*

Proof. Let $(x_\ell)_{\ell=1}^\infty$ and $(x_\ell')_{\ell=1}^\infty$ be two distinct points of X_∞. Then there exists $m \in \mathbb{N}$ such that $x_m \neq x_m'$. Since k_m is one–to–one, $k_m(x_m) \neq k_m(x_m')$. Hence, $k_\infty((x_\ell)_{\ell=1}^\infty) = (k_n \circ f_n((x_\ell)_{\ell=1}^\infty))_{n=1}^\infty = (k_n(x_n))_{n=1}^\infty \neq (k_n(x_n'))_{n=1}^\infty = k_\infty((x_\ell')_{\ell=1}^\infty)$. Therefore, k_∞ is one–to–one.

Q.E.D.

2.1.48. Theorem. *Let $\{X_n, f_n^{n+1}\}$ and $\{Y_n, g_n^{n+1}\}$ be inverse sequences of compacta, with surjective bonding maps, whose inverse limit are X_∞ and Y_∞, respectively. Suppose that for each $n \in \mathbb{N}$, there is a map $k_n \colon X_n \to Y_n$ such that $k_n \circ f_n^{n+1} = g_n^{n+1} \circ k_{n+1}$. If all the maps k_n are surjective, then the induced map k_∞ is surjective.*

Proof. Let $g_m^{-1}(V_m)$ be a basic open set in Y_∞, where V_m is an open subset of Y_m (Proposition 2.1.9). Since k_m is surjective, there exists $z_m \in X_m$ such that $k_m(z_m) \in V_m$. Since for each $n \in \mathbb{N}$, f_n^{n+1} is surjective, the projection maps, f_n, are surjective (Remark 2.1.6). Hence, there exists $(x_n)_{n=1}^\infty \in X_\infty$ such that $f_m((x_n)_{n=1}^\infty) = z_m$. Then $g_m \circ k_\infty((x_n)_{n=1}^\infty) = k_m \circ f_m((x_n)_{n=1}^\infty) = k_m(z_m) \in V_m$. Thus, $k_\infty((x_n)_{n=1}^\infty) \in g_m^{-1}(V_m)$. Therefore, $k_\infty(X_\infty)$ is dense in Y_∞. Since X_∞ is compact, $k_\infty(X_\infty) = Y_\infty$.

Q.E.D.

2.1.49. Theorem. *Let $\{X_n, f_n^{n+1}\}$ and $\{Y_n, g_n^{n+1}\}$ be inverse sequences of compacta, whose inverse limit are X_∞ and Y_∞, respectively. Suppose that for each $n \in \mathbb{N}$, $k_n \colon Y_n \to X_n$ and $h_{n+1} \colon X_{n+1} \to Y_n$ are surjective maps such that $k_n \circ h_{n+1} = f_n^{n+1}$ and $h_{n+1} \circ k_{n+1} = g_n^{n+1}$. Then there exists a map $h_\infty \colon X_\infty \to Y_\infty$ such that $h_\infty = k_\infty^{-1}$, where $k_\infty = \varprojlim\{k_n\}$. In particular, k_∞ is a homeomorphism. Also, X_∞ and Y_∞ are homeomorphic.*

Proof. Let $h_\infty \colon X_\infty \to Y_\infty$ be given by $h_\infty((x_n)_{n=1}^\infty) = (h_n(x_n))_{n=2}^\infty$. By Theorem 1.1.9, h_∞ is continuous.

Now, let $(y_n)_{n=1}^\infty \in Y_\infty$. Then

$$
\begin{aligned}
h_\infty \circ k_\infty((y_n)_{n=1}^\infty) &= h_\infty((k_n(y_n))_{n=1}^\infty) = (h_n k_n(y_n))_{n=2}^\infty \\
&= (g_{n-1}^n(y_n))_{n=2}^\infty = (y_n)_{n=1}^\infty,
\end{aligned}
$$

i.e., $h_\infty \circ k_\infty = 1_{Y_\infty}$.

Next, let $(x_n)_{n=1}^\infty \in X_\infty$. Then

$$
\begin{aligned}
k_\infty \circ h_\infty((x_n)_{n=1}^\infty) &= k_\infty((h_n(x_n))_{n=2}^\infty) = (k_{n-1} h_n(x_n))_{n=2}^\infty \\
&= (f_{n-1}^n(x_n))_{n=2}^\infty = (x_n)_{n=1}^\infty,
\end{aligned}
$$

i.e., $k_\infty \circ h_\infty = 1_{X_\infty}$.

Therefore, $h_\infty = k_\infty^{-1}$.

Q.E.D.

Next, we prove a Theorem due to Anderson and Choquet, which tells us when we can embed an inverse limit in a compactum.

2.1.50. Theorem. *Let X be a compactum, with metric d. Let $\{X_n, f_n^{n+1}\}$ be an inverse sequence of closed subsets of X with surjective bonding maps. Assume:*

(1) For each $\varepsilon > 0$, there exists $k \in \mathbb{N}$ such that for all $x \in X_k$,

$$
\operatorname{diam}\left[\bigcup_{j \geq k} \left(f_k^j\right)^{-1}(x)\right] < \varepsilon;
$$

(2) For each $n \in \mathbb{N}$ and each $\delta > 0$, there exists $\delta' > 0$ such that whenever $j > n$ and $x, y \in X_j$, then $d(x, y) > \delta'$.

Then $\varprojlim\{X_n, f_n^{n+1}\}$ is homeomorphic to $\bigcap_{n=1}^\infty \left[Cl\left(\bigcup_{m \geq n} X_m\right)\right]$. In particular, if $X_n \subset X_{n+1}$ for each $n \in \mathbb{N}$, then $\varprojlim\{X_n, f_n^{n+1}\}$ is homeomorphic to $Cl\left(\bigcup_{n=1}^\infty X_n\right)$.

Proof. Let $X_\infty = \varprojlim\{X_n, f_n^{n+1}\}$. If $x = (x_n)_{n=1}^\infty \in X_\infty$, then, by (1), we have that $(x_n)_{n=1}^\infty$ is a Cauchy sequence in X. Thus, $(x_n)_{n=1}^\infty$

converges to a point in X, which we call $h(x)$. Hence, we have defined a function $h\colon X_\infty \to X$. We see that h is an embedding.

First, we show h is continuous. Let $x = (x_n)_{n=1}^\infty \in X_\infty$, and let $\varepsilon > 0$. Now, choose $k \in \mathbb{N}$ as guaranteed by (1). Let $\ell = k+1$, and let

$$V = f_\ell^{-1}(\mathcal{V}_\varepsilon^d(h(x)) \cap X_\ell).$$

Since $x \in X_\infty$ and k satisfies (1), we have from the definition of h that $d(x_\ell, h(x)) < \varepsilon$. Hence, $x_\ell \in \mathcal{V}_\varepsilon^d(h(x)) \cap X_\ell$. Thus, $x \in V$. Next, observe that, since $\mathcal{V}_\varepsilon^d(h(x)) \cap X_\ell$ is open in X_ℓ, V is open in X_∞. Now, let $y = (y_n)_{n=1}^\infty \in V$. For the same reasons as for x, $d(y_\ell, h(y)) < \varepsilon$. Since $y \in V$, $y_\ell \in \mathcal{V}_\varepsilon^d(h(x))$. Hence,

$$d(h(y), h(x)) \leq d(h(y), y_\ell) + d(y_\ell, h(x)) < 2\varepsilon.$$

Therefore, h is continuous.

To prove h is one–to–one, let $x = (x_n)_{n=1}^\infty$ and $y = (y_n)_{n=1}^\infty$ be two distinct points of X_∞. Then there exists $n \in \mathbb{N}$, such that $x_n \neq y_n$. Let $\delta = \dfrac{d(x_n, y_n)}{2}$. Then, by (2), there exists $\delta' > 0$ such that whenever $j > n$, then, since

$$d(f_n^j(x_j), f_n^j(y_j)) = d(x_n, y_n) = 2\delta > \delta$$

we have that $d(x_j, y_j) > \delta'$. Therefore, $h(x) \neq h(y)$.

It remains to see that $h(X_\infty) = \bigcap_{n=1}^\infty \left[Cl\left(\bigcup_{m \geq n} X_m \right) \right]$. For convenience, let

$$Z = \bigcap_{n=1}^\infty \left[Cl\left(\bigcup_{m \geq n} X_m \right) \right].$$

It is clear, from the definition of h, that $h(X_\infty) \subset Z$. To show $h(X_\infty) = Z$, we prove that $h(X_\infty)$ is dense in Z.

For this purpose, let $z \in Z$ and let $\varepsilon > 0$. Let $k \in \mathbb{N}$ be as guaranteed by (1). Since $z \in Z$, there exists $m \geq k$ such that $d(z, p) < \varepsilon$ for some $p \in X_m$. Since the bonding maps are surjective, by Remark 2.1.6, there exists $x = (x_n)_{n=1}^\infty \in X_\infty$ such that $f_m(x) = f_m((x_n)_{n=1}^\infty) = x_m = p$.

Note that, since $m \geq k$, condition (1) holds for m in place of k. Hence, it follows that $d(z, h(x)) < 2\varepsilon$. Thus, $h(X_\infty)$ is dense in Z. Therefore, since X_∞ is compact, $h(X_\infty) = Z$.

Q.E.D.

We end this section proving that any continuum is homeomorphic to an inverse limit of polyhedra.

2.1.51. Theorem. *Each continuum is homeomorphic to an inverse limit of polyhedra.*

Proof. Let X be a continuum. By Theorem 1.1.16, we assume that X is contained in the Hilbert cube $\mathcal{Q}$.

For each $n \in \mathbb{N}$, let d_n denote the Euclidean metric on $[0,1]^n$, and let $\pi_n \colon \mathcal{Q} \to [0,1]^n$ and $\pi_n^{n+1} \colon [0,1]^{n+1} \to [0,1]^n$ be the projection maps. By Lemma 1.7.31, there exists a polyhedron P_1 in $[0,1]$ such that $\pi_1(X) \subset Int(P_1) \subset P_1 \subset \mathcal{V}_1^{d_1}(\pi_1(X))$. Note that $\pi_2(X) \subset (\pi_1^2)^{-1}(Int(P_1))$. Hence, by Lemma 1.7.31, there exists a polyhedron P_2 in $[0,1]^2$ such that $\pi_2(X) \subset Int(P_2) \subset P_2 \subset \mathcal{V}_{\frac{1}{2}}^{d_2}(\pi_2(X))$ and $\pi_1^2(P_2) \subset Int(P_1)$. Continue in this way to define a sequence, $\{P_n\}_{n=1}^\infty$, of polyhedra such that $\pi_n(X) \subset Int(P_n) \subset P_n \subset \mathcal{V}_{\frac{1}{n}}^{d_n}(\pi_n(X))$ and $\pi_n^{n+1}(P_{n+1}) \subset Int(P_n)$. Hence, $\{P_n, \pi_n^{n+1}|_{P_{n+1}}\}$ is an inverse sequence. Let $P_\infty = \varprojlim\{P_n, \pi_n^{n+1}|_{P_{n+1}}\}$. We show that X is homeomorphic to P_∞.

For each $n \in \mathbb{N}$, let $h_n \colon X \to P_n$ by given by $h_n = \pi_n|_X$. Observe that, by construction, if $x \in X$, then $\pi_n^{n+1}(\pi_{n+1}(x)) = \pi_n(x)$ for each $n \in \mathbb{N}$. Thus, $\pi_n^{n+1}(h_{n+1}(x)) = h_n(x)$ for every $n \in \mathbb{N}$. Hence, the sequence $\{h_n\}_{n=1}^\infty$ induces a map $h_\infty \colon X \to P_\infty$ (Theorem 2.1.31). To see h_∞ is one–to–one, let x and x' be two distinct points of X. Then there exists $n \in \mathbb{N}$, such that $\pi_n(x) \neq \pi_n(x')$, i.e., $h_n(x) \neq h_n(x')$. Thus, $h_\infty(x) \neq h_\infty(x')$. Therefore, h_∞ is one–to–one.

Clearly, if z_1 and z_2 are two points of $[0,1]^{k+1}$, then

$$d_k(\pi_k^{k+1}(z_1), \pi_k^{k+1}(z_2)) \leq d_{k+1}(z_1, z_2).$$

Now, let $\bar{p} = (\bar{p}_k)_{k=1}^\infty \in P_\infty$, and let $n \in \mathbb{N}$. Then, by Proposition 2.1.16, $\bar{p}_n \in \bigcap_{m=n+1}^\infty \pi_n^m(P_m)$. Since $h_n(X) = \pi_n^{n+1} \circ h_{n+1}(X)$,

$d_n(\overline{p}_n, h_n(X)) \leq d_{n+1}(\overline{p}_{n+1}, h_{n+1}(X))$. Hence,

$$d_n(\overline{p}_n, h_n(X)) \leq d_m(\overline{p}_m, h_m(X)) \leq \frac{1}{m}$$

for each $m > n$. Therefore, $\overline{p}_n \in \pi_n(X) = h_n(X)$.

Finally, suppose h_∞ is not surjective. Then there exists a point $\overline{p} = (\overline{p}_k)_{k=1}^\infty \in P_\infty \setminus h_\infty(X)$. By Proposition 2.1.9, there exist $N \in \mathbb{N}$ and an open subset U_N of P_N such that $(\overline{p}_k)_{k=1}^\infty \in (\pi_N|_{P_\infty})^{-1}(U_N) \subset P_\infty \setminus h_\infty(X)$. Hence, $\overline{p}_N \in U_N \setminus h_N(X)$, a contradiction with the preceding paragraph. Therefore, h_∞ is surjective.

Q.E.D.

Note that in Theorem 2.1.51 the bonding maps are not surjective. The following result tells us that each continuum can be written as an inverse limit of polyhedra with surjective bonding maps; a proof of this may be found in Theorem 2 of [12].

2.1.52. Theorem. *Each continuum is homeomorphic to an inverse limit of polyhedra with surjective bonding maps.*

2.2   Inverse Limits and the Cantor Set

The purpose of this section is to characterize the Cantor set $\mathcal{C}$ (Example 1.6.4) and to show that every compactum is a continuous image of $\mathcal{C}$.

We begin by showing that $\mathcal{C}$ is an inverse limit of finite discrete spaces. First, we need the following definition:

2.2.1. Definition. If $x \in \mathbb{R}$, then $[x] = \max\{n \in \mathbb{N} \mid n \leq x\}$ is called the *greatest integer function*.

2.2.2. Theorem. *For each $n \in \mathbb{N}$, let $X_n = \{0, 1, \ldots, 2^n - 1\}$ with the discrete topology, and define $f_n^{n+1} \colon X_{n+1} \to X_n$ by $f_n^{n+1}(x) = \left[\frac{x}{2}\right]$. If $X_\infty = \varprojlim\{X_n, f_n^{n+1}\}$, then X_∞ is homeomorphic to $\mathcal{C}$.*

Proof. For each $n \in \mathbb{N}$, let $h_n \colon \mathcal{C} \to X_n$ be given by $h_n(x) = k$ if $x \in I_{n,k}$ (Example 1.6.4). Clearly, h_n is surjective for each $n \in \mathbb{N}$. To see h_n is continuous, it is enough to observe that for each $k \in X_n$, $h_n^{-1}(k) = I_{n,k} \cap \mathcal{C}$, which is an open subset of $\mathcal{C}$.

Let $n \in \mathbb{N}$. Observe that, by the construction of $\mathcal{C}$, $I_{n+1,2\ell} \cup I_{n+1,2\ell+1} \subset I_{n,\ell}$. Hence, if $x \in \mathcal{C}$ and $h_{n+1}(x) = k$, then $h_n(x) = \left[\frac{k}{2}\right]$. Consequently, if $x \in \mathcal{C}$, then $f_n^{n+1} \circ h_{n+1}(x) = \left[\frac{h_{n+1}(x)}{2}\right] = h_n(x)$. Therefore, $f_n^{n+1} \circ h_{n+1} = h_n$. Thus, by Theorem 2.1.31, there exists the induced map $h_\infty \colon \mathcal{C} \to X_\infty$. By Theorem 2.1.33, h_∞ is surjective.

To see h_∞ is one–to–one, let x and x' be two distinct points of $\mathcal{C}$. Then there exist $n \in \mathbb{N}$ and $j, k \in \{0, \ldots, 2^n - 1\}$ such that $j \neq k$, $x \in I_{n,j}$ and $x' \in I_{n,k}$. Hence, $h_n(x) \neq h_n(x')$. Thus, h_∞ is one–to–one.

Therefore, h_∞ is a homeomorphism.

$$\textbf{Q.E.D.}$$

Now, we present some results about totally disconnected compacta.

2.2.3. Proposition. *If X is a totally disconnected compactum, then the family of all open and closed subsets of X forms a basis for the topology of X.*

Proof. Let U be an open subset of X, and let $x \in U$. Since X is totally disconnected, no connected subcontinuum of X intersects both $\{x\}$ and $X \setminus U$. By Theorem 1.6.8, there exist two disjoint open and closed subsets X_1 and X_2 of X such that $X = X_1 \cup X_2$, $\{x\} \subset X_1$ and $X \setminus U \subset X_2$. Hence, $x \in X_1 \subset U$.

$$\textbf{Q.E.D.}$$

2.2.4. Definition. Let X be a metric space. If $\mathcal{U}$ is a family of subsets of X, then the *mesh of* $\mathcal{U}$, denoted by $\mathrm{mesh}(\mathcal{U})$, is defined by:

$$\mathrm{mesh}(\mathcal{U}) = \max\{\mathrm{diam}(U) \mid U \in \mathcal{U}\}.$$

2.2.5. Definition. Let X be a metric space. If $\mathcal{U}$ and $\mathcal{V}$ are two coverings of X, then $\mathcal{V}$ is a *refinement of* $\mathcal{U}$ provided that for each $V \in \mathcal{V}$, there exists $U \in \mathcal{U}$ such that $V \subset U$. If $\mathcal{V}$ refines $\mathcal{U}$, we write $\mathcal{V} \prec \mathcal{U}$.

2.2.6. Theorem. *If X is a totally disconnected compactum, then there exists a sequence $\{\mathcal{U}_n\}_{n=1}^{\infty}$ of finite coverings of X such that for each $n \in \mathbb{N}$:*

(i) $\mathcal{U}_{n+1} \prec \mathcal{U}_n$;

(ii) $\operatorname{mesh}(\mathcal{U}_n) < \dfrac{1}{n}$;

(iii) *if $U \in \mathcal{U}_n$, then U is open and closed in X; and*

(iv) *if $U, V \in \mathcal{U}_n$ and $U \neq V$, then $U \cap V = \emptyset$.*

Proof. We construct $\mathcal{U}_1$ first. By Proposition 2.2.3, for each $x \in X$, there exists an open and closed subset U_x of X such that $x \in U_x$ and $\operatorname{diam}(U_x) < 1$. Hence, $\{U_x \mid x \in X\}$ is an open cover of X. Since X is compact, there exist $x_{11}, \dots, x_{1k_1} \in X$ such that $X = \bigcup_{j=1}^{k_1} U_{x_{1j}}$. The sets $U_{x_{11}}, \dots, U_{x_{1k_1}}$ are not necessarily pairwise disjoint. But this can be fixed. Let $U_{11} = U_{x_{11}}$ and for $j \in \{2, \dots, k_1\}$, let $U_{1j} = U_{x_{1j}} \setminus \bigcup_{\ell=1}^{j-1} U_{1\ell}$. Then $\mathcal{U}_1 = \{U_{11}, \dots, U_{1k_1}\}$ is a finite family of pairwise disjoint open and closed subsets of X covering X.

Let λ_1 be a Lebesgue number of the open covering $\mathcal{U}_1$ (Theorem 1.6.6). Without loss of generality, we assume that $\lambda_1 < \dfrac{1}{2}$. By Proposition 2.2.3, for each $x \in X$, there exists an open and closed subset V_x of X such that $x \in V_x$ and $\operatorname{diam}(V_x) < \lambda_1$. Hence, the family $\{V_x \mid x \in X\}$ forms an open cover of X. Since X is compact, there exist $x_{21}, \dots, x_{2k_2} \in X$ such that $X = \bigcup_{j=1}^{k_2} V_{x_{2j}}$. Again, the elements of $\{V_{x_{21}}, \dots, V_{x_{2k_2}}\}$ may not be pairwise disjoint. Hence, let $U_{21} = V_{x_{21}}$ and for each $j \in \{2, \dots, k_2\}$, let $U_{2j} = V_{x_{2j}} \setminus \bigcup_{\ell=1}^{j-1} V_{2\ell}$. Then $\mathcal{U}_2 = \{U_{21}, \dots, U_{2k_2}\}$ is a finite family of pairwise disjoint open and closed subsets of X covering X. Note that, by construction, $\mathcal{U}_2 \prec \mathcal{U}_1$.

Proceeding by induction, we obtain the result.

Q.E.D.

2.2.7. Theorem. *If X is a totally disconnected compactum, then there exists an inverse sequence $\{X_n, f_n^{n+1}\}$ of finite discrete spaces such that $\varprojlim\{X_n, f_n^{n+1}\}$ is homeomorphic to X.*

Proof. Let $\{\mathcal{U}_n\}_{n=1}^{\infty}$ be a sequence of finite coverings given by Theorem 2.2.6. For each $n \in \mathbb{N}$, let $X_n = \mathcal{U}_n$. Give X_n the discrete topology. We define the bonding maps f_n^{n+1} as follows: If $U_{n+1} \in X_{n+1}$, then $f_n^{n+1}(U_{n+1}) = U_n$, where U_n is the unique element of $\mathcal{U}_n = X_n$ such that $U_{n+1} \subset U_n$. Hence, $\{X_n, f_n^{n+1}\}$ is an inverse sequence. Let $X_\infty = \varprojlim\{X_n, f_n^{n+1}\}$.

For each $n \in \mathbb{N}$, let $h_n \colon X \to X_n$ be given by $h_n(x) = U$, where U is the unique element of X_n such that $x \in U$. Let $n \in \mathbb{N}$. Note that, by construction, h_n is surjective and $f_n^{n+1} \circ h_{n+1} = h_n$. Also note that if $U \in X_n$, then $h_n^{-1}(U) = U$. Then h_n is continuous.

By Theorem 2.1.31, there exists the induced map $h_\infty \colon X \to X_\infty$. Since X is compact and each h_n is surjective, by Theorem 2.1.33, h_∞ is surjective.

To see h_∞ is one–to–one, let x and y be two distinct points of X. Since $\lim_{n \to \infty} \operatorname{mesh}(\mathcal{U}_n) = 0$, there exist $n \in \mathbb{N}$ and $U, V \in \mathcal{U}_n$ such that $x \in U$ and $y \in V$. Hence, $h_n(x) \neq h_n(y)$, since $U \cap V = \emptyset$. Therefore, h_∞ is one–to–one.

Q.E.D.

2.2.8. Lemma. *If X is a totally disconnected and perfect compactum, then for each $n \in \mathbb{N}$, there exist n nonempty pairwise disjoint open and closed subsets of X whose union is X.*

Proof. The proof is done by induction on n. For $n = 1$, X itself is open and closed. Let $n \geq 2$ and assume that there exist n nonempty pairwise disjoint open and closed subsets, $U_1, \ldots, U_n$, of X such that $X = \bigcup_{j=1}^{n} U_j$. Let $x \in U_n$. Since X is perfect, $U_n \neq \{x\}$. By

Proposition 2.2.3, there exists an open and closed subset of X such that $x \in V \subset U_n$. Hence, $X = U_1 \cup \ldots U_{n-1} \cup V \cup (U_n \setminus V)$.

Q.E.D.

We are ready to prove the characterization of the Cantor set, $\mathcal{C}$, mentioned above.

2.2.9. Theorem. *If X and Y are two totally disconnected and perfect compacta, then X and Y are homeomorphic. In particular, X and Y are homeomorphic to $\mathcal{C}$ (Example 1.6.4).*

Proof. By Proposition 2.2.3 and Lemma 2.2.8, there exist two open covers $\mathcal{U}_1$ and $\mathcal{U}_1'$ of X and Y, respectively, such that the following four conditions hold:

(i) The elements of each cover are open and closed.

(ii) The elements of each cover are pairwise disjoint.

(iii) The mesh of each cover is less than one.

(iv) Both covers have the same cardinality.

Let $h_1 \colon \mathcal{U}_1 \twoheadrightarrow \mathcal{U}_1'$ be any bijection. Apply Proposition 2.2.3 to obtain refinements $\mathcal{U}_2$ and $\mathcal{U}_2'$ of $\mathcal{U}_1$ and $\mathcal{U}_1'$, respectively, of mesh less than $\dfrac{1}{2}$ and satisfying conditions (i) and (ii) above.

By Lemma 2.2.8, we assume that if $U_1 \in \mathcal{U}_1$, then U_1 and $h_1(U_1)$ contain the same number of elements of $\mathcal{U}_2$ and $\mathcal{U}_2'$, respectively. It follows that there is a bijection $h_2 \colon \mathcal{U}_2 \twoheadrightarrow \mathcal{U}_2'$ such that if $U_2 \in \mathcal{U}_2$, $U_1 \in \mathcal{U}_1$ and $U_2 \subset U_1$, then $h_2(U_2) \subset h_1(U_1)$. This process can be repeated inductively to obtain two inverse sequences, $\{\mathcal{U}_n, f_n^{n+1}\}$ and $\{\mathcal{U}_n', (f')_n^{n+1}\}$, and a sequence $\{h_n\}_{n=1}^{\infty}$ of bijections such that we have the following infinite ladder:

$$
\begin{array}{ccccccccc}
\cdots & \longleftarrow & \mathcal{U}_{n-1} & \overset{f_{n-1}^n}{\longleftarrow} & \mathcal{U}_n & \overset{f_n^{n+1}}{\longleftarrow} & \mathcal{U}_{n+1} & \longleftarrow \cdots & : \mathcal{U}_\infty \\
& & \downarrow{\scriptstyle h_{n-1}} & & \downarrow{\scriptstyle h_n} & & \downarrow{\scriptstyle h_{n+1}} & & \downarrow{\scriptstyle h_\infty} \\
\cdots & \longleftarrow & \mathcal{U}_{n-1}' & \underset{(f')_{n-1}^n}{\longleftarrow} & \mathcal{U}_n' & \underset{(f')_n^{n+1}}{\longleftarrow} & \mathcal{U}_{n+1}' & \longleftarrow \cdots & : \mathcal{U}_\infty'
\end{array}
$$

where each $\mathcal{U}_n$ and each $\mathcal{U}_n'$ have the discrete topology, and the functions f_n^{n+1} and $(f')_n^{n+1}$ are defined as in the proof of Theorem 2.2.7. Since each h_n is a homeomorphism, by Theorems 2.1.47

and 2.1.48, h_∞ is a homeomorphism. It follows from the proof of Theorem 2.2.7 that $\mathcal{U}_\infty$ and $\mathcal{U}'_\infty$ are homeomorphic to X and Y, respectively. Therefore, X and Y are homeomorphic.

$$\textbf{Q.E.D.}$$

2.2.10. Lemma. *If X is a totally disconnected compactum, then $X \times \mathcal{C}$ is homeomorphic to $\mathcal{C}$.*

Proof. Let X be a totally disconnected compactum. By Theorem 2.2.9, we only need to show that $X \times \mathcal{C}$ is perfect. Let $(x, t) \in X \times \mathcal{C}$, and let $U \times V$ be a basic open set of $X \times \mathcal{C}$ such that $(x, t) \in U \times V$. Since $\mathcal{C}$ is perfect (Example 1.6.4), there exists $t' \in V \setminus \{t\}$. Hence, $(x, t') \in U \times V$. Therefore, $X \times \mathcal{C}$ is perfect.

$$\textbf{Q.E.D.}$$

The characterization of $\mathcal{C}$ in Theorem 2.2.9 allows us to show that any compactum is a continuous image of $\mathcal{C}$.

2.2.11. Theorem. *If X is a compactum, with metric d, then there exists a surjective map $f\colon \mathcal{C} \twoheadrightarrow X$ of the Cantor set onto X.*

Proof. Since X is compact, using Theorem 1.6.6 (Lebesgue numbers), there exists a sequence $\{\mathcal{U}_n\}_{n=1}^\infty$ of finite covers of X such that, for each $n \in \mathbb{N}$, the following three conditions are satisfied:

(i) each $U \in \mathcal{U}_n$ is the closure of an open subset of X;

(ii) $\mathrm{mesh}(\mathcal{U}_n) < \dfrac{1}{2^n}$; and

(iii) $\mathcal{U}_{n+1} \prec \mathcal{U}_n$.

Let $\mathcal{U}_1 = \{U_{11}, \dots, U_{1k_1}\}$. The elements of $\mathcal{U}_1$ may not be pairwise disjoint. We use the following trick. For each $i \in \{1, \dots, k_1\}$, let $V_{1i} = U_{1i} \times \{i\}$. Then we say that a subset $W \times \{i\}$ is open in V_{1i} if W is an open subset of U_{1i}. Let $\mathcal{V}_1 = \bigcup_{i=1}^{k_1} V_{1i}$. We define a metric d_1 on $\mathcal{V}_1$ as follows:

$$d_1((x, i), (y, j)) = \begin{cases} d(x, y) & \text{if } i = j; \\ 1 & \text{if } i \neq j. \end{cases}$$

Now, let $\mathcal{U}_2 = \{U_{21}, \dots, U_{2k_2}\}$. For each $U_{2j} \in \mathcal{U}_2$, let $i \in \{1, \dots, k_1\}$ such that $U_{2j} \subset U_{1i}$. Then let $V_{2ij} = U_{2j} \times \{i\} \times \{j\}$ (whenever $U_{2j} \subset U_{1i}$). Again, we say that a subset $W \times \{i\} \times \{j\}$ is open in V_{2ij} if W is open in U_{2j}. Let $\mathcal{V}_2 = \bigcup_{i,j} V_{2ij}$. We define a metric d_2 in a similar way as we defined d_1 on $\mathcal{V}_1$. Finally, let $f_1^2 \colon \mathcal{V}_2 \to \mathcal{V}_1$ be given by $f_1^2(u,i,j) = (u,i)$. Clearly, f_1^2 is continuous.

Now, let $\mathcal{U}_3 = \{U_{31}, \dots, U_{3k_3}\}$. For each $U_{3k} \in \mathcal{U}_3$, let $j \in \{1, \dots, k_1\}$ and $i \in \{1, \dots, k_1\}$ such that $U_{3k} \subset U_{2j} \subset U_{1i}$. Then define $V_{3ijk} = U_{3k} \times \{i\} \times \{j\} \times \{k\}$. As before, a subset $W \times \{i\} \times \{j\} \times \{k\}$ is open in V_{3ijk} if W is open in U_{3k}. Let $\mathcal{V}_3 = \bigcup_{i,j,k} V_{3ijk}$. We define a metric on $\mathcal{V}_3$ in the same way we defined d_1 on $\mathcal{V}_1$. Finally, let $f_2^3 \colon \mathcal{V}_3 \to \mathcal{V}_2$ be given by $f_2^3(u,i,j,k) = (u,i,j)$. Clearly, f_2^3 is continuous. Although it is notationally complicated, the general inductive step is clear now. Hence, we obtain an inverse sequence $\{\mathcal{V}_n, f_n^{n+1}\}$ of compacta. Let $\mathcal{V}_\infty = \varprojlim\{\mathcal{V}_n, f_n^{n+1}\}$. Then $\mathcal{V}_\infty$ is a compactum.

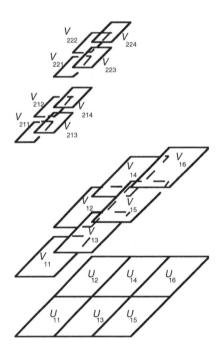

A second inverse sequence, $\{X_n, g_n^{n+1}\}$, may be constructed letting for each $n \in \mathbb{N}$, $X_n = X$ and $g_n^{n+1} = 1_X$. It follows, from

Proposition 2.1.37, that X is homeomorphic to $\varprojlim\{X_n, g_n^{n+1}\}$.

For each $n \in \mathbb{N}$, let $h_n \colon \mathcal{V}_n \to X_n$ be given by $h_n(u, i, j, \ldots, p) = u$. Thus, the following diagram

$$
\begin{array}{ccccccccc}
\cdots & \longleftarrow & \mathcal{V}_{n-1} & \overset{f_{n-1}^n}{\longleftarrow} & \mathcal{V}_n & \overset{f_n^{n+1}}{\longleftarrow} & \mathcal{V}_{n+1} & \longleftarrow & \cdots & : \mathcal{V}_\infty \\
 & & \downarrow h_{n-1} & & \downarrow h_n & & \downarrow h_{n+1} & & & \downarrow h_\infty \\
\cdots & \longleftarrow & X_{n-1} & \underset{g_{n-1}^n}{\longleftarrow} & X_n & \underset{g_n^{n+1}}{\longleftarrow} & X_{n+1} & \longleftarrow & \cdots & : X_\infty
\end{array}
$$

is commutative, where $X_\infty = \varprojlim\{X_n, g_n^{n+1}\}$ and $h_\infty = \varprojlim\{h_n\}$. Since each h_n is surjective and each $\mathcal{V}_n$ is a compactum, by Theorem 2.1.48, h_∞ is surjective.

Now, we show that $\mathcal{V}_\infty$ is totally disconnected. To this end, we prove that for any two distinct points of $\mathcal{V}_\infty$, there exists an open and closed subset of $\mathcal{V}_\infty$ containing one of them and not containing the other.

Let $x = (x_n)_{n=1}^\infty$ and $y = (y_n)_{n=1}^\infty$ be two distinct points of $\mathcal{V}_\infty$. Then there exists $n_0 \in \mathbb{N}$ such that $x_{n_0} \neq y_{n_0}$ and n_0 is the smallest with this property.

First, suppose that $n_0 = 1$, i.e., $x_1 \neq y_1$. Suppose, also, that $x_1 = (u, i)$ and $y_1 = (u', i')$. If $u \neq u'$, then there exist $m \in \mathbb{N}$ and $U_{m\ell}, U_{m\ell'} \in \mathcal{U}_m$ such that $u \in U_{m\ell}$, $u' \in U_{m\ell'}$ and $U_{m\ell} \cap U_{m\ell'} = \emptyset$. Then the corresponding sets $V_{mij\ldots\ell}$ and $V_{mij\ldots\ell'}$ are disjoint and both are open and closed subsets of $\mathcal{V}_m$. Hence, $f_m^{-1}(V_{mij\ldots\ell})$ is an open and closed subset of $\mathcal{V}_\infty$ containing x and not containing y.

If $u = u'$, then $i \neq i'$. Hence, $x_1 \in V_{1i}$ and $y_1 \in V_{1i'}$. Note that V_{1i} and $V_{1i'}$ are disjoint open and closed subsets of $\mathcal{V}_1$. Then $f_1^{-1}(V_{1i})$ is an open and closed subset of $\mathcal{V}_\infty$ containing x and not containing y.

Now, suppose $n_0 \geq 2$. For simplicity, we assume that $n_0 = 3$. Then $x_3 \neq y_3$, $x_3 = (u, i, j, k)$ and $y = (u, i, j, k')$, with $k \neq k'$. Hence, $x_3 \in V_{3ijk}$ and $y_3 \in V_{3ijk'}$. Note that V_{3ijk} and $V_{3ijk'}$ are disjoint open and closed subsets of $\mathcal{V}_3$. Then $f_3^{-1}(V_{ijk})$ is an open and closed subset of $\mathcal{V}_\infty$ containing x and not containing y.

Therefore, $\mathcal{V}_\infty$ is totally disconnected.

Even though $\mathcal{V}_\infty$ may not be perfect, by Lemma 2.2.10, $\mathcal{V}_\infty \times \mathcal{C}$ is perfect. Hence, by Theorem 2.2.9, $\mathcal{V}_\infty \times \mathcal{C}$ is homeomorphic to $\mathcal{C}$. Thus, we have the following maps:

(a) A homeomorphism $\xi \colon \mathcal{C} \to \mathcal{V}_\infty \times \mathcal{C}$;

(b) A surjective map $\zeta\colon \mathcal{V}_\infty \times \mathcal{C} \to \mathcal{V}_\infty$ given by $\zeta(v,c) = v$ (the projection map);

(c) A surjective map $h_\infty\colon \mathcal{V}_\infty \to X_\infty$; and

(d) A homeomorphism $\varphi\colon X_\infty \to X$.

Therefore, $f = \varphi \circ h_\infty \circ \zeta \circ \xi$ is a surjective map from $\mathcal{C}$ onto X.

Q.E.D.

2.3   Inverse Limits and Other Operations

We show that taking inverse limits commute with products, cones and hyperspaces.

2.3.1. Theorem. *Let* $\{X_n, f_n^{n+1}\}$ *and* $\{Y_n, g_n^{n+1}\}$ *be inverse sequences of compacta, with surjective bonding maps, whose inverse limits are* X_∞ *and* Y_∞*, respectively. Then* $\{X_n \times Y_n, f_n^{n+1} \times g_n^{n+1}\}$ *is an inverse sequence and* $\varprojlim\{X_n \times Y_n, f_n^{n+1} \times g_n^{n+1}\}$ *is homeomorphic to* $X_\infty \times Y_\infty$.

Proof. Let $n \in \mathbb{N}$. By Theorem 1.1.10, the function $f_n \times g_n$ is continuous.

For each $n \in \mathbb{N}$, let $h_n\colon X_\infty \times Y_\infty \to X_n \times Y_n$ be given by $h_n = f_n \times g_n$. Then $(f_n^{n+1} \times g_n^{n+1}) \circ h_{n+1} = h_n$. Hence, by Theorem 2.1.31, there exists the induced map $h_\infty\colon X_\infty \times Y_\infty \to \varprojlim\{X_n \times Y_n, f_n^{n+1} \times g_n^{n+1}\}$. Since all the bonding maps are surjective, h_n is surjective for each $n \in \mathbb{N}$. Then h_∞ is surjective by Theorem 2.1.33. Clearly, h_∞ is one–to–one.

Q.E.D.

The next Theorem shows that taking inverse limits and taking cones commute.

2.3.2. Theorem. *Let* $\{X_n, f_n^{n+1}\}$ *be an inverse sequence of compacta whose inverse limit is* X_∞. *Then* $\{K(X_n), K(f_n^{n+1})\}$ *is an* *inverse sequence of cones with induced bonding maps and $K(X_\infty)$ is homeomorphic to $\varprojlim \{K(X_n), K(f_n^{n+1})\}$.*

Proof. By Proposition 1.2.12, $K(f_n^{n+1})$ is a map from $K(X_{n+1})$ into $K(X_n)$ for each $n \in \mathbb{N}$. Hence, $\{K(X_n), K(f_n^{n+1})\}$ is an inverse sequence of continua. Let $Y_\infty = \varprojlim\{K(X_n), K(f_n^{n+1})\}$.

Note that for each $n \in \mathbb{N}$, by Proposition 1.2.12, $K(f_n) \colon K(X_\infty) \to K(X_n)$ is a map. Clearly, $K(f_n^{n+1}) \circ K(f_{n+1}) = K(f_n)$ for every $n \in \mathbb{N}$. Hence, by Theorem 2.1.31, there exists an induced map $K(f)_\infty \colon K(X_\infty) \to Y_\infty$. To see $K(f)_\infty$ is a homeomorphism, we show it is a bijection.

Note that $K(f)_\infty(\nu_{X_\infty}) = (\nu_{X_n})_{n=1}^\infty$. If $((x_n, t))_{n=1}^\infty \in Y_\infty \setminus \{(\nu_{X_n})_{n=1}^\infty\}$, then $((x_n)_{n=1}^\infty, t)$ is a point of $K(X_\infty)$ satisfying that $K(f)_\infty(((x_n)_{n=1}^\infty, t)) = ((x_n, t))_{n=1}^\infty$. Hence, $K(f)_\infty$ is surjective.

To see $K(f)_\infty$ is one–to–one, note that $(K(f)_\infty)^{-1}((\nu_{X_n})_{n=1}^\infty) = \{\nu_{X_\infty}\}$. Now, if $((x_n)_{n=1}^\infty, t)$ and $((x_n')_{n=1}^\infty, t')$ are two points of $K(X_\infty)$ such that

$$K(f)_\infty(((x_n)_{n=1}^\infty, t)) = K(f)_\infty(((x_n')_{n=1}^\infty, t')),$$

then $((x_n, t))_{n=1}^\infty = ((x_n', t'))_{n=1}^\infty$. Hence, $x_n = x_n'$ for each $n \in \mathbb{N}$, and $t = t'$. Thus, $((x_n)_{n=1}^\infty, t) = ((x_n')_{n=1}^\infty, t')$. Consequently, $K(f)_\infty$ is one–to–one.

Therefore, $K(f)_\infty$ is a homeomorphism.

$$\textbf{Q.E.D.}$$

The following Lemma gives a base for the hyperspace of subcompacta of an inverse limit in terms of the open subsets of the factor spaces and the projection maps.

2.3.3. Lemma. *Let $\{X_n, f_n^{n+1}\}$ be an inverse sequence of compacta whose inverse limit is X_∞. For each $j \in \mathbb{N}$, let*

$$\mathcal{B}_j = \{\langle f_j^{-1}(U_1), \dots, f_j^{-1}(U_k)\rangle \mid U_1, \dots, U_k \text{ are open subsets of } X_j\}.$$

If $\mathcal{B} = \bigcup_{j=1}^\infty \mathcal{B}_j$, then $\mathcal{B}$ is a base for the Vietoris topology for 2^{X_∞}.

Proof. By Proposition 2.1.9 and Theorem 1.8.14,

$$\mathcal{B}^* = \{\langle f_{n(1)}^{-1}(U_{n(1)}), \dots, f_{n(m)}^{-1}(U_{n(m)})\rangle \mid U_{n(\ell)} \text{ is an open}$$

$$\text{subset of } X_{n(\ell)} \text{ for each } \ell \in \{1, \dots, m\},\ m \in \mathbb{N}\}$$

is a base for the Vietoris topology for 2^{X_∞}. We show that $\mathcal{B} = \mathcal{B}^*$. Clearly, $\mathcal{B} \subset \mathcal{B}^*$.

Let $\langle f_{n(1)}^{-1}(U_{n(1)}), \dots, f_{n(m)}^{-1}(U_{n(m)})\rangle \in \mathcal{B}^*$, and let

$$k = \max\{n(1), \dots, n(m)\}.$$

Note that, since $f_{n(\ell)}^k \circ f_k = f_{n(\ell)}$, we have that for every $\ell \in \{1, \dots, m\}$, $f_{n(\ell)}^{-1}(U_{n(\ell)}) = f_k^{-1}(f_{n(\ell)}^k)^{-1}(U_{n(\ell)})$.

For each $\ell \in \{1, \dots, m\}$, let $V_{n(\ell)} = (f_{n(\ell)}^k)^{-1}(U_{n(\ell)})$. Then, since the bonding maps are continuous, each $V_{n(\ell)}$ is an open subset of X_k. Hence,

$$\langle f_k^{-1}(V_{n(1)}), \dots, f_k^{-1}(V_{n(m)})\rangle \in \mathcal{B}.$$

Since $f_k^{-1}(V_{n(\ell)}) = f_{n(\ell)}^{-1}(U_{n(\ell)})$ for each $\ell \in \{1, \dots, m\}$,

$$\langle f_k^{-1}(V_{n(1)}), \dots, f_k^{-1}(V_{n(m)})\rangle = \langle f_{n(1)}^{-1}(U_{n(1)}), \dots, f_{n(m)}^{-1}(U_{n(m)})\rangle.$$

Hence, $\langle f_{n(1)}^{-1}(U_{n(1)}), \dots, f_{n(m)}^{-1}(U_{n(m)})\rangle \in \mathcal{B}$.

Therefore, $\mathcal{B} = \mathcal{B}^*$.

$$\text{Q.E.D.}$$

2.3.4. Theorem. *Let $\{X_n, f_n^{n+1}\}$ be an inverse sequence of continua whose inverse limit is X_∞. Then the following hold.*

(1) 2^{X_∞} is homeomorphic to $\varprojlim\{2^{X_n}, 2^{f_n^{n+1}}\}$;

(2) $\mathcal{C}_m(X_\infty)$ is homeomorphic to $\varprojlim\{\mathcal{C}_m(X_n), \mathcal{C}_m(f_n^{n+1})\}$ for each $m \in \mathbb{N}$; and

(3) $\mathcal{F}_m(X_\infty)$ is homeomorphic to $\varprojlim\{\mathcal{F}_m(X_n), \mathcal{F}_m(f_n^{n+1})\}$ for each $m \in \mathbb{N}$.

Furthermore, there is a homeomorphism

$$h \colon \varprojlim\{2^{X_n}, 2^{f_n^{n+1}}\} \twoheadrightarrow 2^{X_\infty}$$

such that for each $m \in \mathbb{N}$,

$$h\left(\varprojlim\{\mathcal{C}_m(X_n), \mathcal{C}_m(f_n^{n+1})\}\right) = \mathcal{C}_m(X_\infty)$$

and

$$h\left(\varprojlim\{\mathcal{F}_m(X_n), \mathcal{F}_m(f_n^{n+1})\}\right) = \mathcal{F}_m(X_\infty).$$

Proof. We define the homeomorphism h after making some observations about the points of $\varprojlim\{2^{X_n}, 2^{f_n^{n+1}}\}$.

Let $(A_n)_{n=1}^\infty \in \varprojlim\{2^{X_n}, 2^{f_n^{n+1}}\}$. Then, by definition, $2^{f_n^{n+1}}(A_{n+1})$ $= A_n$. Thus, $\{A_n, f_n^{n+1}|_{A_{n+1}}\}$ is an inverse sequence with surjective bonding maps.

Since, for each $n \in \mathbb{N}$, $A_n \subset X_n$, we have that $\varprojlim\{A_n, f_n^{n+1}|_{A_{n+1}}\}$ $\subset \varprojlim\{X_n, f_n^{n+1}\} = X_\infty$. Hence, $\varprojlim\{A_n, f_n^{n+1}|_{A_{n+1}}\} \in 2^{X_\infty}$.

Now, let $h\colon \varprojlim\{2^{X_n}, 2^{f_n^{n+1}}\} \twoheadrightarrow 2^{X_\infty}$ be given by

$$h((A_n)_{n=1}^\infty) = \varprojlim\{A_n, f_n^{n+1}|_{A_{n+1}}\}.$$

By the above considerations, h is well defined.

Let $K \in 2^{X_\infty}$. Then K is a closed subset of X_∞. Hence, by $(*)$ of Proposition 2.1.20, $K = \varprojlim\{f_n(K), f_n^{n+1}|_{f_{n+1}(K)}\}$. Note that $f_n(K) \in 2^{X_n}$ and $2^{f_n^{n+1}}(f_{n+1}(K)) = f_n(K)$ for each $n \in \mathbb{N}$. Thus, $(f_n(K))_{n=1}^\infty \in \varprojlim\{2^{X_n}, 2^{f_n^{n+1}}\}$, and $h((f_n(K))_{n=1}^\infty) = K$. Therefore, h is surjective.

Next, let $(A_n)_{n=1}^\infty$ and $(B_n)_{n=1}^\infty$ be two elements of $\varprojlim\{2^{X_n}, 2^{f_n^{n+1}}\}$ such that $h((A_n)_{n=1}^\infty) = h((B_n)_{n=1}^\infty)$. We prove that $A_k = B_k$ for every $k \in \mathbb{N}$.

Let $k \in \mathbb{N}$, and let $p \in A_k$. Since $\{A_n, f_n^{n+1}|_{A_{n+1}}\}$ is an inverse sequence with surjective bonding maps, there exists a point $(x_n)_{n=1}^\infty \in h((A_n)_{n=1}^\infty)$ such that $x_k = p$ (Remark 2.1.6). Since $h((A_n)_{n=1}^\infty) = h((B_n)_{n=1}^\infty)$, $(x_n)_{n=1}^\infty \in h((B_n)_{n=1}^\infty)$. Hence, $x_k \in B_k$, i.e., $p \in B_k$. Therefore, $A_k \subset B_k$. A similar argument shows that $B_k \subset A_k$. Thus, $A_k = B_k$. Therefore, h is one–to–one.

Now, we see that h is continuous. Let $\pi_n\colon \varprojlim\{2^{X_n}, 2^{f_n^{n+1}}\} \to 2^{X_n}$ be the projection map for every $n \in \mathbb{N}$.

Let $\langle f_j^{-1}(U_1), \ldots, f_j^{-1}(U_k)\rangle \in \mathcal{B}$ (Lemma 2.3.3). We show that

$$h^{-1}(\langle f_j^{-1}(U_1), \ldots, f_j^{-1}(U_k)\rangle)$$

is an open subset of $\varprojlim\{2^{X_n}, 2^{f_n^{n+1}}\}$. To this end, it suffices to prove that

$$h^{-1}(\langle f_j^{-1}(U_1), \ldots, f_j^{-1}(U_k)\rangle) = \pi_j^{-1}(\langle U_1, \ldots, U_k\rangle).$$

First observe that if $(A_n)_{n=1}^\infty \in \varprojlim\{2^{X_n}, 2^{f_n^{n+1}}\}$, then $f_j(h((A_n)_{n=1}^\infty)) = A_j$ for each $j \in \mathbb{N}$. Hence,

$$h^{-1}(\langle f_j^{-1}(U_1), \ldots, f_j^{-1}(U_k)\rangle) =$$

$$\left\{ (A_n)_{n=1}^\infty \in \varprojlim\{2^{X_n}, 2^{f_n^{n+1}}\} \ \middle| \ h((A_n)_{n=1}^\infty) \subset \bigcup_{\ell=1}^k f_j^{-1}(U_\ell) \text{ and} \right.$$

$$\left. h((A_n)_{n=1}^\infty) \cap f_j^{-1}(U_\ell) \neq \emptyset, \text{ for each } \ell \in \{1, \ldots, k\} \right\} =$$

$$\left\{ (A_n)_{n=1}^\infty \in \varprojlim\{2^{X_n}, 2^{f_n^{n+1}}\} \ \middle| \ h((A_n)_{n=1}^\infty) \subset f_j^{-1}\left(\bigcup_{\ell=1}^k U_\ell\right) \text{ and} \right.$$

$$\left. h((A_n)_{n=1}^\infty) \cap f_j^{-1}(U_\ell) \neq \emptyset, \text{ for each } \ell \in \{1, \ldots, k\} \right\} =$$

$$\left\{ (A_n)_{n=1}^\infty \in \varprojlim\{2^{X_n}, 2^{f_n^{n+1}}\} \ \middle| \ f_j(h((A_n)_{n=1}^\infty)) \subset \bigcup_{\ell=1}^k U_\ell \text{ and} \right.$$

$$\left. f_j(h((A_n)_{n=1}^\infty)) \cap U_\ell \neq \emptyset, \text{ for each } \ell \in \{1, \ldots, k\} \right\} =$$

$$\left\{ (A_n)_{n=1}^\infty \in \varprojlim\{2^{X_n}, 2^{f_n^{n+1}}\} \ \middle| \ A_j \subset \bigcup_{\ell=1}^k U_\ell \text{ and} \right.$$

$$\left. A_j \cap U_\ell \neq \emptyset, \text{ for each } \ell \in \{1, \ldots, k\} \right\} =$$

$$\pi_j^{-1}(\langle U_1, \ldots, U_k\rangle).$$

Therefore, h is continuous. Hence, h is a homeomorphism.

Observe that by the definition of h and by Theorem 2.1.45, it follows that

$$h\left(\varprojlim\{\mathcal{C}_m(X_n), \mathcal{C}_m(f_n^{n+1})\}\right) = \mathcal{C}_m(X_\infty)$$

and that

$$h\left(\varprojlim\{\mathcal{F}_m(X_n), \mathcal{F}_m(f_n^{n+1})\}\right) = \mathcal{F}_m(X_\infty)$$

for every $m \in \mathbb{N}$.

Q.E.D.

2.4   Chainable Continua

We discuss an important class of continua, namely, chainable continua. Besides presenting some of its properties, we prove that it coincides with the class of arc–like continua.

2.4.1. Definition. Let X be a compactum. A *chain* $\mathcal{U}$ *in* X is a finite sequence, $U_1, \ldots, U_n$, of subsets of X such that $U_i \cap U_j \neq \emptyset$ if and only if $|i - j| \leq 1$ for each $i, j \in \{1, \ldots, n\}$. Each U_j is called a *link* of $\mathcal{U}$. If each link of $\mathcal{U}$ is open, then $\mathcal{U}$ is called an *open chain*. If $\varepsilon > 0$, $\mathcal{U}$ is an open chain and the mesh$(\mathcal{U}) < \varepsilon$, then $\mathcal{U}$ is called an ε–*chain*.

2.4.2. Remark. Let us observe that we do not require the links of a chain to be connected. Hence, a chain may look like in the following picture:

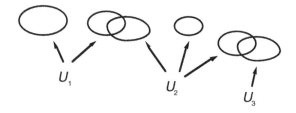

2.4.3. Definition. A continuum X is said to be *chainable* provided that for each $\varepsilon > 0$, there exists an ε–chain covering X. If $x_1, x_2 \in X$, then X *is chainable from x_1 to x_2* if for each $\varepsilon > 0$ there is an ε–chain $\mathcal{U} = \{U_1, \ldots, U_n\}$ covering X such that $x_1 \in U_1$ and $x_2 \in U_n$.

2.4.4. Example. The unit interval $[0, 1]$ is chainable from 0 to 1.

0   1

2.4.5. Example. Let

$$X = \{0\} \times [-1, 1] \cup \left\{ \left(x, \sin\left(\frac{1}{x}\right) \right) \in \mathbb{R}^2 \;\middle|\; x \in \left(0, \frac{2}{\pi} \right] \right\}.$$

Then X is called the *topologist sine curve*. Note that X is a chainable continuum from $(0, -1)$ to $\left(\frac{2}{\pi}, 1 \right)$.

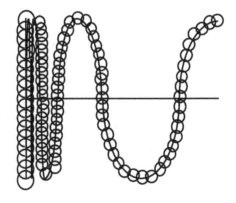

2.4.6. Example. Let X be the topologist sine curve and let X' be the reflection of X in $\mathbb{R}^2$ with respect to the line $x = \frac{2}{\pi}$. Let

$Z = X \cup X'$. Then Z is a chainable continuum from $(0, -1)$ to $\left(\dfrac{4}{\pi}, -1\right)$.

2.4.7. Example. The *Knaster continuum*, $\mathcal{K}$, is defined in the following way. The continuum consists of

(a) all semi–circles in $\mathbb{R}^2$ with nonnegative ordinates, with center at the point $\left(\dfrac{1}{2}, 0\right)$ and passing through every point of the Cantor set $\mathcal{C}$.

(b) all semi–circles in $\mathbb{R}^2$ with nonpositive ordinates, which have for each $n \in \mathbb{N}$, the center at the point $\left(\dfrac{5}{2 \cdot 3^n}, 0\right)$ and passing through each point of the Cantor set $\mathcal{C}$ lying in the interval $\left[\dfrac{2}{3^n}, \dfrac{1}{3^{n-1}}\right]$.

Then $\mathcal{K}$ is a chainable continuum.

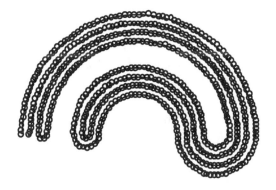

2.4.8. Remark. It is well known that $\mathcal{K}$ is an indecomposable continuum (Remark to Theorem 8 (p. 213) of [11]).

2.4.9. Example. Let $\mathcal{K}'$ be the reflection of $\mathcal{K}$ in $\mathbb{R}^2$ with respect to the origin $(0,0)$. Let $\mathcal{M} = \mathcal{K} \cup \mathcal{K}'$. Then $\mathcal{M}$ is a chainable continuum.

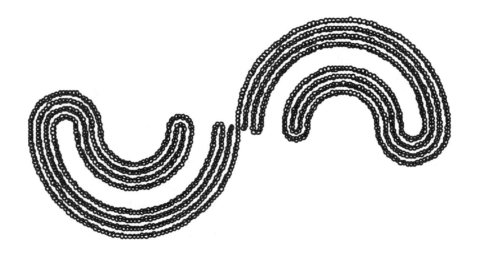

The next Theorem shows that the property of being chainable is hereditary, i.e., each nondegenerate subcontinuum has the property.

2.4.10. Theorem. *If X is a chainable continuum and K is a subcontinuum of X, then K is chainable.*

Proof. Let $\varepsilon > 0$. Since X is chainable, there exists an ε–chain $\mathcal{U} = \{U_1, \ldots, U_n\}$ covering X. Let $i = \min\{k \in \{1, \ldots, n\} \mid K \cap U_k \neq \emptyset\}$, and let $j = \max\{k \in \{1, \ldots, n\} \mid K \cap U_k \neq \emptyset\}$. We show that $\mathcal{U}' = \{U_i \cap K, \ldots, U_j \cap K\}$ is an ε–chain covering K. Clearly, $\text{mesh}(\mathcal{U}') < \varepsilon$.

Suppose $\mathcal{U}'$ is not a chain. Then there exists $\ell \in \{i, \ldots, j-1\}$ such that $(U_\ell \cap K) \cap (U_{\ell+1} \cap K) = \emptyset$. Hence,

$$K \subset \left(\bigcup_{m=1}^{\ell} (U_m \cap K) \right) \bigcup \left(\bigcup_{m=\ell+1}^{j} (U_m \cap K) \right)$$

and

$$\left(\bigcup_{m=i}^{\ell} (U_m \cap K) \right) \bigcap \left(\bigcup_{m=\ell+1}^{j} (U_m \cap K) \right) = \emptyset.$$

This contradicts the fact that K is connected since $K \cap U_i \neq \emptyset$ and $K \cap U_j \neq \emptyset$. Thus, $\mathcal{U}'$ is a chain.

Therefore, K is chainable.

Q.E.D.

The following Lemma is known as the Shrinking Lemma.

2.4.11. Lemma. *Let X be a compactum. If $\mathcal{V} = \{V_1, \dots, V_m\}$ is a finite open cover of X, then there exists an open cover $\mathcal{U} = \{U_1, \dots, U_m\}$ such that $Cl(U_j) \subset V_j$ for each $j \in \{1, \dots, m\}$.*

Proof. The construction of $\mathcal{U}$ is done inductively.

Let $F_1 = X \setminus \left(\bigcup_{j=2}^{m} V_j \right)$. Then $F_1 \subset V_1$, and F_1 is a closed subset of X. Since X is a metric space, there exists an open subset U_1 of X such that $F_1 \subset U_1 \subset Cl(U_1) \subset V_1$.

Suppose U_{k-1} has been defined for each $k < m$. Let

$$F_k = X \setminus \left(\bigcup_{j=1}^{k-1} U_j \cup \bigcup_{j=k+1}^{m} V_j \right).$$

Then $F_k \subset V_k$, and F_k is a closed subset of X. Since X is a metric space, there exists an open subset U_k of X such that $F_k \subset U_k \subset Cl(U_k) \subset V_k$. Thus, we finish the inductive step.

Now, let $\mathcal{U} = \{U_1, \dots, U_m\}$. Then $\mathcal{U}$ is a family of m open subsets of X. We show $\mathcal{U}$ covers X. To this end, let $x \in X$. Then x belongs to finitely many elements of $\mathcal{V}$, say $V_{k_1}, \dots, V_{k_n}$. Let $k = \max\{k_1, \dots, k_n\}$. Now, $x \in X \setminus U_\ell$ for every $\ell > k$, and hence, if $x \in X \setminus U_j$ for every $j < k$, then $x \in F_k \subset U_k$. Thus, in any case, $x \in U_j$ for some $j \in \{1, \dots, m\}$. Therefore, $\mathcal{U}$ covers X.

Q.E.D.

2.4.12. Definition. Let X be a compactum. A cover $\mathcal{U}$ of X is said to be *essential* provided that no proper subfamily of $\mathcal{U}$ covers X.

2.4.13. Lemma. *If X is a chainable continuum, then there exists a sequence $\{\mathcal{U}_n\}_{n=1}^{\infty}$ of essential covers of X such that for each $n \in \mathbb{N}$, the following conditions hold:*

(a) $\mathcal{U}_n$ is a $\dfrac{1}{2^n}$–chain with the property that disjoint links have disjoint closures; and

(b) The closure of the union of any three consecutive links of $\mathcal{U}_{n+1}$ is contained in one link of $\mathcal{U}_n$.

Proof. Note that if $\mathcal{U} = \{U_1, \ldots, U_n\}$ is a chain covering X is not essential, by the definition of chain, then $U_1 \subset U_2$ and/or $U_n \subset U_{n-1}$. Since one can always modify such a chain $\mathcal{U}$ so that its end links contain a point not in any other link (by simply removing U_1 and/or U_n if necessary) we may assume that the chain is essential.

The construction of the coverings is done inductively.

Since X is chainable, there exists a $\dfrac{1}{2}$–chain $\mathcal{V}_1 = \{V_{11}, \ldots, V_{1,m_1}\}$ covering X. By Lemma 2.4.11, there exists an open cover $\mathcal{U}_1 = \{U_{11}, \ldots, U_{1m_1}\}$ such that $Cl(U_{1j}) \subset V_{1j}$ for each $j \in \{1, \ldots, m_1\}$. Then $\mathcal{U}_1$ is a $\dfrac{1}{2}$–chain covering X with the property that disjoint links have disjoint closures.

Suppose that, for some $n \in \mathbb{N}$, we have constructed open covers $\mathcal{U}_1, \ldots, \mathcal{U}_n$ of X satisfying conditions (a) and (b) of the statement above. We construct $\mathcal{U}_{n+1}$ as follows.

Let λ_{n+1} be a Lebesgue number for the cover $\mathcal{U}_n$ (Theorem 1.6.6). Let $\alpha < \min\left\{\dfrac{1}{3}\lambda_{n+1}, \dfrac{1}{2^{n+1}}\right\}$. Since X is chainable, there is an α–chain $\mathcal{V}_{n+1} = \{V_{n+11}, \ldots, V_{n+1m_{n+1}}\}$ covering X. By Lemma 2.4.11, there exists an open cover $\mathcal{U}_{n+1} = \{U_{n+11}, \ldots, U_{n+1m_{n+1}}\}$ such that $Cl(U_{n+1j}) \subset V_{n+1j}$. Then $\mathcal{U}_{n+1}$ is a $\dfrac{1}{2^{n+1}}$–chain covering X with the property that disjoint links have disjoint closures.

We need to see that the closure of the union of three consecutive links of $\mathcal{U}_{n+1}$ is contained in a link of $\mathcal{U}_n$. Let U_{n+1j}, U_{n+1j+1} and

U_{n+1j+2} be three consecutive links of $\mathcal{U}_{n+1}$. Since $\mathrm{diam}(Cl(U_{n+1j}) \cup Cl(U_{n+1j+1}) \cup Cl(U_{n+1j+2})) = \mathrm{diam}(U_{n+1j} \cup U_{n+1j+1} \cup U_{n+1j+2}) \leq 3\alpha < \lambda_{n+1}$, there exists a link U_{nk} of $\mathcal{U}_n$ such that $Cl(U_{n+1j}) \cup Cl(U_{n+1j+1}) \cup Cl(U_{n+1j+2}) \subset U_{nk}$ (since λ_{n+1} is a Lebesgue number for $\mathcal{U}_n$).

In this way, we finish the inductive step. Therefore, the existence of the sequence of open coverings is proven.

Q.E.D.

2.4.14. Definition. Let X be a chainable continuum. A sequence $\{\mathcal{U}_n\}_{n=1}^{\infty}$ of essential covers satisfying the conditions of Lemma 2.4.13 is called a *defining sequence of chains for X*.

2.4.15. Definition. Let X be a metric space, and let $f\colon X \to X$ be a map. We say that *f has a fixed point* if there exists $x \in X$ such that $f(x) = x$.

2.4.16. Definition. Let X be a metric space. We say that *X has the fixed point property* provided that for each map $f\colon X \to X$, f has a fixed point.

It is a well known fact from calculus that the unit interval $[0,1]$ has the fixed point property. Hamilton [6] has shown that this property is shared by all chainable continua. To prove this result, we need the following Lemma:

2.4.17. Lemma. *Let X be a compactum, with metric d, and let $f\colon X \to X$ be a map. If for each $\varepsilon > 0$, there exists a point x_ε in X such that $d(f(x_\varepsilon), x_\varepsilon) < \varepsilon$, then f has a fixed point.*

Proof. By hypothesis, for each $n \in \mathbb{N}$, there exists $x_n \in X$ such that $d(x_n, f(x_n)) < \dfrac{1}{n}$. Since X is compact, without loss of generality, we assume that the sequence $\{x_n\}_{n=1}^{\infty}$ converges to a point $x \in X$. Since f is continuous, the sequence $\{f(x_n)\}_{n=1}^{\infty}$ converges to $f(x)$.

Let $\varepsilon > 0$. Then there exists $N \in \mathbb{N}$ such that

$$d(x_N, x) < \frac{\varepsilon}{3}, \ d(f(x_N), f(x)) < \frac{\varepsilon}{3} \text{ and } \frac{1}{N} < \frac{\varepsilon}{3}.$$

Hence,

$$d(x, f(x)) \leq d(x, x_N) + d(x_N, f(x_N)) + d(f(x_N), f(x)) <$$

$$\frac{\varepsilon}{3} + \frac{1}{N} + \frac{\varepsilon}{3} < \frac{2}{3}\varepsilon + \frac{\varepsilon}{3} = \varepsilon.$$

Since ε was arbitrary, $d(x, f(x)) = 0$. Thus, $f(x) = x$. Therefore, f has a fixed point

Q.E.D.

2.4.18. Theorem. *If X is a chainable continuum, with metric d, then X has the fixed point property.*

Proof. Let $f \colon X \to X$ be a map. Let $\{\mathcal{U}_n\}_{n=1}^{\infty}$ be a defining sequence of chains for X. For each $k \in \mathbb{N}$, we assume that $\mathcal{U}_k = \{U_{k1}, \ldots, U_{kn_k}\}$.

Let $\varepsilon > 0$. By Lemma 2.4.17, we need to find a point $x_\varepsilon \in X$ such that $d(x_\varepsilon, f(x_\varepsilon)) < \varepsilon$.

Let $k \in \mathbb{N}$ such that $\frac{1}{2^k} < \varepsilon$. Consider the chain $\mathcal{U}_k$ and define the following subsets of X:

$$A = \{x \in X \mid \text{if } x \in Cl(U_{kj}) \text{ and } f(x) \in Cl(U_{ki}), \text{ then } j < i\};$$

$$B = \{x \in X \mid x, f(x) \in Cl(U_{ki}) \text{ for some } i \in \{1, \ldots, n_k\}\};$$

$$C = \{x \in X \mid \text{if } x \in Cl(U_{kj}) \text{ and } f(x) \in Cl(U_{ki}), \text{ then } j > i\}.$$

We show that $B \neq \emptyset$. To this end, suppose $B = \emptyset$. Let $x \in X \setminus A$, and let λ be a Lebesgue number of $\mathcal{U}_k$ (Theorem 1.6.6). Since X is compact, f is uniformly continuous. Hence, there exists $\delta > 0$ such that $\delta < \lambda$ and such that if $y, z \in X$ and $d(y, z) < \delta$, then $d(f(y), f(z)) < \lambda$. Let $y \in X$ such that $d(y, x) < \delta$. Since $\delta < \lambda$, there exists $U_{km} \in \mathcal{U}_k$ such that $x, y \in U_{km} \subset Cl(U_{km})$. Since $x \in Cl(U_{km}) \cap (X \setminus A)$, then $f(x) \in Cl(U_{kj})$, where $j < m$. Now, since $d(f(y), f(x)) < \lambda$, $f(y) \in Cl(U_{kn})$, where $n \leq m$. Since $B = \emptyset$, $n < m$. Hence, $y \in X \setminus A$. Consequently, $X \setminus A$ is an open subset of X. Therefore, A is closed. Similarly, C is a closed subset of X. Since $X = A \cup C$ and $A \cap C = \emptyset$, we obtain a contradiction. Therefore, $B \neq \emptyset$.

Q.E.D.

The next concept is used to define classes of continua.

2.4.19. Definition. Let $f\colon X \twoheadrightarrow Y$ be a surjective map between metric spaces, and let $\varepsilon > 0$. We say that f is an ε–*map* provided that for each $y \in Y$, $\operatorname{diam}(f^{-1}(y)) < \varepsilon$.

The following Lemma says that ε–maps between compacta behave in the same way with sets of positive small diameter.

2.4.20. Lemma. *Let X and Y be compacta, with metrics d and d', respectively. Let $\varepsilon > 0$. If $f\colon X \twoheadrightarrow Y$ is an ε–map, then there exists $\delta > 0$ such that $\operatorname{diam}(f^{-1}(U)) < \varepsilon$ for each subset U of Y with $\operatorname{diam}(U) < \delta$.*

Proof. First note that since $\operatorname{diam}(U) = \operatorname{diam}(Cl(U))$ for any set U, it suffices to prove the lemma for closed sets.

Suppose the lemma is not true. Then for each $n \in \mathbb{N}$, there is a closed subset U_n of Y such that $\operatorname{diam}(U_n) < \dfrac{1}{n}$ and $\operatorname{diam}(f^{-1}(U_n)) \geq \varepsilon$. Let $x_n, x'_n \in f^{-1}(U_n)$ such that $d(x_n, x'_n) = \operatorname{diam}(f^{-1}(U_n))$. Since X is compact, without loss of generality, we assume that there exist two points $x, x' \in X$ such that $\lim\limits_{n \to \infty} x_n = x$ and $\lim\limits_{n \to \infty} x'_n = x'$. Note that $d(x, x') \geq \varepsilon$.

Now, by continuity,

$$d'(f(x), f(x')) = \lim_{n \to \infty} d'(f(x_n), f(x'_n)) \leq$$

$$\lim_{n \to \infty} \operatorname{diam}(U_n) \leq \lim_{n \to \infty} \frac{1}{n} = 0,$$

a contradiction to the fact that f is an ε–map. Therefore, the lemma is true.

Q.E.D.

2.4.21. Definition. A continuum X is said to be *like an arc* provided that for each $\varepsilon > 0$, there exists an ε–map $f\colon X \twoheadrightarrow [0, 1]$.

The next Theorem gives us the equality of the class of chainable continua and the class of arc–like continua. The proof we present is due to James T. Rogers, Jr.

2.4.22. Theorem. *If X be a continuum with metric d, then the following are equivalent:*

(1) X is chainable.

(2) X is arc–like.

(3) X is like an arc.

Proof. Suppose X is a chainable continuum. We show X is arc–like.

Let $\{\mathcal{U}_n\}_{n=1}^{\infty}$ be a defining sequence of chains for X (Lemma 2.4.13). For every $n \in \mathbb{N}$, we use the notation $\mathcal{U}_n = \{U_{n,0}, \ldots, U_{n,k(n)}\}$.

For each $n \in \mathbb{N}$, let $X_n = [0, 1]$. Divide X_n into $k(n)$ equal subintervals with vertexes $v_{n,0} = 0, \ldots, v_{n,k(n)} = 1$. Note that there is a one–to–one correspondence between the vertexes of the subintervals of X_n and the links of the chain $\mathcal{U}_n$.

We define the functions $f_n^{n+1}\colon X_{n+1} \twoheadrightarrow X_n$ as follows:

$$f_n^{n+1}(v_{n+1,m}) = \begin{cases} v_{n,j} & \text{if } U_{n,j} \text{ is the only link of } \mathcal{U}_n \\ & \text{containing} \\ & U_{n+1,m}; \\ \dfrac{v_{n,j} + v_{n,j+1}}{2} & \text{if } U_{n+1,m} \subset U_{n,j} \cap U_{n,j+1} \end{cases}$$

and extend f_n^{n+1} linearly over X_{n+1}. Since the chains are essential, the function f_n^{n+1} is well defined for every $n \in \mathbb{N}$. Also, all these functions are continuous and surjective.

Let $X_\infty = \varprojlim\{X_n, f_n^{n+1}\}$. Then X_∞ is an arc–like continuum. We show X_∞ is homeomorphic to X. To this end, let $h_n\colon X \to \mathcal{C}(X_n)$ be given by:

$$h_n(x) = \begin{cases} \{v_{n,j}\} & \text{if } U_{n,j} \text{ is the only link of } \mathcal{U}_n \text{ containing } x; \\ [v_{n,j}, v_{n,j+1}] & \text{if } x \in U_{n,j} \cap U_{n,j+1}. \end{cases}$$

Note that for each $n \in \mathbb{N}$, $f_n^{n+1}(h_{n+1}(x)) \subset h_n(x)$. To see this, we consider six cases.

If $x \in U_{n+1,m} \setminus (U_{n+1,m-1} \cup U_{n+1,m+1})$ and $U_{n+1,m} \subset U_{n,j} \setminus (U_{n,j-1} \cup U_{n,j+1})$, then $f_n^{n+1}(h_{n+1}(x)) = f_n^{n+1}(\{v_{n+1,m}\}) = \{v_{n,j}\} = h_n(x)$.

If $x \in U_{n+1,m} \setminus (U_{n+1,m-1} \cup U_{n+1,m+1})$ and $U_{n+1,m} \subset U_{n,j} \cap U_{n,j+1}$, then $f_n^{n+1}(h_{n+1}(x)) = f_n^{n+1}(\{v_{n+1,m}\}) = \left\{ \dfrac{v_{n,j} + v_{n,j+1}}{2} \right\} \subset [v_{n,j}, v_{n,j+1}] = h_n(x)$.

If $x \in U_{n+1,m} \setminus (U_{n+1,m-1} \cup U_{n+1,m+1})$, $U_{n+1,m} \subset U_{n,j}$ and $U_{n+1,m} \cap U_{n,j+1} \neq \emptyset$, then $f_n^{n+1}(h_{n+1}(x)) = f_n^{n+1}(\{v_{n+1,m}\}) = \{v_{n,j}\} \subset [v_{n,j}, v_{n,j+1}] = h_n(x)$.

If $x \in U_{n+1,m} \cap U_{n+1,m+1}$ and $U_{n+1,m} \cap U_{n+1,m+1} \subset U_{n,j} \setminus (U_{n,j-1} \cup U_{n,j+1})$, then $f_n^{n+1}(h_{n+1}(x)) = f_n^{n+1}([v_{n+1,m}, v_{n+1,m+1}]) = \{v_{n,j}\} = h_n(x)$.

If $x \in U_{n+1,m} \cap U_{n+1,m+1}$ and $U_{n+1,m} \cup U_{n+1,m+1} \subset U_{n,j} \cap U_{n,j+1}$, then $f_n^{n+1}(h_{n+1}(x)) = f_n^{n+1}([v_{n+1,m}, v_{n+1,m+1}]) = \left\{ \dfrac{v_{n,j} + v_{n,j+1}}{2} \right\} \subset [v_{n,j}, v_{n,j+1}] = h_n(x)$.

If $x \in U_{n+1,m} \cap U_{n+1,m+1}$, $U_{n+1,m} \cup U_{n+1,m+1} \subset U_{n,j}$ and $U_{n+1,m+1} \cap U_{n,j+1} \neq \emptyset$, then $f_n^{n+1}(h_{n+1}(x)) = f_n^{n+1}([v_{n+1,m}, v_{n+1,m+1}]) = \left[v_{n,j}, \dfrac{v_{n,j} + v_{n,j+1}}{2} \right] \subset [v_{n,j}, v_{n,j+1}] = h_n(x)$.

Therefore, $f_n^{n+1}(h_{n+1}(x)) \subset h_n(x)$. Note that this implies that $\{f_n^{-1}(h_n(x))\}_{n=1}^{\infty}$ is a decreasing sequence of closed subsets of X_{∞}. Thus, $\bigcap_{n=1}^{\infty} f_n^{-1}(h_n(x)) \neq \emptyset$.

Now, we are ready to define the homeomorphism $g \colon X \twoheadrightarrow X_{\infty}$ as follows: $g(x)$ is the unique point in $\bigcap_{n=1}^{\infty} f_n^{-1}(h_n(x))$.

Since $\lim_{n \to \infty} k(n) = \infty$ and the projection maps are $\dfrac{1}{2^n}$–maps, we have that $\lim_{n \to \infty} \operatorname{diam}(f_n^{-1}(h_n(x))) = 0$. Hence, g is well defined.

Now, we show that g is one–to–one. Let $x, x' \in X$ such that $x \neq x'$. Let $\mathcal{U}_n$ be an element of the defining sequence chains for X with the property that x and x' belong to different links of $\mathcal{U}_n$. In this case, $h_n(x) \cap h(x') = \emptyset$. Hence, $g(x) \neq g(x')$.

To see g is surjective, let $y = (y_n)_{n=1}^{\infty} \in X_{\infty}$. Let $\mathcal{M}_n \colon X_{\infty} \to \mathcal{C}(X_n)$ be given by

$$
\mathcal{M}_n(y) = \begin{cases} \{v_{n,j}\} & \text{if } y_n = v_{n,m}; \\ [v_{n,j}, v_{n,j+1}] & \text{if } y_n \in (v_{n,m}, v_{n,m+1}). \end{cases}
$$

Let R_n be the union of the links of $\mathcal{U}_n$ whose vertexes are in $\mathcal{M}_n(y)$. Then R_n consists of one link or it is the union of two consecutive links of $\mathcal{U}_n$. Let $\{x\} = \bigcap_{n=1}^{\infty} R_n = \bigcap_{n=1}^{\infty} Cl(R_n)$ (recall that $Cl(R_{n+1}) \subset R_n$ for every $n \in \mathbb{N}$). We prove that $g(x) = y$. Let $n \in \mathbb{N}$. Then either $R_n = U_{n,j}$ or $R_n = U_{n,j} \cup U_{n,j+1}$, and $x \in R_n$. Hence, either $h_n(x) = \{v_{n,j}\}$ or $h_n(x) = [v_{n,j}, v_{n,j+1}]$. In either case, $h_n(x) = \mathcal{M}_n(y)$. Thus, $y \in f_n^{-1}(\mathcal{M}_n(y)) = f_n^{-1}(h_n(x))$. Hence, $y \in \bigcap_{n=1}^{\infty} f_n^{-1}(h_n(x))$. Therefore, $g(x) = y$.

Now, we see that g is continuous. To this end, by Theorem 1.1.9, it is enough to prove that for each $n \in \mathbb{N}$, $f_n \circ g$ is continuous. Let $n \in \mathbb{N}$ and let $\varepsilon > 0$. Let $\ell > n$ such that for each $m \in \{1, \dots, k(\ell) - 2\}$, $\mathrm{diam}(f_n^{\ell}([v_{\ell,m}, v_{\ell,m+2}])) < \varepsilon$. Let $\lambda > 0$ be a Lebesgue number for the cover $\mathcal{U}_{\ell}$ (Theorem 1.6.6). Hence, if $x, x' \in X$ and $d(x, x') < \lambda$, then there exists $j \in \{1, \dots, k(\ell)\}$ such that $x, x' \in U_{\ell,j}$.

Note that $f_{\ell} \circ g(U_{\ell,j}) \subset [v_{\ell,m}, v_{\ell,m+2}]$ for some $m \in \{1, \dots, k(\ell) - 2\}$. Hence, $\mathrm{diam}(f_n \circ g(U_{\ell,j})) = \mathrm{diam}(f_n^{\ell} \circ f_{\ell} \circ g(U_{\ell,j})) < \varepsilon$. Thus, $d_n(f_n \circ g(x), f_n \circ g(x')) < \varepsilon$, where d_n is the metric on X_n. Therefore, $f_n \circ g$ is continuous.

Next, suppose X is arc–like. We show that X is like an arc. Since X is arc–like, there exists an inverse sequence $\{X_n, f_n^{n+1}\}$, where $X_n = [0, 1]$ and f_n^{n+1} is surjective for each $n \in \mathbb{N}$, and whose inverse limit is X. Let $\varepsilon > 0$, and let $N \in \mathbb{N}$ such that $\sum_{n=N+1}^{\infty} \frac{1}{2^n} < \varepsilon$. To see f_N is an ε–map, let $z \in X_N$. Let $x = (x_n)_{n=1}^{\infty}$ and $y = (y_n)_{n=1}^{\infty}$ be two points of $f_N^{-1}(z)$. Then

$$
\rho(x, y) = \sum_{n=1}^{\infty} \frac{1}{2^n} d_n(x_n, y_n) = \sum_{n=N+1}^{\infty} \frac{1}{2^n} d_n(x_n, y_n) \le \sum_{n=N+1}^{\infty} \frac{1}{2^n} < \varepsilon.
$$

Therefore, $\mathrm{diam}(f_N^{-1}(z)) < \varepsilon$.

Finally, suppose X is like an arc. To see X is chainable, let $f\colon X \twoheadrightarrow [0,1]$ be an ε–map. By Lemma 2.4.20, there exists $\delta > 0$ such that if U is a subset of $[0,1]$ with $\mathrm{diam}(U) < \delta$, then $\mathrm{diam}(f^{-1}(U)) < \varepsilon$.

Let $n \in \mathbb{N}$ such that $\dfrac{1}{n} < \dfrac{\delta}{2}$. Divide $[0,1]$ into n equal parts and let

$$\mathcal{U} = \left\{ f^{-1}\left(\left[0, \tfrac{1}{n}\right)\right), f^{-1}\left(\left(0, \tfrac{2}{n}\right)\right), f^{-1}\left(\left(\tfrac{1}{n}, \tfrac{3}{n}\right)\right), \dots, \right.$$

$$\left. f^{-1}\left(\left(\tfrac{n-2}{n}, 1\right)\right), f^{-1}\left(\left(\tfrac{n-1}{n}, 1\right]\right) \right\}.$$

Since $\left\{ \left[0, \tfrac{1}{n}\right), \left(0, \tfrac{2}{n}\right), \left(\tfrac{1}{n}, \tfrac{3}{n}\right), \dots, \left(\tfrac{n-2}{n}, 1\right), \left(\tfrac{n-1}{n}, 1\right] \right\}$ is a δ–chain of $[0,1]$, $\mathcal{U}$ is an ε–chain covering X. Since ε was arbitrary, therefore, X is chainable.

Q.E.D.

2.5 Circularly Chainable and $\mathcal{P}$–Like Continua

We present basic facts about circularly chainable and $\mathcal{P}$–like continua.

2.5.1. Definition. Let X be a compactum. A *circular chain* $\mathcal{U}$ in X is a finite sequence, $U_0, \dots, U_n$, of subsets of X such that $U_i \cap U_j \neq \emptyset$ if and only if $|i - j| \leq 1$ or $i, j \in \{1, n\}$. Each U_j is called a *link* of $\mathcal{U}$. If each link of $\mathcal{U}$ is open, then $\mathcal{U}$ is called an *open circular chain*. If $\varepsilon > 0$, $\mathcal{U}$ is an open circular chain and the $\mathrm{mesh}(\mathcal{U}) < \varepsilon$, then $\mathcal{U}$ is called a *circular ε–chain*.

2.5.2. Definition. A continuum X is *circularly chainable* provided that for each $\varepsilon > 0$, there exists a circular ε–chain covering X.

2.5.3. Example. The unit circle $\mathcal{S}^1$ is a circularly chainable continuum.

2.5.4. Example. Let X be the topologist sine curve (see Example 2.4.5). Join the points $(0, -1)$ and $\left(\dfrac{2}{\pi}, 1\right)$ with an arc, Y, in $\mathbb{R}^2$. Then $\mathcal{W} = X \cup Y$ is called the *Warsaw circle*, and it is a circularly chainable continuum.

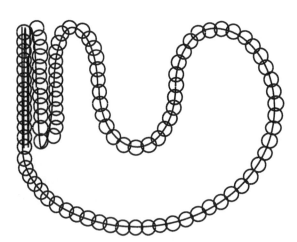

2.5.5. Example. The *double Warsaw circle* consists of two copies of the topologist sine curve joined by two arcs (see picture below). It is also a circularly chainable continuum.

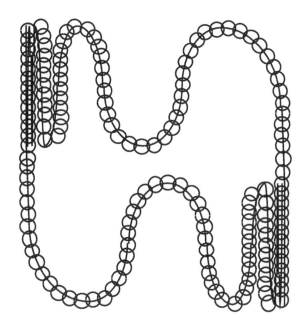

2.5.6. Example. Some chainable continua are circularly chainable. For instance, let X be either $\mathcal{K}$, the Knaster continuum (Example 2.4.7), or $\mathcal{M}$ (Example 2.4.9). Then X is a circularly chainable continuum. If $\varepsilon > 0$ and $\mathcal{U} = \{U_0, \ldots, U_n\}$ is an $\frac{\varepsilon}{2}$–chain, then $\mathcal{C} = \{U_0, \ldots, U_n, U_0 \cup U_n\}$ is a circular ε–chain.

2.5.7. Definition. A continuum X is said to be *like a circle* if for each $\varepsilon > 0$, there exists an ε–map from X onto $\mathcal{S}^1$.

The following Theorem is analogous to Theorem 2.4.22, for circularly chainable continua, and it is a special case of Theorem 2.5.13:

2.5.8. Theorem. *Let X be a continuum. Then the following are equivalent:*
(1) X is a circularly chainable continuum.
(2) X is a circle–like continuum.
(3) X is a like a circle continuum.

The next Theorem says that each circle–like continuum may be written as an inverse limit of circles where the bonding maps have nonnegative degree.

2.5.9. Theorem. *If X is a circle–like continuum, then there exists an inverse sequence $\{X_n, f_n^{n+1}\}$ such that $\varprojlim\{X_n, f_n^{n+1}\}$ is homeomorphic to X, where each $X_n = \mathcal{S}^1$ and such that the bonding maps satisfy one of the following three conditions:*
(a) for each $n \in \mathbb{N}$, $\deg(f_n^{n+1}) = 0$;
(b) for each $n \in \mathbb{N}$, $\deg(f_n^{n+1}) = 1$;
(c) for each $n \in \mathbb{N}$, $\deg(f_n^{n+1}) > 1$.

Proof. Since X is circle–like, there is an inverse sequence $\{Y_n, r_n^{n+1}\}$ of circles with surjective bonding maps such that $\varprojlim\{Y_n, r_n^{n+1}\}$ is homeomorphic to X.

First, we show that X is homeomorphic to an inverse limit of circles where all the bonding maps have nonnegative degrees. To this end, we construct a new inverse sequence $\{Y_n, g_n^{n+1}\}$ and maps $h_n \colon Y_n \twoheadrightarrow Y_n$ such that $h_n \circ r_n^{n+1} = g_n^{n+1} \circ h_{n+1}$ and $\deg(g_n^{n+1}) \geq 0$ for every $n \in \mathbb{N}$.

Let $h_1 = 1_{X_1}$. If $\deg(r_1^2) \geq 0$, then $g_1^2 = r_1^2$, and $h_2 = 1_{X_2}$. If $\deg(r_1^2) < 0$, then $h_2 = \aleph$, where $\aleph$ is the antipodal map, and $g_1^2 = r_1^2 \circ \aleph$. Hence, $h_1 \circ r_1^2 = g_1^2 \circ h_2$, and $\deg(g_1^2) \geq 0$.

Next, suppose inductively that we have constructed $h_1, \dots, h_k$, where either $h_j = 1_{X_j}$ or $h_j = \aleph$, $j \in \{1, \dots, k\}$, and $g_1^2, \dots, g_{k-1}^k$, where $\deg(g_j^{j+1}) \geq 0$, $j \in \{1, \dots, k-1\}$, such that $h_{j-1} \circ r_{j-1}^j = g_{j-1}^j \circ h_j$ for every $j \in \{2, \dots, k\}$. To define h_{k+1} and g_k^{k+1} we consider four cases.

If $h_k = 1_{X_k}$ and $\deg(r_k^{k+1}) \geq 0$, then let $h_{k+1} = 1_{X_{k+1}}$ and let $g_k^{k+1} = r_k^{k+1}$.

If $h_k = 1_{X_k}$ and $\deg(r_k^{k+1}) < 0$, then let $h_{k+1} = \aleph$ and let $g_k^{k+1} = r_k^{k+1} \circ \aleph$.

If $h_k = \aleph$ and $\deg(r_k^{k+1}) \geq 0$, then let $h_{k+1} = \aleph$ and let $g_k^{k+1} = \aleph \circ r_k^{k+1} \circ \aleph$.

If $h_k = \aleph$ and $\deg(r_k^{k+1}) < 0$, then let $h_{k+1} = 1_{x_{k+1}}$ and let $g_k^{k+1} = \aleph \circ r_k^{k+1}$.

In each case, $h_k \circ r_k^{k+1} = g_k^{k+1} \circ h_{k+1}$ and $\deg(g_k^{k+1}) \geq 0$. In this way we finish with the inductive step.

So, we have constructed an inverse sequence $\{Y_n, g_n^{n+1}\}$, where $\deg(g_n^{n+1}) \geq 0$ for all $n \in \mathbb{N}$. By Theorems 2.1.46, 2.1.47 and 2.1.48, X is homeomorphic to $\varprojlim\{Y_n, g_n^{n+1}\}$.

Now, suppose there exists a subsequence $\{m(n)\}_{n=1}^{\infty}$ of $\mathbb{N}$ such that each map $g_{m(n)}^{m(n)+1}$ has degree zero. Then, by Lemma 1.3.39, $\deg\left(g_{m(n)}^{m(n+1)}\right) = 0$ for each $n \in \mathbb{N}$. Thus, each bonding map of the inverse sequence $\left\{Y_{m(n)}, g_{m(n)}^{m(n+1)}\right\}$ has degree zero and, by Theorem 2.1.38, $\varprojlim\left\{Y_{m(n)}, g_{m(n)}^{m(n+1)}\right\}$ is homeomorphic to X. Therefore, taking $X_n = Y_{m(n)}$ and $f_n^{n+1} = g_{m(n)}^{m(n+1)}$, $\{X_n, f_n^{n+1}\}$ is an inverse sequence of circles with surjective bonding maps with degree zero, whose inverse limit is homeomorphic to X.

Next, suppose none of the bonding maps g_n^{n+1} has degree zero. Hence, we assume that for each $n \in \mathbb{N}$, $\deg(g_n^{n+1}) > 0$.

If only finitely many bonding maps have degree greater than one, then taking a subsequence, by Theorem 2.1.38, we assume that all the bonding maps have degree one. Thus, X is homeomorphic to the inverse limit of such subsequence.

Finally, suppose there exists a subsequence $\{m(n)\}_{n=1}^{\infty}$ of $\mathbb{N}$ such that each map $g_{m(n)}^{m(n)+1}$ has degree greater than one. Then, by Lemma 1.3.39, $\deg\left(g_{m(n)}^{m(n+1)}\right) > 1$ for each $n \in \mathbb{N}$. Thus, each bonding map of the inverse sequence $\left\{Y_{m(n)}, g_{m(n)}^{m(n+1)}\right\}$ has degree greater than one and, by Theorem 2.1.38, $\varprojlim\left\{Y_{m(n)}, f_{m(n)}^{m(n+1)}\right\}$ is homeomorphic to X. Therefore, taking $X_n = Y_{m(n)}$ and $f_n^{n+1} = g_{m(n)}^{m(n+1)}$, $\{X_n, f_n^{n+1}\}$ is an inverse sequence of circles with surjective bonding maps with degree greater than one, whose inverse limit is homeomorphic to X.

Q.E.D.

The following Theorem tells us that each circle–like continuum, expressed as an inverse sequence of circles with bonding maps having degree zero, is arc–like too.

2.5.10. Theorem. *Let $\{X_n, f_n^{n+1}\}$ be an inverse sequence of circles with surjective bonding maps having degree zero. If $X_\infty = \lim\{X_n, f_n^{n+1}\}$, then X_∞ is an arc–like continuum.*

Proof. For each $n \in \mathbb{N}$, let $p_n \colon \mathbb{R} \to \mathcal{S}^1$ be given by $p_n(t) = \exp(2\pi t)$. Since $\deg(f_n^{n+1}) = 0$ for every $n \in \mathbb{N}$, each map f_n^{n+1} is homotopic to a constant map (Remark 1.3.38). Hence, by Theorem 1.3.40, there is a map $h_{n+1} \colon X_{n+1} \to \mathbb{R}$ such that $h_{n+1}((1,0)) = 0$ and $f_n^{n+1} = p_n \circ h_{n+1}$ for every $n \in \mathbb{N}$.

For each $n \in \mathbb{N}$, let $Y_n = h_{n+1}(X_{n+1})$. Note that each Y_n is an interval of length at least one (the bonding maps f_n^{n+1} are surjective). For each $n \in \mathbb{N}$, let $g_n^{n+1} \colon Y_{n+1} \twoheadrightarrow Y_n$ be given by $g_n^{n+1} = h_{n+1} \circ (p_n|_{Y_{n+1}})$. Note that g_n^{n+1} is surjective since it is the composition of surjective maps. Hence, $\{Y_n, g_n^{n+1}\}$ is an inverse sequence of arcs with surjective bonding maps whose inverse limit, Y_∞, is homeomorphic to X_∞ (Theorem 2.1.49). Therefore, X_∞ is an arc–like continuum.

Q.E.D.

2.5.11. Remark. It is known that circle–like continua expressed as an inverse limit of circles with bonding maps having degree one are planar continua which separate the plane. If the degree of the bonding maps are greater than one, then such continua are not planar [1].

Next, we present a generalization of arc–like and circle–like continua.

2.5.12. Definition. Let $\mathcal{P}$ be a collection of continua. If the continuum X is homeomorphic to an inverse limit of elements of $\mathcal{P}$ with surjective bonding maps, then X is said to be $\mathcal{P}$–*like*. If $\mathcal{P}$ consists of just one element P, then X is said to be P–*like*.

A proof of the following Theorem may be found in [12]. Note that it generalizes Theorems 2.4.22 and 2.5.8:

2.5.13. Theorem. *Let $\mathcal{P}$ be a collection of polyhedra, and let X be a continuum. Then the following are equivalent:*

(a) For each $\varepsilon > 0$, there exists a finite open cover $\mathcal{U}$ of X such that $\mathrm{mesh}(\mathcal{U}) < \varepsilon$ and the polyhedron, $\mathcal{N}^\star(\mathcal{U})$, associated with the nerve of $\mathcal{U}$, is homeomorphic to an element of $\mathcal{P}$;

(b) X is a $\mathcal{P}$–like continuum; and

(c) For each $\varepsilon > 0$, there exists a ε–map from X onto an element of $\mathcal{P}$.

2.5.14. Corollary. *If X is a continuum, then there exists a countable family $\mathcal{P}$ of polyhedra such that X is a $\mathcal{P}$–like continuum.*

Proof. For each $n \in \mathbb{N}$, let $\mathcal{U}_n$ be a finite open cover of X such that $\mathrm{mesh}(\mathcal{U}_n) < \dfrac{1}{n}$. Let $\mathcal{P} = \{\mathcal{N}^\star(\mathcal{U}_n)\}_{n=1}^\infty$. Then, by Theorem 2.5.13, X is a $\mathcal{P}$–like continuum.

$$\textbf{Q.E.D.}$$

2.5.15. Lemma. *Let $\mathcal{P}$ and $\mathcal{R}$ be two families of continua. If X is a $\mathcal{P}$–like continuum and each element of $\mathcal{P}$ is an $\mathcal{R}$–like continuum, then X is an $\mathcal{R}$–like continuum.*

Proof. Let $\varepsilon > 0$. By Theorem 2.5.13, there exist an element $P \in \mathcal{P}$ and an ε–map $f\colon X \twoheadrightarrow P$. By Lemma 2.4.20, there exists $\delta > 0$ such that if U is a subset of P with $\mathrm{diam}(U) < \delta$, then $\mathrm{diam}(f^{-1}(U)) < \varepsilon$. Since P is an $\mathcal{R}$–like continuum, there exist an element $R \in \mathcal{R}$ and a δ–map $g\colon P \twoheadrightarrow R$. Hence, $g \circ f\colon X \twoheadrightarrow R$ is an ε–map from X onto R. Since the ε was arbitrary, by Theorem 2.5.13, X is $\mathcal{R}$–like.

$$\textbf{Q.E.D.}$$

2.5.16. Theorem. *Let $T = [-1,1] \times \{0\} \cup \{0\} \times [-1,0] \subset \mathbb{R}^2$. Then $[0,1]$ is T–like.*

Proof. By Theorem 2.5.13, it suffices to show that for each $\varepsilon > 0$, there exists an ε–map from $[0,1]$ onto T.

Let $\varepsilon > 0$. Let $f \colon [0,1] \twoheadrightarrow T$ be given by

$$
f(t) = \begin{cases}
\left(0, \dfrac{t+\varepsilon-1}{1-\varepsilon}\right) & \text{if } t \in [0, 1-\varepsilon]; \\[2ex]
\left(-\dfrac{2}{\varepsilon}(t+\varepsilon-1), 0\right) & \text{if } t \in \left[1-\varepsilon, 1-\dfrac{\varepsilon}{2}\right]; \\[2ex]
\left(\dfrac{1}{\varepsilon}(\varepsilon+4(t-1)), 0\right) & \text{if } t \in \left[1-\dfrac{\varepsilon}{2}, 1\right].
\end{cases}
$$

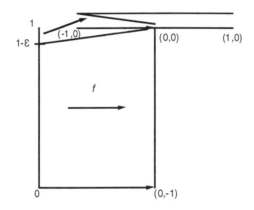

Then f is an ε–map.

Q.E.D.

2.5.17. Corollary. *Let* $T = [-1,1] \times \{0\} \cup \{0\} \times [-1,0] \subset \mathbb{R}^2$. *If* X *is an arc–like continuum, then* X *is* T*–like.*

Proof. It follows easily from Theorem 2.5.16 and Lemma 2.5.15.

Q.E.D.

The following Theorem is a slight modification of the Hahn–Mazurkiewicz Theorem:

2.5.18. Theorem. *Let* X *be a locally connected continuum. If* x *and* y *are two points of* X, *then there exists a surjective map* $f \colon [0,1] \twoheadrightarrow X$ *such that* $f(0) = x$ *and* $f(y) = 1$.

Proof. Since X is locally connected, by Hahn–Mazurkiewicz Theorem (8.14 of [18]), there exists a surjective map $g\colon [0,1] \twoheadrightarrow X$. Let s_x and s_y be points of $[0,1]$ such that $g(s_x) = x$ and $g(s_y) = y$. Let $h\colon [0,1] \twoheadrightarrow [0,1]$ be given by

$$
h(t) = \begin{cases}
s_x - 3s_x t & \text{if } t \in \left[0, \dfrac{1}{3}\right]; \\[2mm]
3t - 1 & \text{if } t \in \left[\dfrac{1}{3}, \dfrac{2}{3}\right]; \\[2mm]
3(s_y - 1)t - 2s_y + 3 & \text{if } t \in \left[\dfrac{2}{3}, 1\right].
\end{cases}
$$

Then h is a surjective map such that $h(0) = s_x$ and $h(1) = s_y$. Hence, $f = g \circ h$ is a surjective map from $[0,1]$ onto X such that $f(0) = x$ and $f(1) = y$.

Q.E.D.

The proof of the following Theorem is due to Ray L. Russo.

2.5.19. Theorem. *Let $n \in \mathbb{N}$. If $\mathcal{B} = \{x \in \mathbb{R}^n \mid \|x\| \leq 1\}$, then each arc–like continuum is $\mathcal{B}$–like.*

Proof. By Lemma 2.5.15, it suffices to show that $[0,1]$ is $\mathcal{B}$–like. Let $\varepsilon > 0$. Let $m \in \mathbb{N}$ such that $\dfrac{1}{m} < \dfrac{\varepsilon}{2}$. For each $k \in \{1, \ldots, m\}$, let

$$
E_k = \left\{ x \in \mathbb{R}^n \ \middle| \ 1 - \frac{k}{m} \leq \|x\| \leq 1 - \frac{k-1}{m} \right\}.
$$

Note that E_m is the closed ball of radius $\dfrac{1}{m}$ and E_k is an annular region, $k \in \{1, \ldots, m-1\}$. Note, also, that $\mathcal{B} = \bigcup\limits_{k=1}^{m} E_k$.

Divide $[0,1]$ into m equal subintervals of length $\dfrac{1}{m}$. By Theorem 2.5.18, there exists a surjective map $g_1\colon \left[0, \dfrac{1}{m}\right] \twoheadrightarrow E_1$ such that $g_1\left(\dfrac{1}{m}\right) \in E_1 \cap E_2$. By the same Theorem, there exists a surjective map $g_2\colon \left[\dfrac{1}{m}, \dfrac{2}{m}\right] \twoheadrightarrow E_2$ such that $g_2\left(\dfrac{1}{m}\right) = g_1\left(\dfrac{1}{m}\right)$ and $g_2\left(\dfrac{2}{m}\right) \in E_2 \cap E_3$. In general, we have a surjective map $g_k\colon \left[\dfrac{k-1}{m}, \dfrac{k}{m}\right] \twoheadrightarrow E_k$ such that $g_{k-1}\left(\dfrac{k-1}{m}\right) = g_k\left(\dfrac{k-1}{m}\right)$ for each $k \in \{2, \ldots, m\}$.

Let $f\colon [0,1] \twoheadrightarrow \mathcal{B}$ be given by $f(t) = g_k(t)$ if $t \in \left[\dfrac{k-1}{m}, \dfrac{k}{m}\right]$. Then f is an ε–map. Hence, by Theorem 2.5.13, $[0,1]$ is $\mathcal{B}$–like.

Q.E.D.

2.6 Universal and A–H Essential Maps

We study universal maps and AH–essential maps. We present some of their properties. In particular, we show that these maps are the same, when the range space is an n–cell.

We begin this section with the following Theorem:

2.6.1. Theorem. *Let $\{X_n, f_n^{n+1}\}$ be an inverse of continua with the fixed point property, and surjective bonding maps. For each $n \in \mathbb{N}$, let $h_n\colon X_n \to X_n$ be a map such that $f_n^{n+1} \circ h_{n+1} = h_n \circ f_n^{n+1}$. If $X_\infty = \varprojlim\{X_n, f_n^{n+1}\}$, then $h_\infty = \varprojlim\{h_n\}$ has a fixed point.*

Proof. Let $\varepsilon > 0$. Let $m \in \mathbb{N}$ such that $\dfrac{1}{2^m} < \varepsilon$.

Since each X_m has the fixed point property, there exists $x_m \in X_m$ such that $h_m(x_m) = x_m$. Since all the bonding maps are surjective, by Remark 2.1.6, the projection maps are surjective too. Hence, there exists a point $z = (z_n)_{n=1}^\infty \in X_\infty$ such that $f_m(z) = z_m = x_m$.

Since $f_n^{n+1} \circ h_{n+1} = h_n \circ f_n^{n+1}$ for each $n \in \mathbb{N}$,

$$h_\infty(z) = (z_1, z_2, \ldots, z_{m-1}, x_m, h_{m+1}(z_{m+1}), \ldots).$$

Hence,

$$\rho\left(h_\infty(z), z\right) = \sum_{n=m+1}^{\infty} \frac{1}{2^n} d(h_n(z_n), z_n) \leq \sum_{n=m+1}^{\infty} \frac{1}{2^n} = \frac{1}{2^m} < \varepsilon.$$

Therefore, by Lemma 2.4.17, h_∞ has a fixed point.

Q.E.D.

2.6.2. Remark. Let us observe that Theorem 2.6.1 does not imply that the inverse limit of continua with the fixed point property has the fixed point property; compare with Theorem 2.6.14.

2.6.3. Definition. A map $f\colon X \to Y$ between metric spaces is called *universal* provided that for any other map $g\colon X \to Y$, there exists a point $x \in X$ such that $f(x) = g(x)$.

As an easy consequence of the definition of a universal map we have the following two Propositions:

2.6.4. Proposition. *Let $f\colon X \to Y$ be a universal map between metric spaces. Then the following hold:*

(1) f is surjective.

(2) Y has the fixed point property.

(3) If X_0 is a subspace of X and $f|_{X_0}\colon X_0 \to Y$ is universal, then f is universal.

(4) If $g\colon Y \to Z$ is a map between metric spaces and $g\circ f$ is universal, then g is universal.

2.6.5. Proposition. *A metric space X has the fixed point property if and only if the identity map, 1_X, is universal.*

2.6.6. Lemma. *Let X and Y be metric spaces, and let $f\colon X \twoheadrightarrow Y$ be a universal map. If $g\colon W \twoheadrightarrow X$ and $h\colon Y \twoheadrightarrow Z$ are homeomorphisms, then $f \circ g$ and $h \circ f$ are universal.*

Proof. We show $f \circ g$ is universal. The proof for $h \circ f$ is similar.

Let $\ell \colon W \to Y$ be a map. Note that $\ell \circ g^{-1} \colon X \to Y$. Since f is universal, there exists $x \in X$ such that $(\ell \circ g^{-1})(x) = f(x)$. Hence, $\ell(g^{-1}(x)) = (f \circ g)(g^{-1}(x))$. Therefore, $f \circ g$ is universal.

Q.E.D.

2.6.7. Proposition. *For each $n \in \mathbb{N}$, let $f_n \colon X \to Y$ be a map between compacta. Suppose the sequence $\{f_n\}_{n=1}^{\infty}$ converges uniformly to a map $f \colon X \to Y$. If each f_n is universal, then f is universal.*

Proof. Let $g \colon X \to Y$ be a map. Since for each $n \in \mathbb{N}$, f_n is universal, there exists $x_n \in X$ such that $f_n(x_n) = g(x_n)$ for every $n \in \mathbb{N}$. Since X is a compactum, the sequence $\{x_n\}_{n=1}^{\infty}$ has a convergent subsequence $\{x_{n_k}\}_{k=1}^{\infty}$. Let x be the limit point of $\{x_{n_k}\}_{k=1}^{\infty}$. Then $f(x) = g(x)$. Therefore, f is universal.

Q.E.D.

2.6.8. Theorem. *Let X be a continuum. If $f \colon X \twoheadrightarrow [0,1]$ is a surjective map, then f is universal.*

Proof. Let $g \colon X \to [0,1]$ be a map. Note that the sets

$$\{x \in X \mid f(x) \leq g(x)\} \text{ and } \{x \in X \mid f(x) \geq g(x)\}$$

are nonempty closed subsets of X whose union is X. Hence, these sets have a point x in common. Thus, $f(x) = g(x)$. Therefore, f is universal.

Q.E.D.

2.6.9. Theorem. *Let $\{Y_n, g_n^{n+1}\}$ be an inverse sequence of compacta whose inverse limit is Y_∞, and let X be a compactum. For each $n \in \mathbb{N}$, let $h_n \colon X \to Y_n$ be a map such that $g_n^{n+1} \circ h_{n+1} = h_n$. If each h_n is universal, then the induced map $h_\infty = \varprojlim\{h_n\}$ is universal.*

Proof. Let $k\colon X \to Y_\infty$ be a map. Since each h_n is universal, for each $n \in \mathbb{N}$, there exists a point $x_n \in X_n$ such that $(g_n \circ k)(x_n) = h_n(x_n)$. Since $h_n(x_n) = (g_n \circ h_\infty)(x_n)$,

$$
\begin{aligned}
\rho(h_\infty(x_n), k(x_n)) &= \sum_{\ell=n+1}^{\infty} \frac{1}{2^\ell} d_\ell((g_\ell \circ h_\infty)(x_n), (g_\ell \circ k)(x_n)) \\
&\le \sum_{\ell=n+1}^{\infty} \frac{1}{2^\ell} = \frac{1}{2^n}
\end{aligned}
$$

for every $n \in \mathbb{N}$.

Since X is a compactum, without loss of generality, we assume that the sequence $\{x_n\}_{n=1}^{\infty}$ converges to a point $x \in X$. Hence, $\rho(h_\infty(x), k(x)) = \lim_{n\to\infty} \rho(h_\infty(x_n), k(x_n)) \le \lim_{n\to\infty} \frac{1}{2^n} = 0$. Thus, $h_\infty(x) = k(x)$. Therefore, h_∞ is universal.

Q.E.D.

2.6.10. Corollary. *Let $\{Y_n, g_n^{n+1}\}$ be an inverse sequence of arcs, with surjective bonding maps, whose inverse limit is Y_∞. If X is a continuum and $f\colon X \twoheadrightarrow Y_\infty$ is a surjective map, then f is universal.*

Proof. Note that $f = \varprojlim\{g_n \circ f\}$. Since each Y_n is an arc, by Theorem 2.6.8, each $g_n \circ f$ is universal. Hence, by Theorem 2.6.9, f is universal.

Q.E.D.

2.6.11. Corollary. *If X is an arc–like continuum, then X has the fixed point property.*

2.6.12. Definition. Let Y be a metric space. We say that Y is an *absolute retract* (*absolute neighborhood retract*), provided that for any metric space X, and for any closed subset A of X, every map $f\colon A \to Y$ can be extended over Y (over a neighborhood (depending on f) of A in X).

2.6.13. Remark. The concepts defined in Definition 2.6.12 are *absolute extensor* and *absolute neighborhood extensor*. It is known that these concepts are equivalent to the original definition of an absolute retract and an absolute neighborhood retract (1.5.2 of [15]).

The following Theorem gives us a sufficient condition to obtain an inverse limit with the fixed point property.

2.6.14. Theorem. *Let $\{X_n, f_n^{n+1}\}$ be an inverse sequence of absolute neighborhood retracts, where all the bonding maps $f_n^m\colon X_m \twoheadrightarrow X_n$ are universal. Then $X_\infty = \varprojlim\{X_n, f_n^{n+1}\}$ has the fixed point property.*

Proof. First, we prove each projection map $f_n\colon X_\infty \twoheadrightarrow X_n$ is universal. To this end, let $n \in \mathbb{N}$. Let $(a_\ell)_{\ell=1}^\infty \in \prod_{\ell=1}^\infty X_\ell$. For each $m \in \mathbb{N}$, let $i_m\colon X_m \to \prod_{\ell=1}^\infty X_\ell$ be given by

$$i_m(x) = (f_1^m(x), f_2^m(x), \dots, f_{m-1}^m(x), x, a_{m+1}, a_{m+2}, \dots).$$

Note that $i_m\colon X_m \twoheadrightarrow i_m(X_m) \subset S_m$ (Definition 2.1.7) is a homeomorphism. Also note that given $z \in X_m$, $i_m(z)$ is not necessarily an element of X_∞.

Let $k\colon X_\infty \to X_n$ be a map. We show there exists $x = (x_\ell)_{\ell=1}^\infty \in X_\infty$ such that $k(x) = f_n(x)$. Since X_n is an absolute neighborhood retract, there exist an open set G of $\prod_{\ell=1}^\infty X_\ell$ containing X_∞ and a map $\hat{k}\colon G \to X_n$ extending k.

Since $X_\infty = \bigcap_{m=1}^\infty S_m$ (Proposition 2.1.8), by Lemma 1.6.7, there exists $M \in \mathbb{N}$ such that $i_m(X_m) \subset S_m \subset G$ for each $m \geq M$. Without loss of generality, we assume that $M \geq n$.

Let $m \geq M$. Then, by Lemma 2.6.6, $f_n^m \circ i_m^{-1}$ is universal. Hence, since $i_m(X_m) \subset G$, there exists $x_m \in X_m$ such that $f_n^m \circ i_m^{-1}(i_m(x_m)) = f_n^m(x_m) = \hat{k} \circ i_m(x_m)$. Note that this is true for each $m \geq M$. Since $\prod\limits_{\ell=1}^{\infty} X_\ell$ is compact, without loss of generality, we assume that the sequence $\{i_m(x_m)\}_{m=M}^{\infty}$ converges to a point $x' = (x'_\ell)_{\ell=1}^{\infty} \in X_\infty$.

Note that the bonding maps are surjective, since they are universal (Proposition 2.6.4 (1)). Hence, the projection maps are surjective (Remark 2.1.6). Thus, for each $m \geq M$, there exists $z^m = (z^m_\ell)_{\ell=1}^{\infty} \in X_\infty$ such that $f_m(z^m) = x_m$. Note that $\lim\limits_{m\to\infty} z^m = x'$. In consequence:

$$k(x') = \hat{k}(x') = \hat{k}(\lim_{m\to\infty} i_m(x_m)) = \lim_{m\to\infty} \hat{k}\circ i_m(x_m) = \lim_{m\to\infty} f_n^m(x_m) =$$

$$\lim_{m\to\infty} f_n^m f_m(z^m) = \lim_{m\to\infty} f_n(z^m) = f_n\left(\lim_{m\to\infty} z^m\right) = f_n(x').$$

Therefore, f_n is universal.

Now, observe that $\lim\limits_{\longleftarrow}\{f_n\} = 1_{X_\infty}$. Hence, by Theorem 2.6.9, 1_{X_∞} is universal. Therefore, by Proposition 2.6.5, X_∞ has the fixed point property.

Q.E.D.

A proof of the following Theorem may be found in 3.6.11 of [15].

2.6.15. Theorem. *Each polyhedron is an absolute neighborhood retract.*

2.6.16. Corollary. *Let $\{X_n, f_n^{n+1}\}$ be an inverse sequence of polyhedra, where all the bonding maps $f_n^m\colon X_m \twoheadrightarrow X_n$ are universal. Then $X_\infty = \lim\limits_{\longleftarrow}\{X_n, f_n^{n+1}\}$ has the fixed point property.*

2.6.17. Definition. Let X be a metric space and let A be a subset of X. We say that A is a *retract* of X if there is a map $r\colon X \twoheadrightarrow A$ such that $r(a) = a$ for each $a \in A$. The map r is called a *retraction*.

2.6.18. Corollary. *Let $\{X_n, f_n^{n+1}\}$ be an inverse sequence of absolute neighborhood retracts with the fixed point property. If each bonding map is a retraction, then $\varprojlim\{X_n, f_n^{n+1}\}$ has the fixed point property.*

Proof. By Theorem 2.6.14, it suffices to show that each bonding map $f_n^m \colon X_m \to X_n$ is universal.

Since each bonding map f_n^m is a retraction, $f_n^m|_{X_n} \colon X_n \twoheadrightarrow X_n$ is the identity map 1_{X_n}. Since every X_n has the fixed point property, by Proposition 2.6.5, 1_{X_n} is universal. Hence, $f_n^m|_{X_n} = 1_{X_n}$ is universal. Therefore, by (3) of Proposition 2.6.4, f_n^m is universal.

$$\textbf{Q.E.D.}$$

From now on, we identify the manifold boundary, $\partial\left([0,1]^n\right)$, of the n-cell $[0,1]^n$ with the unit $(n-1)$–dimensional sphere $\mathcal{S}^{n-1}$.

2.6.19. Theorem. *Let X be a separable metric space such that every map $g \colon X \to \mathcal{S}^{n-1}$ is homotopic to a constant map. Then a map $f \colon X \twoheadrightarrow [0,1]^n$ is universal if and only if the restriction $f|_{f^{-1}(\mathcal{S}^{n-1})} \colon f^{-1}(\mathcal{S}^{n-1}) \twoheadrightarrow \mathcal{S}^{n-1}$ is not homotopic to a constant map.*

Proof. Let $f \colon X \twoheadrightarrow [0,1]^n$ be a map and suppose that $f|_{f^{-1}(\mathcal{S}^{n-1})}$ is homotopic to a constant map. Hence, by Theorem 1.3.5, there exists a map $f' \colon X \twoheadrightarrow \mathcal{S}^{n-1}$ such that $f'|_{f^{-1}(\mathcal{S}^{n-1})} = f|_{f^{-1}(\mathcal{S}^{n-1})}$. Let $g \colon X \twoheadrightarrow \mathcal{S}^{n-1} \subset [0,1]^n$ be given by $g(x) = -f'(x)$. Then $f(x) \neq g(x)$ for any $x \in X$. Therefore, f is not universal.

Next, suppose $f \colon X \twoheadrightarrow [0,1]^n$ is not universal. Then there exists a map $g \colon X \to [0,1]^n$ such that $f(x) \neq g(x)$ for any $x \in X$. Let $f' \colon X \twoheadrightarrow \mathcal{S}^{n-1}$ be the map satisfying the following equation:

$$||f'(x) - f(x)|| + ||f(x) - g(x)|| = ||f'(x) - g(x)||.$$

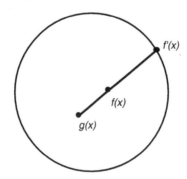

Then f' is map such that $f'|_{f^{-1}(\mathcal{S}^{n-1})} = f|_{f^{-1}(\mathcal{S}^{n-1})}$. Note that, by hypothesis, f' is homotopic to a constant map.

Q.E.D.

2.6.20. Definition. Let $\mathbf{n} = \{n_k\}_{k=1}^{\infty}$ be a sequence of positive integers. For each $k \in \mathbb{N}$, let $f_k^{k+1} \colon \mathcal{S}^1 \twoheadrightarrow \mathcal{S}^1$ be given by $f_k^{k+1}(z) = z^{n_k}$. The $\mathbf{n}$–*solenoid*, denoted by $\Sigma_{\mathbf{n}}$, is $\varprojlim\{X_k, f_k^{k+1}\}$, where $X_k = \mathcal{S}^1$ for every $k \in \mathbb{N}$. Note that, by Theorem 2.1.19, $\Sigma_{\mathbf{n}}$ is an indecomposable continuum.

2.6.21. Theorem. *Let* $\mathbf{n} = \{n_k\}_{k=1}^{\infty}$ *be a sequence of positive integers. Then the cone over the* $\mathbf{n}$–*solenoid,* $\Sigma_{\mathbf{n}}$, *has the fixed point property.*

Proof. Let $\Sigma_{\mathbf{n}} = \varprojlim\{X_k, f_k^{k+1}\}$, where every $X_k = \mathcal{S}^1$. By Theorem 2.3.2, $K(\Sigma_{\mathbf{n}})$ is homeomorphic to $\varprojlim\{K(X_k), K(f_k^{k+1})\}$. It is well known that $K(\mathcal{S}^1)$ is homeomorphic to $\{x \in \mathbb{R}^2 \mid \|x\| \leq 1\}$ (which is, in turn, homeomorphic to $[0,1]^2$). Since $K(\Sigma_{\mathbf{n}})$ is contractible, then every map from $K(\Sigma_{\mathbf{n}})$ into $\mathcal{S}^1$ is homotopic to a constant map (Theorem 1.3.12).

Observe that for each $k \in \mathbb{N}$, $K(f_k)^{-1}(\mathcal{S}^1) = \Sigma_{\mathbf{n}} \times \{0\}$. Hence, each $K(f_k)|_{K(f_k)^{-1}(\mathcal{S}^1)} \colon K(f_k)^{-1}(\mathcal{S}^1) \to \mathcal{S}^1$ is not homotopic to a constant map. Thus, by Theorem 2.6.19, every $K(f_k)$ is universal. Since the induced map $\varprojlim\{K(f_k)\} = 1_{K(\Sigma_{\mathbf{n}})}$, by Theorem 2.6.9, $1_{K(\Sigma_{\mathbf{n}})}$ is universal. Therefore, by Proposition 2.6.5, $K(\Sigma_{\mathbf{n}})$ has the fixed point property.

Q.E.D.

2.6.22. Definition. Let X be a metric space. A map $f \colon X \to [0,1]^n$ is said to be *AH–inessential* provided that there exists a map $g \colon X \to \mathcal{S}^{n-1}$ such that $f|_{f^{-1}(\mathcal{S}^{n-1})} = g|_{f^{-1}(\mathcal{S}^{n-1})}$. The map f is *AH–essential* if it is not AH–inessential.

2.6.23. Lemma. *Let* X *be a separable metric space. If* $f \colon X \twoheadrightarrow [0,1]^n$ *is AH-essential, then* $f|_{f^{-1}(\mathcal{S}^{n-1})} \colon f^{-1}(\mathcal{S}^{n-1}) \twoheadrightarrow \mathcal{S}^{n-1}$ *is not homotopic to a constant map.*

Proof. Suppose $f|_{f^{-1}(\mathcal{S}^{n-1})}$ is homotopic to a constant map g'. Then, clearly, g' can be extended to a constant map g defined on X. Hence, by Theorem 1.3.5, there exists a map $\hat{f}\colon X \to \mathcal{S}^{n-1}$ such that $\hat{f}|_{f^{-1}(\mathcal{S}^{n-1})} = f|_{f^{-1}(\mathcal{S}^{n-1})}$ and $\hat{f}$ is homotopic to g. Hence, f is AH–inessential.

Q.E.D.

2.6.24. Corollary. *Let X be a separable metric space. If $f\colon X \twoheadrightarrow [0,1]^2$ is AH-essential, then there exists a component K of $f^{-1}(\mathcal{S}^1)$ such that $f|_K\colon K \to \mathcal{S}^1$ is not homotopic to a constant map.*

Proof. By Lemma 2.6.23, $f|_{f^{-1}(\mathcal{S}^1)}$ is not homotopic to a constant map. Hence, by Theorem 1.3.7, there exists a subcontinuum C of $f^{-1}(\mathcal{S}^1)$ such that $f|_C$ is not homotopic to a constant map. Let K be the component of $f^{-1}(\mathcal{S}^1)$ such that $C \subset K$. Then $f|_K\colon K \twoheadrightarrow \mathcal{S}^1$ is not homotopic to a constant map.

Q.E.D.

2.6.25. Theorem. *Let X be a metric space. Then the map $f\colon X \twoheadrightarrow [0,1]^n$ is AH–essential if and only if f is universal.*

Proof. Suppose f is not universal. Then there exists a map $g\colon X \to [0,1]^n$ such that $f(x) \neq g(x)$ for any $x \in X$. Let $f'\colon X \to \mathcal{S}^{n-1}$ be the map satisfying the equation

$$||f'(x) - f(x)|| + ||f(x) - g(x)|| = ||f'(x) - g(x)||.$$

Then f' is an extension of $f|_{f^{-1}(\mathcal{S}^{n-1})}$ to X. Therefore, f is AH–inessential.

Next, suppose f is AH–inessential. Then there exists a map $g\colon X \to \mathcal{S}^{n-1}$ such that $g|_{f^{-1}(\mathcal{S}^{n-1})} = f|_{f^{-1}(\mathcal{S}^{n-1})}$. Let $h\colon X \to \mathcal{S}^{n-1}$ be given by $h(x) = -g(x)$. Hence, $f(x) \neq h(x)$ for any $x \in X$. Therefore, f is not universal.

Q.E.D.

Now, our goal is to prove Mazurkiewicz's Theorem about the existence of indecomposable subcontinua in continua of dimension at least two.

A proof of the following Theorem may be found in 18.6 of [19].

2.6.26. Theorem. *Let X be a separable metric space and let $n \in \mathbb{N}$. Then $\dim(X) \geq n$ if and only if there is an AH-essential map of X to $[0,1]^n$.*

2.6.27. Lemma. *Let X be a continuum, let $f \colon X \twoheadrightarrow [0,1]^2$ be an AH–essential map, and let J be a simple closed curve contained in $[0,1]^2$. If H is the bounded complementary domain (in $\mathbb{R}^2$) of J, then $f|_{f^{-1}(H \cup J)} \colon f^{-1}(H \cup J) \twoheadrightarrow H \cup J$ is an AH–essential map.*

Proof. Since $H \cup J$ is homeomorphic to $[0,1]^2$, by Theorem 2.6.25, it suffices to show that $f|_{f^{-1}(H \cup J)} \colon f^{-1}(H \cup J) \to H \cup J$ is universal.

Let $g \colon f^{-1}(H \cup J) \to H \cup J$ be a map. Since $H \cup J$ is an absolute retract (see 1.5.5 of [15]), there exists a map $\hat{g} \colon X \to H \cup J$ such that $\hat{g}|_{f^{-1}(H \cup J)} = g$. Since f is a universal map (Theorem 2.6.25), there exists $x \in X$ such that $f(x) = \hat{g}(x)$. Note that $f(x) \in H \cup J$. Hence, $x \in f^{-1}(H \cup J)$. Thus, $\hat{g}(x) = g(x)$. Therefore, $f|_{f^{-1}(H \cup J)}$ is a universal map.

Q.E.D.

2.6.28. Corollary. *Let X be a continuum, and let $f \colon X \twoheadrightarrow [0,1]^2$ be an AH–essential map. If J is a simple closed curve contained in $[0,1]^2$, then there exists a subcontinuum K of X such that $f(K) = J$.*

Proof. Note that, by Lemmas 2.6.27 and 2.6.23, $f|_{f^{-1}(J)} \colon f^{-1}(J) \twoheadrightarrow J$ is not homotopic to a constant map. Hence, by Corollary 2.6.24, there exists a component K of $f^{-1}(J)$ such that $f|_K \colon K \twoheadrightarrow J$ is not homotopic to a constant map. In particular, $f(K) = J$.

Q.E.D.

A proof of the following Theorem may be found on p. 157 of [13].

2.6.29. Theorem. *If C is a subcontinuum of $[0,1]^2$, then there exists a sequence $\{J_n\}_{n=1}^{\infty}$ of simple closed curves such that $\lim_{n \to \infty} J_n = C$.*

2.6.30. Definition. A map $f\colon X \to Y$ between continua is said to be *weakly confluent* provided that for every subcontinuum Z of Y, there exists a subcontinuum W of X such that $f(W) = Z$.

2.6.31. Theorem. *Let X be a continuum. If $f\colon X \twoheadrightarrow [0,1]^2$ is an AH–essential map, then f is weakly confluent.*

Proof. Let C be a subcontinuum of $[0,1]^2$. Then, by Theorem 2.6.29, there exists a sequence $\{J_n\}_{n=1}^{\infty}$ of simple closed curves converging to C. By Corollary 2.6.28, for each $n \in \mathbb{N}$, there exists a subcontinuum K_n of X such that $f(K_n) = J_n$. Since $\mathcal{C}(X)$ is compact (Theorem 1.8.5), without loss of generality, we assume that the sequence $\{K_n\}_{n=1}^{\infty}$ converges to a subcontinuum K of X. Therefore, by Corollary 1.8.23, $f(K) = C$.

Q.E.D.

As a consequence of Theorems 2.6.26 and 2.6.31, we have the following:

2.6.32. Corollary. *If X is a continuum of dimension greater than one, then there exists a weakly confluent map $f\colon X \twoheadrightarrow [0,1]^2$.*

Now, we are ready to show Mazurkiewicz's Theorem.

2.6.33. Theorem. *If X is a continuum of dimension greater than one, then X contains an indecomposable continuum.*

Proof. Since the dimension of X is greater than one, by Corollary 2.6.32, there exists a weakly confluent map $f\colon X \to [0,1]^2$. Let K be an indecomposable subcontinuum (like Knaster's continuum, Example 2.4.7) of $[0,1]^2$. Since f is weakly confluent, there exists a subcontinuum Y' of X such that $f(Y') = K$. Using Kuratowski–Zorn Lemma, there exists a subcontinuum Y of Y' such that $f(Y) = K$ and for each proper subcontinuum C of Y, $f(C) \neq K$.

We assert that Y is indecomposable. Suppose this is not true. Then there are two proper subcontinua A and B of Y such that

$Y = A \cup B$. Since A and B are proper subcontinua of Y, $f(A) \neq K$ and $f(B) \neq K$. Note that $K = f(Y) = f(A \cup B) = f(A) \cup f(B)$. Hence, K is decomposable, a contradiction. Therefore, Y is an indecomposable subcontinuum of X.

Q.E.D.

2.6.34. Corollary. *Each hereditarily decomposable continuum is of dimension one.*

To finish this section, we describe the Waraszkiewicz spirals [23] and present some applications.

Let $\mathcal{M}$ be the following subset of $\mathbb{R}^2$, defined in polar coordinates (e denotes the real exponential map):

$$\mathcal{M} = \{(1, \theta) \mid \theta \geq 0\} \cup \{(1 + e^{-\theta}, \theta) \mid \theta \geq 0\} \cup \{(1 + e^{-\theta}, -\theta) \mid \theta \geq 0\}.$$

Hence, $\mathcal{M}$ is the union of the unit circle $\mathcal{S}^1$, a ray R^+ spiraling in a counterclockwise direction onto $\mathcal{S}^1$, and a ray R^- spiraling in a clockwise direction onto $\mathcal{S}^1$.

2.6.35. Definition. The *Waraszkiewicz spirals* are subcontinua of $\mathcal{M}$. Each spiral is homeomorphic to a compactification of $[0, 1)$ with remainder $\mathcal{S}^1$. To determine a Waraszkiewicz spiral, begin at the point with Cartesian coordinates $(2, 0)$ and follow R^+ or R^-. At any time that the positive x–axis is crossed, feel free to change rays. This procedure determines the uncountable collection of continua called the Waraszkiewicz spirals.

Waraszkiewicz gave a proof of the following result in [23]. Ray L. Russo has improved the proof [22]:

2.6.36. Theorem. *No continuum can be mapped onto all Waraszkiewicz spirals.*

2.6.37. Corollary. *Given a countable collection of continua, there exists a Waraszkiewicz spiral W such that no member of this collection can be mapped onto W.*

Proof. Suppose the result is not true. Then there exists a countable family $\{X_n\}_{n=1}^{\infty}$ of continua such that for each Waraszkiewicz spiral W, there is $n \in \mathbb{N}$ such that X_n can be mapped onto W.

Without loss of generality, we assume that the members of the family $\{X_n\}_{n=1}^{\infty}$ are pairwise disjoint and $\lim_{n \to \infty} \text{diam}(X_n) = 0$. For each $n \in \mathbb{N}$, let $x_n \in X_n$. Identify all the points of the sequence $\{x_n\}_{n=1}^{\infty}$ to a point ω. In this way we obtain an infinite wedge. We call this quotient space X. Note that, by construction, X is a continuum.

Observe that, for each $n \in \mathbb{N}$, there exists a retraction $r_n \colon X \twoheadrightarrow X_n$ given by

$$r_n(x) = \begin{cases} \omega & \text{if } x \in X_m \text{ and } n \neq m; \\ x & \text{if } x \in X_n. \end{cases}$$

Hence, X can be mapped onto all the Waraszkiewicz spirals, a contradiction to Theorem 2.6.36.

$$\textbf{Q.E.D.}$$

2.6.38. Definition. A continuum X is *hereditarily equivalent* provided that X is homeomorphic to each of its nondegenerate subcontinuum.

2.6.39. Corollary. *Each hereditarily equivalent continuum has dimension one.*

Proof. Let X be a continuum of dimension greater than one. By Corollary 2.6.32, there exists a weakly confluent map $f\colon X \twoheadrightarrow [0,1]^2$. By Theorem 2.6.36, there exists a Waraszkiewicz spiral W such that X cannot be mapped onto W. We assume that $W \subset [0,1]^2$. Since f is weakly confluent, there exists a subcontinuum Y of X such that $f(Y) = W$. Hence, Y and X are not homeomorphic. Therefore, X is not hereditarily equivalent.

Q.E.D.

2.6.40. Corollary. *If X is a continuum of dimension greater than one, then X contains uncountably many nonhomeomorphic subcontinua.*

Proof. Suppose X has countably many nonhomeomorphic subcontinua $\{X_n\}_{n=1}^{\infty}$. By Corollary 2.6.37, there exists a Waraszkiewicz spiral W such that X_n cannot be mapped onto W for any $n \in \mathbb{N}$. We assume that $W \subset [0,1]^2$.

Since X is a continuum of dimension greater than one, by Corollary 2.6.32, there exists a weakly confluent map $f\colon X \twoheadrightarrow [0,1]^2$. Hence, there exists a subcontinuum Y of X such that $f(Y) = W$. This implies that Y is not homeomorphic to X_n for any $n \in \mathbb{N}$, a contradiction. Therefore, X contains uncountably many nonhomeomorphic subcontinua.

Q.E.D.

An argument for the following result may be found on p. 484 of [21].

2.6.41. Theorem. *For each Waraszkiewicz spiral W, there exists a plane continuum $\widehat{W}$ such that each nondegenerate subcontinuum of $\widehat{W}$ can be mapped onto W.*

2.6.42. Definition. A metric space X is *homogeneous* provided that for each pair of points $x, y \in X$, there exists a homeomorphism $h\colon X \twoheadrightarrow X$ such that $h(x) = y$.

We end this chapter showing that homogeneous hereditarily indecomposable continua are one–dimensional.

2.6.43. Theorem. *If X is a hereditarily indecomposable continuum of dimension greater than one, then X is not homogeneous.*

Proof. Let x be a point of X, and let $[\![x, X]\!]$ be the family of all subcontinua of X containing x, with the topology generated by the Hausdorff metric (Theorem 1.8.3). Note that $[\![x, X]\!]$ is the image of an order arc. Hence, $[\![x, X]\!]$ is homeomorphic to $[0, 1]$. Let $\{X_n\}_{n=1}^{\infty}$ be a countable dense subset of $[\![x, X]\!]$.

By Corollary 2.6.37, there exists a Waraszkiewicz spiral W such that X_n cannot be mapped onto W for every $n \in \mathbb{N}$. By Theorem 2.6.41, there exists a continuum $\widehat{W}$ in $[0, 1]^2$ such that each map from X_n into $\widehat{W}$ is constant for each $n \in \mathbb{N}$.

Suppose X is homogeneous. Since X is of dimension greater than one, by Corollary 2.6.32, there exists a weakly confluent map $f\colon X \twoheadrightarrow [0, 1]^2$. Hence, there exists a subcontinuum Z of X such that $f(Z) = \widehat{W}$. Since X is homogeneous, we assume that $x \in Z$, i.e., $Z \in [\![x, X]\!]$. Since $\{X_n\}_{n=1}^{\infty}$ is dense in $[\![x, X]\!]$, there exists a subsequence $\{X_{n_k}\}_{k=1}^{\infty}$ of $\{X_n\}_{n=1}^{\infty}$ such that $\lim_{k\to\infty} X_{n_k} = Z$ and $X_{n_k} \subset Z$ for every $k \in \mathbb{N}$. Since for each $k \in \mathbb{N}$, $x \in X_{n_k}$, $f(X_{n_k}) = \{f(x)\}$. This implies the contradiction that $f(Z) = \{f(x)\}$. Therefore, X is not homogeneous.

Q.E.D.

REFERENCES

[1] R. H. Bing, Embedding Circle–like Continua in the Plane, Canad. J. Math., 14 (1962), 113–128.

[2] C. E. Capel, Inverse Limit Spaces, Duke Math. J. 21 (1954), 233–245.

[3] C. O. Christenson and W. L. Voxman, *Aspects of Topology*, Monographs and Textbooks in Pure and Applied Math., Vol. 39, Marcel Dekker, New York, Basel, 1977.

[4] J. Dugundji, *Topology*, Allyn and Bacon, Inc., Boston, 1966.

[5] J. Grispolakis and E. D. Tymchatyn, On Confluent Mappings and Essential Mappings–A Survey, Rocky Mountain J. Math., 11 (1981), 131–153.

[6] O. H. Hamilton, A Fixed Point Theorem for Pseudo–arcs and Certain Other Metric Continua, Proc. Amer. Math. Soc., 2 (1951), 173–174.

[7] J. G. Hocking and G. S. Young, *Topology*, Dover Publications, Inc., New York, 1988.

[8] W. Holsztyński, Universal Mappings and Fixed Point Theorems, Bull. Acad. Polon. Sci. Sér. Sci. Math. Astronom. Phys., 15 (1967), 433–438

[9] W. Holsztyński, A Remark on the Universal Mappings of 1–dimensional Continua, Bull. Acad. Polon. Sci. Sér. Sci. Math. Astronom. Phys., 15 (1967), 547–549.

[10] W. T. Ingram, *Inverse Limits*, Aportaciones Matemáticas, Textos # 15, Sociedad Matemática Mexicana, 2000.

[11] K. Kuratowski, *Topology*, Vol. II, Academic Press, New York, 1968.

[12] S. Mardešić and J. Segal, ε–mappings onto Polyhedra, Trans. Amer. Math. Soc., 109 (1963), 146–164.

[13] S. Mazurkiewicz, Sur les Continus Absolutment Indécomposables, Fund. Math., 16 (1930), 151–159.

[14] S. Mazurkiewicz, Sur L'existence des Continus Indéscomposables, Fund. Math., 25 (1935), 327–328.

[15] J. van Mill, *Infinite–Dimensional Topology*, North Holland, Amsterdam, 1989.

[16] S. B. Nadler, Jr., Multicoherence Techniques Applied to Inverse Limits, Trans. Amer. Math. Soc., 157 (1971), 227–234.

[17] S. B. Nadler, Jr., *Hyperspaces of Sets,* Monographs and Textbooks in Pure and Applied Math., Vol. 49, Marcel Dekker, New York, Basel, 1978.

[18] S. B. Nadler, Jr., *Continuum Theory: An Introduction,* Monographs and Textbooks in Pure and Applied Math., Vol. 158, Marcel Dekker, New York, Basel, Hong Kong, 1992.

[19] S. B. Nadler, Jr., *Dimension Theory: An Introduction with Exercises,* Aportaciones Matemáticas, Textos # 18, Sociedad Matemática Mexicana, 2002.

[20] J. T. Rogers, Jr., The Pseudo–circle is not Homogeneous, Trans. Amer. Math. Soc. 148 (1970), 417–428.

[21] J. T. Rogers, Jr., Orbits of Higher–Dimensional Hereditarily Indecomposable Continua, Proc. Amer. Math. Soc., 95 (1985), 483–486.

[22] R. L. Russo, Universal Continua, Fund. Math., 105 (1979), 41–60.

[23] Z. Waraszkiewicz, Sur un Probléme de M. H. Hahn, Fund. Math., 18 (1932), 118–137.

Chapter 3

JONES'S SET FUNCTION $\mathcal{T}$

We prove basic results about the set function $\mathcal{T}$ defined by F. Burton Jones [17]. We define this function on compacta and then we concentrate on continua. In particular, we present some of the well known properties (such as connectedness im kleinen, local connectedness, semi–local connectedness, etc.) using the set function $\mathcal{T}$. The notion of aposyndesis was the main motivation of Jones to define this function. We present some properties of a continuum assuming the continuity of the set function $\mathcal{T}$. We also give some applications.

3.1 The Set Function $\mathcal{T}$

3.1.1. Definition. Given a compactum X, the *power set of X*, denoted by $\mathcal{P}(X)$, is:

$$\mathcal{P}(X) = \{A \mid A \subset X\}.$$

143

3.1.2. Remark. Let X be a metric space. Let us note that if $A \in \mathcal{P}(X)$, then its closure satisfies:

$$Cl(A) = \{x \in X \mid \text{for each open subset } U \text{ of } X \text{ such that}$$

$$x \in U, \text{ we have that } U \cap A \neq \emptyset\}.$$

As we see below (Definition 3.1.3), this property is similar to the definition of the function $\mathcal{T}$.

3.1.3. Definition. Let X be a compactum. Define

$$\mathcal{T} \colon \mathcal{P}(X) \to \mathcal{P}(X)$$

by

$$\mathcal{T}(A) = \{x \in X \mid \text{for each subcontinuum } W \text{ of } X \text{ such that}$$

$$x \in Int(W), \text{ we have that } W \cap A \neq \emptyset\},$$

for each $A \in \mathcal{P}(X)$. The function $\mathcal{T}$ is called *Jones's set function* $\mathcal{T}$.

3.1.4. Remark. In general, when working with the set function $\mathcal{T}$, we usually work with complements. Hence, for any compactum X and any $A \in \mathcal{P}(X)$, we have that:

$$\mathcal{T}(A) = X \setminus \{x \in X \mid \text{there exists a subcontinuum } W \text{ of } X$$

$$\text{such that } x \in Int(W) \subset W \subset X \setminus A\}.$$

3.1.5. Remark. Let X be a compactum. If $A \in \mathcal{P}(X)$, then $A \subset \mathcal{T}(A)$, and $\mathcal{T}(A)$ is closed in X. Hence, the range of $\mathcal{T}$ is 2^X (see Definition 1.8.1), i.e.:

$$\mathcal{T} \colon \mathcal{P}(X) \to 2^X.$$

The following proposition gives a relation between aposyndesis and $\mathcal{T}$.

3.1.6. Proposition. *Let X be a compactum, and let $A \in \mathcal{P}(X)$. If $x \in X \setminus \mathcal{T}(A)$, then X is aposyndetic at x with respect to each point of A.*

Proof. Let $x \in X \setminus \mathcal{T}(A)$, and let $a \in A$. Since $x \in X \setminus \mathcal{T}(A)$, by Definition 3.1.3, there exists a subcontinuum W of X such that $x \in Int(W) \subset W \subset X \setminus A$. In particular, $W \subset X \setminus \{a\}$. Therefore, X is aposyndetic at x with respect to a.

Q.E.D.

3.1.7. Proposition. *Let X be a compactum. If $A, B \in \mathcal{P}(X)$ and $A \subset B$, then $\mathcal{T}(A) \subset \mathcal{T}(B)$.*

Proof. Let $x \in X \setminus \mathcal{T}(B)$. Then there exists a subcontinuum W of X such that $x \in Int(W) \subset W \subset X \setminus B$. Since $X \setminus B \subset X \setminus A$, $x \in Int(W) \subset W \subset X \setminus A$. Therefore, $x \in X \setminus \mathcal{T}(A)$.

Q.E.D.

3.1.8. Corollary. *Let X be a compactum. If $A, B \in \mathcal{P}(X)$, then $\mathcal{T}(A) \cup \mathcal{T}(B) \subset \mathcal{T}(A \cup B)$.*

3.1.9. Remark. Let us note that the reverse inclusion of Corollary 3.1.8 is, in general, not true (see Example 3.1.18). It is an open question to characterize continua X such that $\mathcal{T}(A \cup B) = \mathcal{T}(A) \cup \mathcal{T}(B)$ for each $A, B \in \mathcal{P}(X)$.

Let us see a couple of examples.

3.1.10. Example. Let X be the Cantor set. Then $\mathcal{T}(\emptyset) = X$. To see this, suppose there is a point $x \in X \setminus \mathcal{T}(\emptyset)$. Then there exists a subcontinuum W of X such that $x \in Int(W) \subset W \subset X \setminus \emptyset = X$. Since X is the Cantor set, X is totally disconnected and perfect. Hence, no subcontinuum of X has interior. Therefore, W cannot exist. Hence, $\mathcal{T}(\emptyset) = X$. Note that, by Proposition 3.1.7, $\mathcal{T}(A) = X$ for each $A \in \mathcal{P}(X)$.

3.1.11. Example. Let $X = \{0\} \cup \left\{\dfrac{1}{n}\right\}_{n=1}^{\infty}$. Then $\mathcal{T}(\emptyset) = \{0\}$. This follows from the fact that the only subcontinuum of X with empty interior is $\{0\}$. Hence, by Remark 3.1.5 and Proposition 3.1.7, $\mathcal{T}(A) = \{0\} \cup A$ for each $A \in \mathcal{P}(X)$.

3.1.12. Remark. Note that if X is as in Example 3.1.10 or 3.1.11, then $\mathcal{T}|_{2^X} : 2^X \rightarrow 2^X$ is continuous, where 2^X has the topology given by the Hausdorff metric. In Section 3.2, we discuss the continuity of $\mathcal{T}$ on continua.

The following Theorem gives us a characterization of compacta X for which $\mathcal{T}(\emptyset) = \emptyset$.

3.1.13. Theorem. *Let X be a compactum. Then $\mathcal{T}(\emptyset) = \emptyset$ if and only if X has finitely many components.*

Proof. Suppose X has finitely many components. Let $x \in X$, and let C be the component of X such that $x \in C$. Hence, C is a subcontinuum of X and, by Lemma 1.6.2, $x \in Int(C)$. Thus, each point of X is contained in the interior of a proper subcontinuum of X. Therefore, $\mathcal{T}(\emptyset) = \emptyset$.

Now, suppose $\mathcal{T}(\emptyset) = \emptyset$. Then for each point $x \in X$, there exists a subcontinuum W_x of X such that $x \in Int(W_x) \subset W_x \subset X \setminus \emptyset$. Hence, $\{Int(W_x) \mid x \in X\}$ is an open cover of X. Since X is compact, there exist $x_1, \ldots, x_n \in X$ such that $X = \bigcup_{j=1}^{n} Int(W_{x_j}) \subset$ $\bigcup_{j=1}^{n} W_{x_j} \subset X$. Thus, X is the union of finitely many continua. Therefore, X has finitely many components.

$$\text{Q.E.D.}$$

3.1.14. Corollary. *If X is a continuum, then $\mathcal{T}(\emptyset) = \emptyset$.*

Let us see more examples.

3.1.15. Example. Let X be the cone over the Cantor set, with vertex ν_X and base B. Since the Cantor set is totally disconnected and perfect, it follows that the subcontinua of X having nonempty interior must contain ν_X. Hence, $\mathcal{T}(\{\nu_X\}) = X$. If $r \in X \setminus \{\nu_X\}$, then $\mathcal{T}(\{r\})$ is the line segment from r to the base B. Also, $\mathcal{T}(B) = B$.

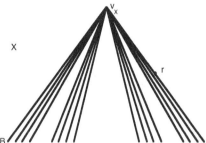

3.1.16. Example. Let $Y = X \cup X'$, where X is the cone over the Cantor set, with vertex ν_X and base B, and X' is another copy of the cone over the Cantor set, with vertex $\nu_{X'} \in B$. In this case, $\mathcal{T}(\{\nu_X\}) = X$, $\mathcal{T}(\{\nu_{X'}\}) = X'$ and $\mathcal{T}(\mathcal{T}(\{\nu_X\})) = \mathcal{T}^2(\{\nu_X\}) = Y$. Note that this implies, in general, that the function $\mathcal{T}$ is not idempotent (see Definition 3.1.51).

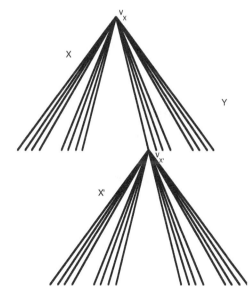

3.1.17. Example. Let X be the topologist sine curve, see Example 2.4.5. Let $J = X \setminus \left\{ \left(x, \sin\left(\dfrac{1}{x}\right) \right) \;\middle|\; 0 < x \le \dfrac{2}{\pi} \right\}$. Since X is not locally connected at any point of J, if $p \in J$, then $\mathcal{T}(\{p\}) = J$. Also, $\mathcal{T}(\{q\}) = \{q\}$ for each $q \in X \setminus J$.

The following example shows that, in general, $\mathcal{T}(A \cup B) = \mathcal{T}(A) \cup \mathcal{T}(B)$ does not hold.

3.1.18. Example. Let X be the suspension over the harmonic sequence $\{0\} \cup \left\{ \dfrac{1}{n} \right\}_{n=1}^{\infty}$, with vertices a and b. Note that for each $x \in X$, $\mathcal{T}(\{x\}) = \{x\}$. In particular, $\mathcal{T}(\{a\}) = \{a\}$ and $\mathcal{T}(\{b\}) = \{b\}$; however $\mathcal{T}(\{a, b\})$ consists of the limit segment from a to b.

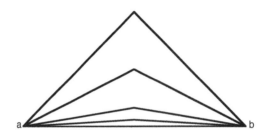

According to VandenBoss (p. 18 of [24]), the following result is due to H. S. Davis.

3.1.19. Theorem. *Let X be a compactum, and let A and B be nonempty closed subsets of X. Then the following are equivalent:*

(1) $\mathcal{T}(A) \cap B = \emptyset$.

(2) There exist two closed subsets M and N of X such that $A \subset Int(M)$, $B \subset Int(N)$ and $\mathcal{T}(M) \cap N = \emptyset$.

Proof. Assume (1), we show (2). For each $b \in B$, there exists a subcontinuum W_b of X such that $b \in Int(W_b) \subset W_b \subset X \setminus A$. Since B is compact, there exist $b_1, \ldots, b_n \in B$ such that $B \subset \displaystyle\bigcup_{j=1}^{n} Int(W_{b_j}) \subset \bigcup_{j=1}^{n} W_{b_j} \subset X \setminus A$. Since X is a metric space, there exist two open sets U and V of X such that $A \subset U \subset Cl(U) \subset$ $\left(X \setminus \bigcup_{j=1}^{n} W_{b_j}\right)$ and $B \subset V \subset Cl(V) \subset \bigcup_{j=1}^{n} Int(W_{b_j})$.

Let $M = Cl(U)$ and let $N = Cl(V)$. Then M and N are closed subsets of X such that $A \subset Int(M)$, $B \subset Int(N)$ and $\mathcal{T}(M) \cap N = \emptyset$.

The other implication follows easily from Proposition 3.1.7.

Q.E.D.

The following Corollary is an easy consequence of proof of Theorem 3.1.19.

3.1.20. Corollary. *Let X be a continuum, and let A be a closed subset of X. If $x \in X \setminus \mathcal{T}(A)$, then there exists an open subset U of X such that $A \subset U$ and $x \in X \setminus \mathcal{T}(Cl(U))$.*

3.1.21. Theorem. *Let X be a continuum. If W is a subcontinuum of X, then $\mathcal{T}(W)$ is also a subcontinuum of X.*

Proof. By Remark 3.1.5, $\mathcal{T}(W)$ is a closed subset of X.

Suppose $\mathcal{T}(W)$ is not connected. Then there exist two disjoint closed subsets A and B of X such that $\mathcal{T}(W) = A \cup B$. Since W is connected, without loss of generality, we assume that $W \subset A$. Since X is a metric space, there exists an open subset U of X such that $A \subset U$ and $Cl(U) \cap B = \emptyset$.

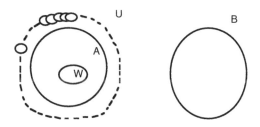

Note that $Bd(U) \cap \mathcal{T}(W) = \emptyset$. Hence, for each $z \in Bd(U)$, there exists a subcontinuum K_z of X such that $z \in Int(K_z) \subset K_z \subset X \setminus A$. Since $Bd(U)$ is compact, there exist $z_1, \ldots, z_n \in Bd(U)$ such that $Bd(U) \subset \bigcup_{j=1}^{n} Int(K_{z_j}) \subset \bigcup_{j=1}^{n} K_{z_j}$. Let $V = U \setminus \left(\bigcup_{j=1}^{n} K_{z_j}\right)$.

Let $Y = X \setminus V = (X \setminus U) \cup \left(\bigcup\limits_{j=1}^{n} K_{z_j} \right)$. By Theorem 1.7.22, Y has
a finite number of components. Observe that $B \subset X \setminus Cl(U) \subset X \setminus U \subset Y$. Hence, $B \subset Int(Y)$. Let $b \in B$, and let C be the component of Y such that $b \in C$. By Lemma 1.6.2, $b \in Int(C)$. By construction, $C \cap W = \emptyset$. Thus, $b \in X \setminus \mathcal{T}(W)$, a contradiction. Therefore, $\mathcal{T}(W)$ is connected.

Q.E.D.

The set function $\mathcal{T}$ may be used to define some of the properties of continua like connectedness im kleinen or local connectedness. In the following Theorems we show the equivalence between both types of definitions.

The following Theorem gives us a local definition of almost connectedness im kleinen using $\mathcal{T}$.

3.1.22. Theorem. *Let X be a continuum, with metric d. If $p \in X$, then X is almost connected im kleinen at p if and only if for each $A \in \mathcal{P}(X)$ such that $p \in Int(\mathcal{T}(A))$, $p \in Cl(A)$.*

Proof. Suppose X is almost connected im kleinen at p. Let $A \in \mathcal{P}(X)$ such that $p \in Int(\mathcal{T}(A))$. Then there exists $N \in \mathbb{N}$ such that $\mathcal{V}_{\frac{1}{n}}^{d}(p) \subset Int(\mathcal{T}(A))$ for each $n \geq N$. Since X is almost connected im kleinen at p, for each $n \geq N$, there exists a subcontinuum W_n of X such that $Int(W_n) \neq \emptyset$ and $W_n \subset \mathcal{V}_{\frac{1}{n}}^{d}(p) \subset Cl(\mathcal{V}_{\frac{1}{n}}^{d}(p)) \subset \mathcal{T}(A)$. Hence, $W_n \cap A \neq \emptyset$ for each $n \geq N$. Let $x_n \in W_n \cap A$ for each $n \geq N$. Note that, by construction, the sequence $\{x_n\}_{n=N}^{\infty}$ converges to p and $\{x_n\}_{n=N}^{\infty} \subset A$. Therefore, $p \in Cl(A)$.

Now suppose that $p \in X$ satisfies that for each $A \in \mathcal{P}(X)$ such that $p \in Int(\mathcal{T}(A))$, $p \in Cl(A)$. Let U be an open subset of X such that $p \in U$. Let V an open subset of X such that $p \in V \subset Cl(V) \subset U$. If some component of $Cl(V)$ has nonempty interior, then X is almost connected im kleinen at p. Suppose, then, that all the components of $Cl(V)$ have empty interior. Let $A = Bd(V)$. Then A is a closed subset of X and $p \in X \setminus A$. We show that $V \subset \mathcal{T}(A)$. To this end, suppose there exists $x \in V \setminus \mathcal{T}(A)$. Thus, there exists a subcontinuum W of X such that

$x \in Int(W) \subset W \subset X \setminus A$. Since all the components of $Cl(V)$ have empty interior, and W is a subcontinuum with nonempty interior, we have that $W \cap (X \setminus V) \neq \emptyset$ and, of course, $W \cap V \neq \emptyset$. Since W is connected, $W \cap Bd(V) = W \cap A \neq \emptyset$, a contradiction. Hence, $V \subset \mathcal{T}(A)$. Since $p \in V \subset \mathcal{T}(A)$, by hypothesis, $p \in Cl(A) = A$, a contradiction. Therefore, $Cl(V)$ has component with nonempty interior, and X is almost connected im kleinen at p.

Q.E.D.

The next Theorem gives us a global definition of almost connectedness im kleinen using $\mathcal{T}$.

3.1.23. Theorem. *A continuum X is almost connected im kleinen if and only if for each closed subset, F, of X, $Int(F) = Int(\mathcal{T}(F))$.*

Proof. Suppose X is almost connected im kleinen. Let F be a closed subset of X. Since $F \subset \mathcal{T}(F)$ (Remark 3.1.5), $Int(F) \subset Int(\mathcal{T}(F))$. Let $x \in Int(\mathcal{T}(F))$. Then, by Theorem 3.1.22, $x \in Cl(F) = F$. Hence, $Int(\mathcal{T}(F)) \subset F$. Therefore, $Int(\mathcal{T}(F)) \subset Int(F)$ and $Int(\mathcal{T}(F)) = Int(F)$.

Now, suppose that $Int(F) = Int(\mathcal{T}(F))$ for each closed subset F of X. Let $x \in X$, and let U be an open subset of X such that $x \in U$. Since $Int(X \setminus U) \cap U = \emptyset$, by hypothesis, $Int(\mathcal{T}(X \setminus U)) \cap U = \emptyset$. Hence, there exists $y \in U$ such that $y \in X \setminus \mathcal{T}(X \setminus U)$. Thus, there is a subcontinuum K of X such that $y \in Int(K) \subset K \subset X \setminus (X \setminus U) = U$. Therefore, X is almost connected im kleinen at x.

Q.E.D.

3.1.24. Theorem. *Let X be continuum. If $p \in X$, then X is connected im kleinen at p if and only if for each $A \in \mathcal{P}(X)$ such that $p \in \mathcal{T}(A)$, $p \in Cl(A)$.*

Proof. Let $p \in X$. Assume X is connected im kleinen at p. Let $A \in \mathcal{P}(X)$ and suppose $p \in X \setminus Cl(A)$. Hence, $Cl(A)$ is a closed subset of $X \setminus \{p\}$. Since X is connected im kleinen at p, there exists a subcontinuum W of X such that $p \in Int(W) \subset W \subset X \setminus A$. Therefore, $p \in X \setminus \mathcal{T}(A)$.

Now, suppose that $p \in X$ satisfies that for each $A \in \mathcal{P}(X)$ such that $p \in \mathcal{T}(A)$, $p \in Cl(A)$. Let D be a closed subset of X such that $D \subset X \setminus \{p\}$. Since $p \in X \setminus D$, by hypothesis, $p \in X \setminus \mathcal{T}(D)$. Then there exists a subcontinuum K of X such that $p \in Int(K) \subset K \subset X \setminus D$. Therefore, X is connected im kleinen at p.

$$\textbf{Q.E.D.}$$

3.1.25. Corollary. *Let X be continuum. If $p \in X$, then X is connected im kleinen at p if and only if for each closed subset A of X such that $p \in \mathcal{T}(A)$, $p \in A$.*

The following Theorem follows from Definition 1.7.13:

3.1.26. Theorem. *Let X be a continuum, and let $p, q \in X$. Then X is semi–aposyndetic at p and q if and only if either $p \in X \setminus \mathcal{T}(\{q\})$ or $q \in X \setminus \mathcal{T}(\{p\})$.*

The next Theorem follows from the definition of aposyndesis:

3.1.27. Theorem. *Let X be a continuum, and let $p, q \in X$. Then X is aposyndetic at p with respect to q if and only if $p \in X \setminus \mathcal{T}(\{q\})$.*

The following Theorem is a global version of Theorem 3.1.27.

3.1.28. Theorem. *A continuum X is aposyndetic if and only if $\mathcal{T}(\{p\}) = \{p\}$ for each $p \in X$.*

Proof. Suppose X is aposyndetic. Let $p \in X$. Then for each $q \in X \setminus \{p\}$, X is aposyndetic at q with respect to p. Hence, by Remark 3.1.5 and Theorem 3.1.27, $q \in X \setminus \mathcal{T}(\{p\})$. Therefore, $\mathcal{T}(\{p\}) = \{p\}$.

Now, suppose $\mathcal{T}(\{x\}) = \{x\}$ for each $x \in X$. Let $p, q \in X$ such that $p \neq q$. Since $\mathcal{T}(\{q\}) = \{q\}$, $p \in X \setminus \mathcal{T}(\{q\})$. Thus, by Theorem 3.1.27, X is aposyndetic at p with respect to q. Since p and q were arbitrary points of X, X is aposyndetic.

$$\textbf{Q.E.D.}$$

3.1.29. Theorem. *Let X be a continuum. If $p \in X$, then X is semi–locally connected at p if and only if $\mathcal{T}(\{p\}) = \{p\}$.*

Proof. Suppose X is semi–locally connected at p. By Remark 3.1.5, $\{p\} \subset \mathcal{T}(\{p\})$. Let $q \in X \setminus \{p\}$, and let U be an open subset of X such that $p \in U$ and $q \in X \setminus Cl(U)$. Since X is semi–locally connected at p, there exists an open subset V of X such that $p \in V \subset U$ and $X \setminus V$ has finitely many components. By Lemma 1.6.2, q belongs to the interior of the component of $X \setminus V$ containing q. Hence, $q \in X \setminus \mathcal{T}(\{p\})$. Therefore, $\mathcal{T}(\{p\}) = \{p\}$.

Now, suppose $\mathcal{T}(\{p\}) = \{p\}$. Let U be an open subset of X such that $p \in U$. Since $\mathcal{T}(\{p\}) = \{p\}$, for each $q \in X \setminus U$, there exists a subcontinuum W_q of X such that $q \in Int(W_q) \subset W_q \subset X \setminus \{p\}$. Note that $\{Int(W_q) \mid q \in X \setminus U\}$ is an open cover of $X \setminus U$. Since this set is compact, there exist $q_1, \ldots, q_n \in X \setminus U$ such that $X \setminus U \subset \bigcup_{j=1}^{n} Int(W_{q_j})$. Let $V = X \setminus \left(\bigcup_{j=1}^{n} W_{q_j} \right)$. Then V is an open subset of X such that $p \in V \subset U$ and $X \setminus V$ has finitely many components. Therefore, X is semi–locally connected at p.

$$\textbf{Q.E.D.}$$

As a consequence of Theorems 3.1.28 and 3.1.29, we have:

3.1.30. Corollary. *A continuum X is aposyndetic if and only if it is semi–locally connected.*

The next Theorem gives us a characterization of local connectedness in terms of the set function $\mathcal{T}$.

3.1.31. Theorem. *A continuum X is locally connected if and only if for each closed subset A of X, $\mathcal{T}(A) = A$.*

Proof. Suppose X is locally connected. Let A be a closed subset of X. Then $X \setminus A$ is an open subset of X. Let $p \in X \setminus A$. Then there exists an open subset U of X such that $p \in U \subset Cl(U) \subset X \setminus A$. Since X is locally connected, there exists an open connected subset V of X such that $p \in V \subset U$. Hence, $Cl(V)$ is a subcontinuum of X such that $p \in V \subset Cl(V) \subset Cl(U) \subset X \setminus A$. Consequently, $p \in X \setminus \mathcal{T}(A)$. Thus $\mathcal{T}(A) \subset A$. By Remark 3.1.5, $A \subset \mathcal{T}(A)$. Therefore, $\mathcal{T}(A) = A$.

Now, suppose $\mathcal{T}(A) = A$ for each closed subset of X. Let $p \in X$, and let U be an open subset of X such that $p \in U$. Then $X \setminus U$ is closed subset of X such that $p \notin X \setminus U$. Since $X \setminus U = \mathcal{T}(X \setminus U)$, there exists a subcontinuum W of X such that $p \in Int(W) \subset W \subset X\setminus(X\setminus U) = U$. Thus, by Theorem 1.7.9, X is connected im kleinen at p. Since p was an arbitrary point of X, by Theorem 1.7.12, X is locally connected.

Q.E.D.

Theorem 3.1.31 may be strengthened as follows:

3.1.32. Theorem. *Let X be a continuum. Then X is locally connected if and only if $\mathcal{T}(Y) = Y$ for each subcontinuum Y of X.*

Proof. If X is locally connected, by Theorem 3.1.31, $\mathcal{T}(Y) = Y$ for each subcontinuum Y of X.

Now suppose that for every subcontinuum Y of X, $\mathcal{T}(Y) = Y$. We show that X is locally connected. To this end, by Lemma 1.7.11, it is enough to see the components of open subsets of X are open. Let V be an open subset of X. Let $p \in V$. Since $\mathcal{T}(\{p\}) = \{p\}$ and $X \setminus V$ is compact, there exist finitely many subcontinua contained in $X \setminus \{p\}$ whose interiors cover $X \setminus V$. Let $\mathcal{W} = \{W_1, \ldots, W_m\}$ be such a collection of smallest possible cardinality. Let $U = X \setminus \left(\bigcup_{j=1}^{m} W_j \right)$, and let P be the component of $Cl(U)$ such that $p \in P$. We show that $p \in Int(P) \subset P \subset V$. By Theorem 1.7.22, each component of $Cl(U)$ intersects $\bigcup_{j=1}^{m} W_j$. By the minimality of m, we assert that no component of $Cl(U)$ except P can intersect more than one of the W_j's. To see this, let P' be a component of $Cl(U)$, different from P, and suppose that P' intersects both W_j and W_k, where $j, k \in \{1, \ldots, m\}$ and $j \neq k$. Note that, in this case, $P' \cup W_j \cup W_j$ is a continuum. Let $\mathcal{W}' = \{W_1, ..., W_m, P' \cup W_j \cup W_k\} \setminus \{W_j, W_k\}$.

Then $\mathcal{W}'$ has $m-1$ elements and $X \setminus V \subset \left(\bigcup_{\substack{l \neq j \\ l \neq k}} W_l \right) \cup (P' \cup W_j \cup W_k)$,

a contradiction to the minimality of m.

For each $j \in \{1, \ldots, m\}$, let A_j be the union of all the components of $Cl(U)$ which intersect W_j. Then $A_j \cap A_k \subset P$, for each $j, k \in \{1, \ldots, m\}$ such that $j \neq k$.

Now, since $p \notin \mathcal{T}(W_j) = W_j$, for each $j \in \{1, \ldots, m\}$, there exists a subcontinuum K_j of X such that $p \in Int(K_j) \subset K_j \subset X \setminus W_j$. Then $K_j \cap (A_j \setminus P) = \emptyset$. To see this, suppose there exists $x \in K_j \cap (A_j \setminus P)$. Let L be the component of $K_j \cap Cl(U)$ such that $x \in L$. By Theorem 1.7.22, L intersects the boundary of $K_j \cap Cl(U)$ in K_j. This boundary is contained in $\bigcup_{\ell=1}^{m} W_\ell$. (If $y \in Bd_{K_j}(K_j \cap Cl(U))$, then $y \in Cl(U)$. If $y \in U$, then $y \in K_j \cap U \subset K_j \cap Cl(U)$ and $(K_j \cap U) \cap (K_j \setminus (K_j \cap Cl(U))) = \emptyset$, a contradiction. Therefore, $y \in Cl(U) \setminus U \subset \bigcup_{\ell=1}^{m} W_\ell$.) Hence, there exists $k \in \{1, \ldots, m\} \setminus \{j\}$ such that $L \cap W_k \neq \emptyset$. But $L \subset M$, for some component M of $Cl(U)$, $M \neq P$ and $M \subset A_j$, so that $M \cap W_k = \emptyset$, a contradiction.

Thus, $\left(\bigcap_{j=1}^{m} K_j \right) \cap \left(\bigcup_{j=1}^{m} (A_j \setminus P) \right) = \emptyset$, and it follows that $\bigcap_{j=1}^{m} K_j \subset P$. Since $p \in Int \left(\bigcap_{j=1}^{m} K_j \right)$, $p \in Int(P)$. Therefore, P is a connected neighborhood of p in X contained in V. Since p was an arbitrary point of V, every component of V is open.

$$\textbf{Q.E.D.}$$

3.1.33. Remark. Note that Theorem 3.1.32 gives a partial answer to Question 7.2.11.

3.1.34. Theorem. *Let X be a continuum. Then X is indecomposable if and only if $\mathcal{T}(\{p\}) = X$ for each $p \in X$. Hence, $\mathcal{T}(A) = X$ for each closed subset A of X.*

Proof. Suppose X is indecomposable. By Corollary 1.7.21, each proper subcontinuum of X has empty interior. Therefore, $\mathcal{T}(\{p\}) = X$ for each $p \in X$.

Now, suppose X is decomposable. Then there exist two proper subcontinua A and B of X such that $X = A \cup B$. Let $a \in A \setminus B$, and let $b \in B \setminus A$. Then $b \in Int(B) \subset B \subset X \setminus \{a\}$. Hence, $b \in X \setminus \mathcal{T}(\{a\})$. Therefore, $\mathcal{T}(\{a\}) \neq X$.

$$\textbf{Q.E.D.}$$

3.1.35. Definition. A continuum X *is $\mathcal{T}$–symmetric* if for each pair, A and B, of closed subsets of X, $A \cap \mathcal{T}(B) = \emptyset$ if and only if $B \cap \mathcal{T}(A) = \emptyset$. We say that X *is point $\mathcal{T}$–symmetric* if for each pair, p and q, of points of X, $p \notin \mathcal{T}(\{q\})$ if and only if $q \notin \mathcal{T}(\{p\})$.

3.1.36. Theorem. *If X is a weakly irreducible continuum, then X is $\mathcal{T}$–symmetric.*

Proof. Let A and B be two closed subsets of X. Suppose $A \cap \mathcal{T}(B) = \emptyset$. Hence, for each $a \in A$, there exists a subcontinuum W_a of X such that $a \in Int(W_a) \subset W_a \subset X \setminus B$. Since A is compact, there exist $a_1, \dots, a_n \in A$ such that $A \subset \bigcup_{j=1}^{n} Int(W_{a_j}) \subset \bigcup_{j=1}^{n} W_{a_j} \subset X \setminus B$. This implies that $B \subset X \setminus \left(\bigcup_{j=1}^{n} W_{a_j} \right) \subset X \setminus \left(\bigcup_{j=1}^{n} Int(W_{a_j}) \right) \subset X \setminus A$.

Let $b \in B$. Since X is weakly irreducible, $X \setminus \left(\bigcup_{j=1}^{n} W_{a_j} \right)$ has a finite number of components which are open subsets of X. Let C be the component of $X \setminus \left(\bigcup_{j=1}^{n} W_{a_j} \right)$ such that $b \in C$. Then $b \in C \subset Cl(C) \subset X \setminus \left(\bigcup_{j=1}^{n} Int(W_{a_j}) \right) \subset X \setminus A$. Thus, $b \in X \setminus \mathcal{T}(A)$. Therefore, $B \cap \mathcal{T}(A) = \emptyset$.

Similarly, if $B \cap \mathcal{T}(A) = \emptyset$, then $A \cap \mathcal{T}(B) = \emptyset$.
Therefore, X is $\mathcal{T}$–symmetric.

Q.E.D.

3.1.37. Corollary. *If X is an irreducible continuum, then X is* $\mathcal{T}$*–symmetric.*

Proof. Let X be an irreducible continuum. By Theorem 1.7.29, X is weakly irreducible. Hence, by Theorem 3.1.36, X is $\mathcal{T}$–symmetric.

Q.E.D.

3.1.38. Theorem. *Let X be a $\mathcal{T}$–symmetric continuum, and let p be a point of X. Then X is connected im kleinen at p if and only if X is semi–locally connected at p.*

Proof. Suppose X is connected im kleinen at p. Let $q \in \mathcal{T}(\{p\})$. Since X is $\mathcal{T}$–symmetric, $p \in \mathcal{T}(\{q\})$. Thus, $p \in \{q\}$ (Corollary 3.1.25). Hence, $p = q$ and $\mathcal{T}(\{p\}) = \{p\}$. Therefore, by Theorem 3.1.29, X is semi–locally connected at p.

Now, suppose X is semi–locally connected at p. Then $\mathcal{T}(\{p\}) = \{p\}$ (Theorem 3.1.29). Let A be a closed subset of X such that $p \in \mathcal{T}(A)$. Hence, since X is $\mathcal{T}$–symmetric, $A \cap \mathcal{T}(\{p\}) \neq \emptyset$. Thus, $p \in A$. Therefore, by Corollary 3.1.25, X is connected im kleinen at p.

Q.E.D.

3.1.39. Definition. A continuum X *is* $\mathcal{T}$*–additive* if for each pair, A and B, of closed subsets of X, $\mathcal{T}(A \cup B) = \mathcal{T}(A) \cup \mathcal{T}(B)$.

The following Theorem gives us a sufficient condition for a continuum X to be $\mathcal{T}$–additive.

3.1.40. Theorem. *Let X be a continuum. If for each point $x \in X$ and any two subcontinua W_1 and W_2 of X such that $x \in Int(W_j)$, $j \in \{1,2\}$, there exists a subcontinuum W_3 of X such that $x \in Int(W_3)$ and $W_3 \subset W_1 \cap W_2$, then X is $\mathcal{T}$–additive.*

Proof. Let A_1 and A_2 be closed subsets of X, and let $x \in X \setminus \mathcal{T}(A_1) \cup \mathcal{T}(A_2)$. Hence, there exist two subcontinua W_1 and W_2 of X such that $x \in Int(W_j) \subset W_j \subset X \setminus A_j, j \in \{1, 2\}$. By hypothesis, there exists a subcontinuum W_3 of X such that $x \in Int(W_3)$ and $W_3 \subset W_1 \cap W_2$. Thus, $x \in Int(W_3) \subset W_3 \subset X \setminus (A_1 \cup A_2)$. Hence, $x \in X \setminus \mathcal{T}(A_1 \cup A_2)$. Therefore, by Corollary 3.1.8, $\mathcal{T}(A_1 \cup A_2) = \mathcal{T}(A_1) \cup \mathcal{T}(A_2)$.

Q.E.D.

We need the following definition to give a characterization of $\mathcal{T}$–additive continua.

3.1.41. Definition. Let X be a continuum. A *filterbase* $\aleph$ *in* X is a family $\aleph = \{A_\omega\}_{\omega \in \Omega}$ of subsets of X having two properties:

(a) For each $\omega \in \Omega$, $A_\omega \neq \emptyset$, and

(b) For each $\omega_1, \omega_2 \in \Omega$, there exists $\omega_3 \in \Omega$ such that $A_{\omega_3} \subset A_{\omega_1} \cap A_{\omega_2}$.

3.1.42. Lemma. Let X be a compactum. If Γ is a filterbase of closed subsets of X, then $\mathcal{T}\left(\bigcap\{G \mid G \in \Gamma\}\right) = \bigcap\{\mathcal{T}(G) \mid G \in \Gamma\}$.

Proof. By Proposition 3.1.7, $\mathcal{T}\left(\bigcap\{G \mid G \in \Gamma\}\right) \subset \bigcap\{\mathcal{T}(G) \mid G \in \Gamma\}$.

Let $p \in X \setminus \mathcal{T}\left(\bigcap\{G \mid G \in \Gamma\}\right)$. Then there exists a subcontinuum W of X such that $p \in Int(W) \subset W \subset X \setminus \left(\bigcap\{G \mid G \in \Gamma\}\right)$. Since W is compact, there exist $G_1, \ldots, G_n \in \Gamma$ such that $W \cap \left(\bigcap_{j=1}^{n} G_j\right) = \emptyset$. Since Γ is a filterbase, there exists $G \in \Gamma$ such that $G \subset \bigcap_{j=1}^{n} G_j$. Note that $W \cap G = \emptyset$. Thus, $p \in X \setminus \mathcal{T}(G)$. Hence, $p \in X \setminus \bigcap\{\mathcal{T}(G) \mid G \in \Gamma\}$. Therefore, $\mathcal{T}\left(\bigcap\{G \mid G \in \Gamma\}\right) = \bigcap\{\mathcal{T}(G) \mid G \in \Gamma\}$.

Q.E.D.

3.1.43. Theorem. *Let X be a continuum. Then X is $\mathcal{T}$–additive if and only if for each family Λ of closed subsets of X whose union is closed, $\mathcal{T}\left(\bigcup\{L \mid L \in \Lambda\}\right) = \bigcup\{\mathcal{T}(L) \mid L \in \Lambda\}$.*

Proof. Suppose X is $\mathcal{T}$–additive. By Proposition 3.1.7,

$$\bigcup\{\mathcal{T}(L) \mid L \in \Lambda\} \subset \mathcal{T}\left(\bigcup\{L \mid L \in \Lambda\}\right).$$

Now, suppose $x \in X \setminus \bigcup\{\mathcal{T}(L) \mid L \in \Lambda\}$. Then for each $L \in \Lambda$, let $F(L) = \{A \subset X \mid A \text{ is closed and } L \subset Int(A)\}$. If $L = \emptyset$, then $\mathcal{T}(L) = \bigcap\{\mathcal{T}(A) \mid A \in F(L)\}$. (Clearly $\mathcal{T}(L) \subset \bigcap\{\mathcal{T}(A) \mid A \in F(L)\}$. Let $z \in X \setminus \mathcal{T}(L)$. Then there exists a subcontinuum W of X such that $z \in Int(W)$. Let A be any proper closed subset of $X \setminus W$. Then $A \in F(L)$ and $z \in X \setminus \mathcal{T}(A)$. Hence, $z \in X \setminus \bigcap\{\mathcal{T}(A) \mid A \in F(L)\}$.) If $L \neq \emptyset$, then $F(L)$ is a filterbase of closed subsets of X. Since $\bigcap\{A \mid A \in F(L)\} = L$, $\mathcal{T}(L) = \bigcap\{\mathcal{T}(A) \mid A \in F(L)\}$, by Lemma 3.1.42.

Hence, for each $L \in \Lambda$, $x \in X \setminus \bigcap\{\mathcal{T}(A) \mid A \in F(L)\}$, and thus, there exists, for each $L \in \Lambda$, $A_L \in F(L)$ such that $x \in X \setminus \mathcal{T}(A_L)$. Note that $\{Int(A_L) \mid L \in \Lambda\}$ is an open cover of $\bigcup\{L \mid L \in \Lambda\}$. Since this set is compact, there exist $L_1, \ldots, L_m \in \Lambda$ such that

$$\bigcup\{L \mid L \in \Lambda\} \subset \bigcup_{j=1}^{m} Int(A_{L_j}).$$

Since, by hypothesis and mathematical induction, $\mathcal{T}\left(\bigcup_{j=1}^{m} A_{L_j}\right) = \bigcup_{j=1}^{m} \mathcal{T}(A_{L_j})$, $\mathcal{T}\left(\bigcup\{L \mid L \in \Lambda\}\right) \subset \bigcup_{j=1}^{m} \mathcal{T}(A_{L_j})$. Now, since for every $j \in \{1, \ldots, m\}$, $x \in X \setminus \mathcal{T}(A_{L_j})$, it follows that $x \in X \setminus \mathcal{T}\left(\bigcup\{L \mid L \in \Lambda\}\right)$. Thus, we have that $\mathcal{T}\left(\bigcup\{L \mid L \in \Lambda\}\right) \subset \bigcup\{\mathcal{T}(L) \mid L \in \Lambda\}$.

The other implication is obvious.

Q.E.D.

3.1.44. Theorem. *Each $\mathcal{T}$–symmetric continuum is $\mathcal{T}$–additive.*

Proof. Let X be a $\mathcal{T}$–symmetric continuum, and let A and B be two closed subsets of X. By Corollary 3.1.8, $\mathcal{T}(A) \cup \mathcal{T}(B) \subset \mathcal{T}(A \cup B)$.

Let $x \in \mathcal{T}(A \cup B)$. Then $\{x\} \cap \mathcal{T}(A \cup B) \neq \emptyset$. Since X is $\mathcal{T}$–symmetric, $\mathcal{T}(\{x\}) \cap (A \cup B) \neq \emptyset$. Hence, either $\mathcal{T}(\{x\}) \cap A \neq \emptyset$ or $\mathcal{T}(\{x\}) \cap B \neq \emptyset$. Thus, since X is $\mathcal{T}$–symmetric, $\{x\} \cap \mathcal{T}(A) \neq \emptyset$ or $\{x\} \cap \mathcal{T}(B) \neq \emptyset$, i.e., $x \in \mathcal{T}(A)$ or $x \in \mathcal{T}(B)$. Then $x \in \mathcal{T}(A) \cup \mathcal{T}(B)$. Therefore, X is $\mathcal{T}$–additive.

Q.E.D.

3.1.45. Theorem. *If X is a hereditarily unicoherent continuum, then X is $\mathcal{T}$–additive.*

Proof. Let X be a hereditarily unicoherent continuum, and let A and B be two closed subsets of X. By Corollary 3.1.8, $\mathcal{T}(A) \cup \mathcal{T}(B) \subset \mathcal{T}(A \cup B)$.

Let $x \in X \setminus (\mathcal{T}(A) \cup \mathcal{T}(B))$. Then $x \notin \mathcal{T}(A)$ and $x \notin \mathcal{T}(B)$. Hence, there exist subcontinua W_A and W_B of X such that $x \in Int(W_A) \cap Int(W_B) \subset W_A \cap W_B \subset X \setminus (A \cup B)$. Since X is hereditarily unicoherent, $W_A \cap W_B$ is a subcontinuum of X. Therefore, $x \in X \setminus \mathcal{T}(A \cup B)$.

Q.E.D.

The following Corollary is a consequence of Theorem 3.1.43; however, we present a different proof based on Corollary 3.1.20.

3.1.46. Corollary. *If X is a $\mathcal{T}$–additive continuum and if A is a closed subset of X, then $\mathcal{T}(A) = \bigcup\limits_{a \in A} \mathcal{T}(\{a\})$.*

Proof. Let A be a closed subset of the $\mathcal{T}$–additive continuum X. If $a \in A$, then $\{a\} \subset A$ and $\mathcal{T}(\{a\}) \subset \mathcal{T}(A)$, by Remark 3.1.5. Hence, $\bigcup\limits_{a \in A} \mathcal{T}(\{a\}) \subseteq \mathcal{T}(A)$.

Now, let $x \in X$ such that $x \in X \setminus \mathcal{T}(\{a\})$ for each $a \in A$. Then, by Corollary 3.1.20, for each $a \in A$, there exists an open subset U_a

of X such that $a \in U_a$ and $x \in X \setminus \mathcal{T}(Cl(U_a))$. Since A is compact and $\{U_a\}_{a \in A}$ is an open cover of A, there exist $a_1, \ldots, a_n \in A$ such that $A \subset \bigcup_{j=1}^{n} U_{a_j}$. Since for each $j \in \{1, \ldots, n\}$, $x \in X \setminus \mathcal{T}(Cl(U_{a_j}))$, $x \in X \setminus \bigcup_{j=1}^{n} \mathcal{T}(Cl(U_{a_j}))$. Thus, $x \in X \setminus \mathcal{T}\left(\bigcup_{j=1}^{n} Cl(U_{a_j})\right)$ (since X is $\mathcal{T}$–additive). Hence, $x \in X \setminus \mathcal{T}(A)$.

Q.E.D.

3.1.47. Theorem. *A continuum X is locally connected if and only if X is aposyndetic and $\mathcal{T}$–additive.*

Proof. Suppose X is a locally connected continuum. Then, by Theorem 3.1.31, $\mathcal{T}(A) = A$ for each closed subset A of X. Hence, $\mathcal{T}(\{p\}) = \{p\}$ for each point $p \in X$ and $\mathcal{T}(A \cup B) = A \cup B = \mathcal{T}(A) \cup \mathcal{T}(B)$ for each pair of closed subsets A and B of X. Therefore, X is aposyndetic (Theorem 3.1.28) and $\mathcal{T}$–additive.

Now, suppose X is aposyndetic and $\mathcal{T}$–additive. Let A be a closed subset of X. Since X is $\mathcal{T}$–additive, by Corollary 3.1.46, $\mathcal{T}(A) = \bigcup_{a \in A} \mathcal{T}(\{a\})$. Since X is aposyndetic, by Theorem 3.1.28, $\mathcal{T}(\{a\}) = \{a\}$ for each $a \in A$. Thus, $\mathcal{T}(A) = \bigcup_{a \in A} \mathcal{T}(\{a\}) = \bigcup_{a \in A} \{a\} = A$. Therefore, by Theorem 3.1.31, X is locally connected.

Q.E.D.

3.1.48. Corollary. *If X is an aposyndetic hereditarily unicoherent continuum, then X is locally connected.*

Proof. Let X be an aposyndetic hereditarily unicoherent continuum. Since X is hereditarily unicoherent, by Theorem 3.1.45, X is $\mathcal{T}$–additive. Then, since X is an aposyndetic $\mathcal{T}$–additive continuum, by Theorem 3.1.47, X is locally connected.

Q.E.D.

3.1.49. Theorem. *A continuum X is $\mathcal{T}$–symmetric if and only if X is point $\mathcal{T}$–symmetric and $\mathcal{T}$–additive.*

Proof. Suppose X is $\mathcal{T}$ symmetric. By Theorem 3.1.44, X is $\mathcal{T}$–additive. Clearly, X is point $\mathcal{T}$–symmetric.

Now, suppose X is point $\mathcal{T}$–symmetric and $\mathcal{T}$–additive. Let A and B be two closed subsets of X. Suppose that $A \cap \mathcal{T}(B) = \emptyset$ and $B \cap \mathcal{T}(A) \neq \emptyset$. Let $x \in B \cap \mathcal{T}(A)$. Then $x \in B$ and $x \in \mathcal{T}(A)$. Note that $x \in X \setminus A$ (if $x \in A$, then $x \in A \cap B \subset A \cap \mathcal{T}(B)$, a contradiction). Since X is $\mathcal{T}$–additive, $\mathcal{T}(A) = \bigcup_{a \in A} \mathcal{T}(\{a\})$. Thus, there exists $a \in A$ such that $x \in \mathcal{T}(\{a\})$. This implies that $a \in \mathcal{T}(\{x\})$ (since X is point $\mathcal{T}$–symmetric). Consequently, since $x \in B$, $a \in \mathcal{T}(\{x\}) \subset \mathcal{T}(B)$ (Proposition 3.1.7). Hence, $a \in A \cap \mathcal{T}(B)$, a contradiction to our assumption. Therefore, $B \cap \mathcal{T}(A) = \emptyset$, and X is $\mathcal{T}$–symmetric.

Q.E.D.

3.1.50. Theorem. *Let X be a semi–aposyndetic irreducible continuum. Then X is homeomorphic to $[0,1]$.*

Proof. Since X is an irreducible continuum, by Corollary 3.1.37, X is $\mathcal{T}$–symmetric. Let $x, y \in X$. Since X is semi–aposyndetic, by Theorem 3.1.26, either $x \in X \setminus \mathcal{T}(\{y\})$ or $y \in X \setminus \mathcal{T}(\{x\})$. This implies that $x \in X \setminus \mathcal{T}(\{y\})$ if and only if $y \in X \setminus \mathcal{T}(\{x\})$, by the $\mathcal{T}$–symmetry of X. Hence, $\mathcal{T}(\{x\}) = \{x\}$ for each $x \in X$. Therefore, X is aposyndetic (Theorem 3.1.28).

Since X is $\mathcal{T}$–symmetric, X is $\mathcal{T}$–additive (Theorem 3.1.44).

Since X is an aposyndetic $\mathcal{T}$–additive continuum, X is locally connected (Theorem 3.1.47). Thus, X is an irreducible locally connected continuum. Consequently, X is an irreducible arcwise connected continuum (Theorem 3–15 (p. 116) of [14]). Therefore, X is homeomorphic to $[0,1]$.

Q.E.D.

3.1.51. Definition. Let X be a continuum. We say that $\mathcal{T}$ *is idempotent on X* provided that $\mathcal{T}^2(A) = \mathcal{T}(A)$ for each closed subset A of X.

3.1.52. Proposition. *Let X be a continuum for which $\mathcal{T}$ is idempotent, and let Z be a nonempty closed subset of X. If $\mathcal{T}(Z) = A \cup B$, where A and B are nonempty closed subsets of X, then $\mathcal{T}(A \cup B) = \mathcal{T}(A) \cup \mathcal{T}(B) = A \cup B$.*

Proof. By the idempotency of $\mathcal{T}$, Remark 3.1.5 and Proposition 3.1.7, we have

$$\mathcal{T}(Z) = A \cup B \subset \mathcal{T}(A) \cup \mathcal{T}(B) \subset \mathcal{T}(A \cup B) = \mathcal{T}^2(Z) = \mathcal{T}(Z).$$

Q.E.D.

3.1.53. Proposition. *Let X be a continuum for which $\mathcal{T}$ is idempotent. If Z is a nonempty closed subset of X, then $\mathcal{T}(Z) = \bigcup \{\mathcal{T}(\{w\}) \mid w \in \mathcal{T}(Z)\}$.*

Proof. Let $x \in \bigcup \{\mathcal{T}(\{w\}) \mid w \in \mathcal{T}(Z)\}$. Then there exists $w \in \mathcal{T}(Z)$ such that $x \in \mathcal{T}(\{w\})$. Since $\mathcal{T}$ is idempotent, $\mathcal{T}(\{w\}) \subset \mathcal{T}^2(Z) = \mathcal{T}(Z)$. Hence, $x \in \mathcal{T}(Z)$, and $\bigcup \{\mathcal{T}(\{w\}) \mid w \in \mathcal{T}(Z)\} \subset \mathcal{T}(Z)$.

The other inclusion is clear.

Q.E.D.

3.1.54. Theorem. *Let X be a continuum. Then $\mathcal{T}$ is idempotent on X if and only if for each subcontinuum W of X and each point $x \in Int(W)$, there exists a subcontinuum M of X such that $x \in Int(M) \subset M \subset Int(W)$.*

Proof. Suppose $\mathcal{T}$ is idempotent on X. Let W be a subcontinuum of X, and let $x \in Int(W)$. Hence, $x \in X \setminus \mathcal{T}(X \setminus W)$. Since $\mathcal{T}$ is idempotent, $x \in X \setminus \mathcal{T}^2(X \setminus W)$. Thus, there exists a subcontinuum M of X such that $x \in Int(M) \subset M \subset X \setminus \mathcal{T}(X \setminus W) \subset X \setminus (X \setminus W) = W$. Therefore, $x \in Int(M) \subset M \subset Int(W)$.

Now, suppose the condition stated holds. We show $\mathcal{T}$ is idempotent. By Remark 3.1.5, for each $A \in \mathcal{P}(X)$, $\mathcal{T}(A) \subset \mathcal{T}^2(A)$. Let $B \in \mathcal{P}(X)$, and let $x \in X \setminus \mathcal{T}(B)$. Then there exists a subcontinuum W of X such that $x \in Int(W) \subset W \subset X \setminus B$. By hypothesis, there exists a subcontinuum M such that $x \in Int(M) \subset M \subset Int(W)$.

Since $Int(W) \subset X \setminus \mathcal{T}(B)$, $x \in Int(M) \subset M \subset X \setminus \mathcal{T}(B)$. Hence, $x \in X \setminus \mathcal{T}^2(B)$. Therefore, $\mathcal{T}$ is idempotent.

Q.E.D.

3.1.55. Corollary. *Let X be a continuum. If $\mathcal{T}$ is idempotent on X, W is a subcontinuum of X and K is a component of $Int(W)$, then K is open.*

Proof. Let W be a subcontinuum of X, and let K be a component of $Int(W)$. If $x \in K$, then, by Theorem 3.1.54, there exists a subcontinuum M of X such that $x \in Int(M) \subset Int(W)$. Hence, $M \subset K$ and x is an interior point of K. Therefore, K is open.

Q.E.D.

3.1.56. Corollary. *Let X be a continuum. If $\mathcal{T}$ is idempotent on X, $x \in X$ and W is a subcontinuum of X such that $x \in Int(W)$, then there exists a subcontinuum M of X such that $x \in Int(M) \subset M \subset W$ and $M = Cl(Int(M))$ (i.e., M is a continuum domain).*

Proof. Let W be a subcontinuum of X, and let $x \in Int(W)$. By Corollary 3.1.55, the component, K, of W containing x is open. Let $M = Cl(K)$. Then $M = Cl(Int(M))$ and $x \in Int(M) \subset M \subset W$.

Q.E.D.

3.1.57. Theorem. *Let X be a continuum. If $\mathcal{T}$ is idempotent on X and $\mathcal{T}(\{p, q\})$ is a continuum for all $p, q \in X$, then X is indecomposable.*

Proof. Suppose X is decomposable. Then, by Corollary 3.1.56, there exists a subcontinuum W of X such that $W = Cl(Int(W))$. Let p_0 and q_0 be any two points in $X \setminus Int(W)$. Note that, by Theorem 3.1.54, $\mathcal{T}(\{p_0, q_0\}) \cap Int(W) = \emptyset$.

Since $\mathcal{T}(\{p_0, q_0\})$ is connected and $\mathcal{T}(\{p_0, q_0\}) \cap Int(W) = \emptyset$, p_0 and q_0 belong to the same component of $X \setminus Int(W)$. Thus, $X \setminus Int(W)$ is a continuum. By Theorem 3.1.54, there exists a subcontinuum M, with nonempty interior, such that $M \subset Int(X \setminus Int(W)) \subset X \setminus W$. Then let $K = X \setminus (Int(W) \cup Int(M))$, and let p_1 and q_1 be any two points of K. By Theorem 3.1.54, $\mathcal{T}(\{p_1, q_1\}) \subset$

K. Hence, K is a continuum also. Now, let $p \in Int(M)$ and let $q \in Int(W)$. Observe that, by Theorem 3.1.54, $\mathcal{T}(\{p,q\}) \cap Int(K) = \emptyset$. Then $\mathcal{T}(\{p,q\}) \subset X \setminus Int(K) \subset W \cup M$. Hence, $\mathcal{T}(\{p,q\})$ is not a continuum, a contradiction.

Q.E.D.

3.1.58. Lemma. *Let X be a compactum. If A is a closed subset of X, then $p \in X \setminus \mathcal{T}(A)$ if and only if there exist a subcontinuum W and an open subset Q of X such that $p \in Int(W) \cap Q$, $Bd(Q) \cap \mathcal{T}(A) = \emptyset$ and $W \cap A \cap Q = \emptyset$.*

Proof. Let $p \in X \setminus \mathcal{T}(A)$. Then there exists a subcontinuum W of X such that $p \in Int(W) \subset W \subset X \setminus A$. Since X is a metric space, there exists an open subset Q of X such that $p \in Q \subset Cl(Q) \subset Int(W)$. Hence, $Bd(Q) \cap \mathcal{T}(A) = \emptyset$ and $W \cap A \cap Q = \emptyset$.

Now, suppose there exist a continuum W and an open subset Q of X such that $p \in Int(W) \cap Q$, $Bd(Q) \cap \mathcal{T}(A) = \emptyset$ and $W \cap A \cap Q = \emptyset$. Since $Bd(Q)$ is a compact set and $Bd(Q) \cap \mathcal{T}(A) = \emptyset$, there exist a finite collection, $W_1, \ldots, W_n$, of subcontinua of X such that $Bd(Q) \subset \bigcup_{j=1}^{n} Int(W_j) \subset \bigcup_{j=1}^{m} W_j \subset X \setminus A$. If $W \subset Q$, then, clearly, $p \in X \setminus \mathcal{T}(A)$.

Assume $W \setminus Q \neq \emptyset$. By Theorem 1.7.22, the closure of each component of $W \cap Q$ must intersect at least one of the W_j's, since $Bd(Q) \subset \bigcup_{j=1}^{m} W_j$. Hence, $H = (W \cap Q) \cup \left(\bigcup_{j=1}^{m} W_j \right)$ has only a finite number of components. Let K be the component of H such that $p \in K$. Since $p \in Int(W) \cap Q$, by Lemma 1.6.2, $p \in Int(K)$, and, of course, $K \cap A \subset H \cap A = \emptyset$. Thus, $p \in X \setminus \mathcal{T}(A)$.

Q.E.D.

3.1.59. Lemma. *Let X be a compactum, and let A be a subset of X. If $\mathcal{T}(A) = M \cup N$, where M and N are disjoint closed sets, then $\mathcal{T}(A \cap M) = M \cup \mathcal{T}(\emptyset)$.*

Proof. Suppose $p \in \mathcal{T}(A \cap M) \setminus (M \cup \mathcal{T}(\emptyset))$. Since $p \in X \setminus \mathcal{T}(\emptyset)$, there exists a subcontinuum W of X such that $p \in Int(W)$. Since

X is a metric space, there exists an open subset Q of X such that $N \subset Q$ and $Cl(Q) \cap M = \emptyset$. Note that $p \in Int(W) \cap Q$ and $Bd(Q) \cap \mathcal{T}(A \cap M) \subset Bd(Q) \cap \mathcal{T}(A) = \emptyset$, and $W \cap (A \cap M) \cap Q \subset Q \cap M = \emptyset$. Then, by Lemma 3.1.58, $p \in X \setminus \mathcal{T}(A \cap M)$, thus contradicting the assumption.

Now, suppose $p \in (M \cup \mathcal{T}(\emptyset)) \setminus \mathcal{T}(A \cap M)$. Since $p \in X \setminus \mathcal{T}(A \cap M)$ and $\emptyset \subset A \cap M$, $p \in X \setminus \mathcal{T}(\emptyset)$. Hence, $p \in M$. Since X is a metric space, there exists an open subset Q of X such that $M \subset Q$ and $Cl(Q) \cap N = \emptyset$. Since $p \in X \setminus \mathcal{T}(A \cap M)$, there exists a subcontinuum W of X such that $p \in Int(W) \subset W \subset X \setminus (A \cap M)$. Observe that $p \in Int(W) \cap Q$ and $Bd(Q) \cap \mathcal{T}(A) = \emptyset$. Since $Q \cap N = \emptyset$, $W \cap A \cap Q = W \cap (A \cap M) = \emptyset$. Hence, by Lemma 3.1.58, $p \in X \setminus \mathcal{T}(A) \subset X \setminus M$, thus contradicting our assumption.

Q.E.D.

3.1.60. Theorem. *Let X be a compactum, and let A be a closed subset of X. If K is a component of $\mathcal{T}(A)$, then $\mathcal{T}(A \cap K) = K \cup \mathcal{T}(\emptyset)$.*

Proof. Let K be a component of $\mathcal{T}(A)$. Let $\Gamma(K) = \{L \subset \mathcal{T}(A) \mid K \subset L$ and L is open and closed in $\mathcal{T}(A)\}$. Note that the collection of $\{A \cap L \mid L \in \Gamma(K)\}$ fails to be a filterbase if for some $L \in \Gamma(K)$, $A \cap L = \emptyset$. In this case the conclusion of Lemma 3.1.42 holds (an argument similar to the one given in the proof of Theorem 3.1.43 shows this). Note that Lemma 3.1.59 remains true even if $A \cap M = \emptyset$. Hence, by Lemma 3.1.59, for each $L \in \Gamma(K)$, $\mathcal{T}(A \cap L) = L \cup \mathcal{T}(\emptyset)$. Therefore, by Lemma 3.1.42, the following sequence of equalities establishes the theorem:

$$\mathcal{T}(A \cap K) = \mathcal{T}\left(\bigcap\{A \cap L \mid L \in \Gamma(K)\}\right) = \bigcap\{\mathcal{T}(A \cap L) \mid L \in$$

$$\Gamma(K)\} = \bigcap\{L \cup \mathcal{T}(\emptyset) \mid L \in \Gamma(K)\} = \bigcap\{L \mid L \in \Gamma(K)\} \cup \mathcal{T}(\emptyset) =$$

$$K \cup \mathcal{T}(\emptyset).$$

(The fact that $\bigcap\{L \mid L \in \Gamma(K)\} = K$ follows from Theorem 2–14 of [14].)

Q.E.D.

The following Corollary says that for each nonempty closed subset A of a continuum X, the components of $\mathcal{T}(A)$ are also in the image of $\mathcal{T}$.

3.1.61. Corollary. *Let X be a continuum, and let A be a closed subset of X. If K is a component of $\mathcal{T}(A)$, then $K = \mathcal{T}(A \cap K)$.*

Proof. Let A be a closed subset of X and let K be a component of $\mathcal{T}(A)$. By Theorem 3.1.60, $\mathcal{T}(A \cap K) = K \cup \mathcal{T}(\emptyset)$. Since $\mathcal{T}(\emptyset) = \emptyset$, by Corollary 3.1.14, $K = \mathcal{T}(A \cap K)$.

$\qquad\qquad\qquad\qquad\qquad\qquad\qquad\qquad\qquad\qquad$ **Q.E.D.**

3.1.62. Corollary. *Let X be a continuum, and let W_1 and W_2 be subcontinua of X. If $\mathcal{T}(W_1 \cup W_2) \neq \mathcal{T}(W_1) \cup \mathcal{T}(W_2)$, then $\mathcal{T}(W_1 \cup W_2)$ is a continuum.*

Proof. Suppose $\mathcal{T}(W_1 \cup W_2)$ is not connected. Then there exist two disjoint closed subsets A and B of X such that $\mathcal{T}(W_1 \cup W_2) = A \cup B$. By Lemma 3.1.59 and Corollary 3.1.14, $\mathcal{T}((W_1 \cup W_2) \cap A) = A$ and $\mathcal{T}((W_1 \cup W_2) \cap B) = B$. Suppose $W_1 \subset A$. If $W_2 \subset A$, then $A = \mathcal{T}((W_1 \cup W_2) \cap A) = \mathcal{T}(W_1 \cup W_2)$, a contradiction. Thus, $W_2 \subset B$. Hence $\mathcal{T}(W_1) = A$ and $\mathcal{T}(W_2) = B$, which implies that $\mathcal{T}(W_1 \cup W_2) = \mathcal{T}(W_1) \cup \mathcal{T}(W_2)$, a contradiction. Therefore, $\mathcal{T}(W_1 \cup W_2)$ is connected.

$\qquad\qquad\qquad\qquad\qquad\qquad\qquad\qquad\qquad\qquad$ **Q.E.D.**

3.1.63. Corollary. *Let X be a continuum. If $p, q \in X$ are such that $\mathcal{T}(\{p, q\}) \neq \mathcal{T}(\{p\}) \cup \mathcal{T}(\{q\})$, then $\mathcal{T}(\{p, q\})$ is a continuum.*

The following Theorem gives relationships between the images of maps and the images of $\mathcal{T}$. We subscript $\mathcal{T}$ to differentiate the continua on which $\mathcal{T}$ is defined.

3.1.64. Theorem. *Let X and Y be continua, and let $f \colon X \twoheadrightarrow Y$ be a surjective map. If $A \in \mathcal{P}(X)$ and $B \in \mathcal{P}(Y)$, then the following hold:*

(a) $\mathcal{T}_Y(B) \subset f\mathcal{T}_X f^{-1}(B)$.
(b) If f is monotone, then $f\mathcal{T}_X(A) \subset \mathcal{T}_Y f(A)$ and $\mathcal{T}_X f^{-1}(B) \subset f^{-1}\mathcal{T}_Y(B)$.
(c) If f is monotone, then $\mathcal{T}_Y(B) = f\mathcal{T}_X f^{-1}(B)$.
(d) If f is open, then $f^{-1}\mathcal{T}_Y(B) \subset \mathcal{T}_X f^{-1}(B)$.
(e) If f is monotone and open, then $f^{-1}\mathcal{T}_Y(B) = \mathcal{T}_X f^{-1}(B)$.

Proof. We show (a). Let $y \in Y \setminus f\mathcal{T}_X f^{-1}(B)$. Then $f^{-1}(y) \cap \mathcal{T}_X f^{-1}(B) = \emptyset$. Thus, for each $x \in f^{-1}(y)$, there exists a subcontinuum W_x of X such that $x \in Int(W_x) \subset W_x \subset X \setminus f^{-1}(B)$. Since $f^{-1}(y)$ is compact, there exist $x_1, ..., x_n \in f^{-1}(y)$ such that $f^{-1}(y) \subset \bigcup_{j=1}^{n} Int(W_{x_j})$. Note that $\left(\bigcup_{j=1}^{n} W_{x_j} \right) \cap f^{-1}(B) = \emptyset$, and for each $j \in \{1, ..., n\}$, $f^{-1}(y) \cap W_{x_j} \neq \emptyset$. Hence, $f\left(\bigcup_{j=1}^{n} W_{x_j} \right) \cap B = \emptyset$ and $f\left(\bigcup_{j=1}^{n} W_{x_j} \right)$ is a continuum ($y \in f(W_{x_j}) \cap f(W_{x_k})$ for every $j, k \in \{1, 2, ..., n\}$).

Observe that $Y \setminus f\left(X \setminus \bigcup_{j=1}^{n} Int(W_{x_j}) \right)$ is an open set of Y and it is contained in $f\left(\bigcup_{j=1}^{n} W_{x_j} \right)$. To see this, note that, since $\bigcup_{j=1}^{n} Int(W_{x_j}) \subset \bigcup_{j=1}^{n} W_{x_j}$, $X \setminus \bigcup_{j=1}^{n} W_{x_j} \subset X \setminus \bigcup_{j=1}^{n} Int(W_{x_j})$. Then $f\left(X \setminus \bigcup_{j=1}^{n} W_{x_j} \right) \subset f\left(X \setminus \bigcup_{j=1}^{n} Int(W_{x_j}) \right)$ and, consequently, $Y \setminus f\left(X \setminus \bigcup_{j=1}^{n} Int(W_{x_j}) \right) \subset Y \setminus f\left(X \setminus \bigcup_{j=1}^{n} W_{x_j} \right)$. Since f is a surjection, $Y \setminus f\left(X \setminus \bigcup_{j=1}^{n} Int(W_{x_j}) \right) \subset f\left(X \setminus \left(X \setminus \bigcup_{j=1}^{n} W_{x_j} \right) \right) = f\left(\bigcup_{j=1}^{n} W_{x_j} \right)$. Therefore, $Y \setminus f\left(X \setminus \bigcup_{j=1}^{n} Int(W_{x_j}) \right) \subset f\left(\bigcup_{j=1}^{n} W_{x_j} \right)$.

Also, since $f^{-1}(y) \subset \bigcup_{j=1}^{n} Int(W_{x_j})$, $\{y\} \cap f\left(X \setminus \bigcup_{j=1}^{n} Int(W_{x_j}) \right) = \emptyset$. Hence, $y \in Y \setminus f\left(X \setminus \bigcup_{j=1}^{n} Int(W_{x_j}) \right)$. Therefore, $y \in Y \setminus \mathcal{T}_Y(B)$.

We prove (b). First, we see $f\mathcal{T}_X(A) \subset \mathcal{T}_Y f(A)$. Let $y \in Y \setminus \mathcal{T}_Y f(A)$. Then there exists a subcontinuum W of Y such that $y \in$

$Int(W) \subset W \subset Y \setminus f(A)$. It follows that $f^{-1}(y) \subset f^{-1}(Int(W)) \subset f^{-1}(W) \subset f^{-1}(Y) \setminus f^{-1}f(A) \subset X \setminus A$. Hence, $f^{-1}(y) \subset X \setminus A$. Since f is monotone, $f^{-1}(W)$ is a subcontinuum of X (Lemma 2.1.12). Thus, since $f^{-1}(y) \subset f^{-1}(Int(W)) \subset f^{-1}(W) \subset X \setminus A$, $f^{-1}(y) \cap \mathcal{T}_X(A) = \emptyset$. Therefore, $y \in Y \setminus f\mathcal{T}_X(A)$.

Now, we show $\mathcal{T}_X f^{-1}(B) \subset f^{-1}\mathcal{T}_Y(B)$. Let $x \in X \setminus f^{-1}\mathcal{T}_Y(B)$. Then $f(x) \in Y \setminus \mathcal{T}_Y(B)$. Hence, there exists a subcontinuum W of Y such that $f(x) \in Int(W) \subset W \subset Y \setminus B$. This implies that $x \in f^{-1}(f(x)) \subset f^{-1}(Int(W)) \subset f^{-1}(W) \subset X \setminus f^{-1}(B)$. Since f is monotone, $f^{-1}(W)$ is a subcontinuum of X. Therefore, $x \in X \setminus \mathcal{T}_X f^{-1}(B)$.

We prove (c). By (a), $\mathcal{T}_Y(B) \subset f\mathcal{T}_X f^{-1}(B)$. Since $\mathcal{T}_X f^{-1}(B) \subset f^{-1}\mathcal{T}_Y(B)$, by (b), $f\mathcal{T}_X f^{-1}(B) \subset ff^{-1}\mathcal{T}_Y(B) = \mathcal{T}_Y(B)$. Therefore, $f\mathcal{T}_X f^{-1}(B) = \mathcal{T}_Y(B)$.

We show (d). Let $x \in X \setminus \mathcal{T}_X f^{-1}(B)$. Then there exists a subcontinuum W of X such that $x \in Int(W) \subset W \subset X \setminus f^{-1}(B)$. Hence, $f(x) \in f(Int(W)) \subset f(W) \subset Y \setminus B$. Since f is open, $f(Int(W))$ is an open subset of Y. Therefore, $f(x) \in Y \setminus \mathcal{T}_Y(B)$.

Note that (e) follows directly from (b) and (d).

Q.E.D.

The following four Theorems are applications of Theorem 3.1.64.

3.1.65. Theorem. *Let X and Y be continua, and let $f: X \twoheadrightarrow Y$ be a surjective map. If X is locally connected, then Y is locally connected.*

Proof. By Theorem 3.1.31, it suffices to show that $\mathcal{T}_Y(B) = B$ for each closed subset B of Y. Let B be a closed subset of Y. By Remark 3.1.5, $B \subset \mathcal{T}_Y(B)$. By Theorem 3.1.64 (a), $\mathcal{T}_Y(B) \subseteq f\mathcal{T}_X f^{-1}(B)$. Since X is locally connected, $\mathcal{T}_X f^{-1}(B) = f^{-1}(B)$. Hence, $\mathcal{T}_Y(B) \subseteq ff^{-1}(B) = B$. Therefore, $\mathcal{T}_Y(B) = B$, and Y is locally connected.

Q.E.D.

3.1.66. Theorem. *The monotone image of an indecomposable continuum is an indecomposable continuum.*

Proof. Let $f\colon X \twoheadrightarrow Y$ be a monotone map, where X is an inde-composable continuum. Let $y \in Y$. Then, by Theorem 3.1.64 (b), $\mathcal{T}_X f^{-1}(\{y\}) \subset f^{-1}\mathcal{T}_Y(\{y\})$. Since X is indecomposable, by Theorem 3.1.34, $\mathcal{T}_X f^{-1}(\{y\}) = X$. Hence, $Y = f(X) \subset f f^{-1}\mathcal{T}_Y(\{y\}) \subset Y$. Thus, $\mathcal{T}_Y(\{y\}) = Y$. Therefore, by Theorem 3.1.34, Y is inde-composable.

Q.E.D.

3.1.67. Theorem. *The monotone image of a $\mathcal{T}$–symmetric con-tinuum is $\mathcal{T}$–symmetric.*

Proof. Let X be a $\mathcal{T}$–symmetric continuum, and let $f\colon X \twoheadrightarrow Y$ be a monotone map. Let A and B be two closed subsets of Y such that $A \cap \mathcal{T}_Y(B) = \emptyset$. Then, by Theorem 3.1.64 (c), $A \cap f\mathcal{T}_X f^{-1}(B) = \emptyset$. Thus, $f^{-1}(A) \cap \mathcal{T}_X f^{-1}(B) = \emptyset$. Since X is $\mathcal{T}$–symmetric, $\mathcal{T}_X f^{-1}(A) \cap f^{-1}(B) = \emptyset$. Hence, $f(\mathcal{T}_X f^{-1}(A) \cap f^{-1}(B)) = f\mathcal{T}_X f^{-1}(A) \cap B = \mathcal{T}_Y(A) \cap B = \emptyset$. Therefore, Y is $\mathcal{T}$–symmetric.

Q.E.D.

3.1.68. Theorem. *The monotone image of a $\mathcal{T}$–additive contin-uum is $\mathcal{T}$–additive.*

Proof. Let X be a $\mathcal{T}$–additive continuum, and let $f\colon X \twoheadrightarrow Y$ be a monotone map. Let A and B be two closed subsets of Y. Then, by Theorem 3.1.64 (c), $\mathcal{T}_Y(A \cup B) = f\mathcal{T}_X f^{-1}(A \cup B) = f\mathcal{T}_X(f^{-1}(A) \cup f^{-1}(B)) = f(\mathcal{T}_X f^{-1}(A) \cup \mathcal{T}_X f^{-1}(B)) = f\mathcal{T}_X f^{-1}(A) \cup f\mathcal{T}_X f^{-1}(B) = \mathcal{T}_Y(A) \cup \mathcal{T}_Y(B)$.

Q.E.D.

Now, we present some relations between $\mathcal{T}$ and inverse limits due to H. S. Davis [10].

3.1.69. Lemma. *Let $\{X_n, f_n^{n+1}\}$ be an inverse sequence of con-tinua whose inverse limit is X. Let $p = (p_n)_{n=1}^{\infty} \in X$, and let A be a closed subset of X. If there exist $N \in \mathbb{N}$, two open subsets U_N and V_N of X_N and an inverse sequence $\{W_n, f_n^{n+1}|_{W_{n+1}}\}$ of subcontinua such that $p_N \in U_N$, $f_N(A) \subset V_N$ and for each $n \geq N$,*

$$(f_N^n)^{-1}(U_N) \subset W_n \subset X_n \setminus (f_N^n)^{-1}(V_N),$$

then $p \in X \setminus \mathcal{T}(A)$.

Proof. Let $W = \varprojlim\{W_n, f_n^{n+1}|_{W_{n+1}}\}$. Then W is a subcontinuum of X, by Proposition 2.1.8. Note that $f_N^{-1}(U_N) \subset W$ since, for each $n \geq N$, $(f_N^n)^{-1}(U_N) \subset Cl((f_N^n)^{-1}(U_N)) \subset Cl(f_n^{-1}(f_N^n)^{-1}(U_N)) \subset W_n$ and $W = \bigcap_{n=N}^{\infty} f_n^{-1}(W_n)$ (Corollary 2.1.21). Since $p \in f_N^{-1}(U_N)$, $p \in Int(W)$.

Since, for every $n \geq N$, $f_n^{-1}(W_n) \subset X \setminus f_n^{-1}(f_N^n)^{-1}(V_N) = X \setminus f_N^{-1}(V_N) = f_N^{-1}(X_N \setminus V_N) \subset X \setminus A$, $\bigcap_{n=N}^{\infty} f_n^{-1}(W_n) \subset X \setminus A$. Thus, $W \cap A = \emptyset$. Therefore, $p \in X \setminus \mathcal{T}(A)$.

Q.E.D.

3.1.70. Lemma. *Let $\{X_n, f_n^{n+1}\}$ be an inverse sequence of continua whose inverse limit is X. Let $p = (p_n)_{n=1}^{\infty} \in X$, and let A be a closed subset of X. If $p \in X \setminus \mathcal{T}(A)$, then there exist $N \in \mathbb{N}$, two open subsets U_N and V_N of X_N and an inverse sequence $\{W_n, f_n^{n+1}|_{W_{n+1}}\}$ of subcontinua such that $p_N \in U_N$, $f_N(A) \subset V_N$ and for every $n \geq N$,*

$$(f_N^n)^{-1}(U_N) \cap f_n(X) \subset W_n \subset f_n(X) \setminus (f_N^n)^{-1}(V_N).$$

Proof. Since $p \in X \setminus \mathcal{T}(A)$, there exists a subcontinuum W of X such that $p \in Int(W) \subset W \subset X \setminus A$. By Proposition 2.1.9, there exist $N(p) \in \mathbb{N}$ and an open subset $U_{N(p)}$ of $X_{N(p)}$ such that $p \in f_{N(p)}^{-1}(U_{N(p)}) \subset Int(W)$. Similarly, for each $a = (a_n)_{n=1}^{\infty} \in A$, there exist $N(a) \in \mathbb{N}$ and an open subset $U_{N(a)}$ of $X_{N(a)}$ such that $a \in f_{N(a)}^{-1}(U_{N(a)}) \subset X \setminus W$. Since A is compact, there exist $a^1, \ldots a^m \in A$ such that $A \subset \bigcup_{j=1}^{m} f_{N(a^j)}^{-1}(U_{N(a^j)})$. Let $N = \max\{N(p), N(a^1), \ldots, N(a^m)\}$. Define $U_N = (f_{N(p)}^N)^{-1}(U_{N(p)})$ and $V_N = \bigcup_{j=1}^{m} (f_{N(a^j)}^N)^{-1}(U_{N(a^j)})$.

For each $n \in \mathbb{N}$, let $W_n = f_n(W)$. Let $n \geq N$, and let $z_n \in (f_N^n)^{-1}(U_N) \cap f_n(X)$. Since $z_n \in f_n(X)$, there exists $z \in X$ such that

$f_n(z) = z_n$. Since $f_{N(p)}^n(z_n) \in U_{N(p)}$, $f_{N(p)}(z) \in U_{N(p)}$. Therefore, $z \in f_{N(p)}^{-1}(U_{N(p)}) \subset Int(W) \subset W$. Thus, $z_n = f_n(z) \in f_n(W) = W_n$. Hence, $(f_N^n)^{-1}(U_N) \cap f_n(X) \subset W_n$.

Now, let $w_n \in W_n = f_n(W)$. Then there exists $w \in W$ such that $f_n(w) = w_n$. Clearly, $w_n \in f_n(X)$. Suppose $w_n \in (f_N^n)^{-1}(V_N)$. Then there exists $j \in \{1, \dots, m\}$ such that $f_{N(a^j)}^n(w_n) \in U_{N(a^j)}$. Hence, $f_{N(a^j)}(w) \in U_{N(a^j)}$. Therefore, $w \in X \setminus W$. This contradiction establishes that $W_n \subset f_n(X) \setminus (f_N^n)^{-1}(V_N)$.

Q.E.D.

3.1.71. Theorem. *Let $\{X_n, f_n^{n+1}\}$ be an inverse sequence of continua whose inverse limit is X. Let $p = (p_n)_{n=1}^\infty \in X$, and let A be a closed subset of X. Then the following are equivalent:*

(a) $p \in X \setminus \mathcal{T}(A)$,

(b) there exist $N \in \mathbb{N}$, two open subsets U_N and V_N of X_N and an inverse sequence $\{W_n, f_n^{n+1}|_{W_{n+1}}\}$ of subcontinua such that $p_N \in U_N$, $f_N(A) \subset V_N$ and for every $n \geq N$,

$$(f_N^n)^{-1}(U_N) \cap f_n(X) \subset W_n \subset f_n(X) \setminus (f_N^n)^{-1}(V_N).$$

Proof. Suppose (a). Then Lemma 3.1.70 shows (b).

Now, suppose (b). Note that if $W = \varprojlim\{W_n, f_n^{n+1}|_{W_{n+1}}\}$, then for each $n \in \mathbb{N}$, $f_n(W) = f_n(X) \cap W_n$, $f_n(W)$ is a continuum and $W = \varprojlim\{f_n(W), f_n^{n+1}|_{f_{n+1}(W)}\}$ (Proposition 2.1.20).

The relations in Lemma 3.1.69 imply that $(f_N^n)^{-1}(U_N) \cap f_n(X) \subset W_n \subset f_n(X) \setminus (f_N^n)^{-1}(V_N)$.

Q.E.D.

3.1.72. Theorem. *Let $\{X_n, f_n^{n+1}\}$ be an inverse sequence of continua, with surjective bonding maps, whose inverse limit is X. Let $p = (p_n)_{n=1}^\infty \in X$, and let A be a closed subset of X. Then the following are equivalent:*

(a) $p \in X \setminus \mathcal{T}(A)$,

(b) there exist $N \in \mathbb{N}$, two open subsets U_N and V_N of X_N and an inverse sequence $\{W_n, f_n^{n+1}|_{W_{n+1}}\}$ of subcontinua such that $p_N \in U_N$, $f_N(A) \subset V_N$ and for every $n \geq N$,

$$(f_N^n)^{-1}(U_N) \subset W_n \subset X_n \setminus (f_N^n)^{-1}(V_N).$$

Proof. Note that, since the bonding maps are surjective, the projection maps are surjective also (Remark 2.1.6). Thus, the Theorem follows directly from Theorem 3.1.71

Q.E.D.

3.1.73. Corollary. *Let $\{X_n, f_n^{n+1}\}$ be an inverse sequence of continua, with surjective bonding maps, whose inverse limit is X. Let A be a closed subset of X, and let $N_0 \in \mathbb{N}$. Then*

$$\bigcap_{n=N_0}^{\infty} f_n^{-1}(\mathcal{T}_{X_n}(f_n(A))) \subset \mathcal{T}_X(A).$$

Proof. Let $p = (p_n)_{n=1}^{\infty} \in X \setminus \mathcal{T}_X(A)$. By Theorem 3.1.72, there exist $N \in \mathbb{N}$, two open subsets U_N and V_N of X_N and an inverse sequence $\{W_n, f_n^{n+1}|_{W_{n+1}}\}$ of subcontinua such that $p_N \in U_N$, $f_N(A) \subset V_N$ and for every $n \geq N$,

$$(f_N^n)^{-1}(U_N) \subset W_n \subset X_n \setminus (f_N^n)^{-1}(V_N).$$

Let $m \geq \max\{N, N_0\}$. Then $p_m \in (f_N^m)^{-1}(U_N) \subset W_m \subset X_m \setminus (f_N^m)^{-1}(V_N) \subset X_m \setminus f_m(A)$. Thus, $p_m \in X_m \setminus \mathcal{T}_{X_m}(f_m(A))$. Hence, $p \in X \setminus f_m^{-1}(\mathcal{T}_{X_m}(f_m(A)))$. Therefore, $p \in X \setminus \bigcap_{n=N_0}^{\infty} f_n^{-1}(\mathcal{T}_{X_n}(f_n(A)))$.

Q.E.D.

3.1.74. Corollary. *Let $\{X_n, f_n^{n+1}\}$ be an inverse sequence of continua, with surjective bonding maps, whose inverse limit is X. Let A be a closed subset of X, and let $N_0 \in \mathbb{N}$. If the bonding maps are monotone, then*

$$\bigcap_{n=N_0}^{\infty} f_n^{-1}(\mathcal{T}_{X_n}(f_n(A))) = \mathcal{T}_X(A).$$

Proof. Let $p = (p_n)_{n=1}^{\infty} \in X \setminus \bigcap_{n=N_0}^{\infty} f_n^{-1}(\mathcal{T}_{X_n}(f_n(A)))$. Then there exist $N \geq N_0$ such that $p_N \in X_N \setminus \mathcal{T}_{X_N}(f_N(A))$. Thus, there exist a subcontinuum W_N of X_N and two open subsets U_N and V_N of X_N such that $p_N \in U_N \subset W_N \subset X_N \setminus V_N \subset X_N \setminus f_N(A)$.

Now, for each $n \in \mathbb{N}$, let

$$W_n = \begin{cases} (f_N^n)^{-1}(W_N) & \text{if } n \geq N; \\ f_n^N(W_N) & \text{if } n \leq N. \end{cases}$$

Since the bonding maps are monotone, W_n is a continuum for each $n \in \mathbb{N}$ (Lemma 2.1.12). By construction, $\{W_n, f_n^{n+1}|_{W_{n+1}}\}$ is an inverse sequence of subcontinua and $(f_N^n)^{-1}(U_N) \subset W_n \subset X_n \setminus (f_N^n)^{-1}(V_N)$. Hence, by Theorem 3.1.72, $p \in X \setminus \mathcal{T}_X(A)$. The other inclusion is given by Corollary 3.1.73.

Q.E.D.

The next Corollary is a consequence of Davis's work:

3.1.75. Corollary. *Let $\{X_n, f_n^{n+1}\}$ be an inverse sequence of a-posyndetic continua, with surjective bonding maps, whose inverse limit is X. If the bonding maps are monotone, then X is aposyndetic.*

Proof. Let $p = (p_n)_{n=1}^\infty \in X$. By Corollary 3.1.74, we have that $\mathcal{T}_X(\{p\}) = \bigcap_{n=1}^\infty f_n^{-1}(\mathcal{T}_{X_n}(\{p_n\}))$. Since each factor space is aposyndetic, by Theorem 3.1.28, $\bigcap_{n=1}^\infty f_n^{-1}(\mathcal{T}_{X_n}(\{p_n\})) = \bigcap_{n=1}^\infty f_n^{-1}(\{p_n\}) = \{p\}$ (Corollary 2.1.21). Therefore, $\mathcal{T}_X(\{p\}) = \{p\}$. Since p was arbitrary, by Theorem 3.1.28, X is aposyndetic.

Q.E.D.

3.2   Continuity of $\mathcal{T}$

Let X be a continuum. By Remark 3.1.5, the image of any subset of X under $\mathcal{T}$ is a closed subset of X. Then we may restrict the domain of $\mathcal{T}$ to the hyperspace, 2^X, of closed subsets of X. Since 2^X has a topology, we may ask if $\mathcal{T}: 2^X \to 2^X$ is continuous. The answer to this question is negative, as can be easily seen from Example 3.1.15. On the other hand, by Theorems 3.1.31 and 3.1.34, $\mathcal{T}$ is continuous for locally connected continua and for indecomposable continua, respectively. In this section we present some results related to the continuity of $\mathcal{T}$. In particular, we show that if a continuum X is almost connected im kleinen at each of its points and $\mathcal{T}$ is continuous, then X is locally connected. Most of the results of this section are taken from [1].

We begin with the following Theorem which says that $\mathcal{T}$ is always upper semicontinuous.

3.2.1. Theorem. *Let X be a continuum. If U is an open subset of X, then $\mathcal{U} = \{A \in 2^X \mid \mathcal{T}(A) \subset U\}$ is open in 2^X.*

Proof. Let U be an open subset of X and let $\mathcal{U} = \{A \in 2^X \mid \mathcal{T}(A) \subset U\}$. We show $\mathcal{U}$ is open in 2^X. Let $B \in Cl_{2^X}(2^X \setminus \mathcal{U})$. Then there exists a sequence $\{B_n\}_{n=1}^{\infty}$ of elements of $2^X \setminus \mathcal{U}$ converging to B. Note that for each $n \in \mathbb{N}$, $\mathcal{T}(B_n) \cap (X \setminus U) \neq \emptyset$. Let $x_n \in \mathcal{T}(B_n) \cap (X \setminus U)$. Without loss of generality, we assume that $\{x_n\}_{n=1}^{\infty}$ converges to a point $x \in X$. Note that $x \in X \setminus U$. We assert that $x \in \mathcal{T}(B)$. Suppose $x \in X \setminus \mathcal{T}(B)$. Then there exists a subcontinuum W of X such that $x \in Int(W) \subset W \subset X \setminus B$. Since $\{B_n\}_{n=1}^{\infty}$ converges to B and $\{x_n\}_{n=1}^{\infty}$ converges to x, there exists $N \in \mathbb{N}$ such that for each $n \geq N$, $B_n \subset X \setminus W$ and $x_n \in Int(W)$. Let $n \geq N$. Then $x \in Int(W) \subset W \subset X \setminus B_n$. This implies that $x_n \in X \setminus \mathcal{T}(B_n)$, a contradiction. Thus, $x \in \mathcal{T}(B) \cap (X \setminus U)$. Hence, $B \in 2^X \setminus \mathcal{U}$. Therefore, $2^X \setminus \mathcal{U}$ is closed in 2^X.

Q.E.D.

The next two Theorems give sufficient conditions for $\mathcal{T}$ to be continuous for a continuum.

3.2.2. Theorem. *Let X and Z be continua, where Z is locally connected. If $f \colon X \twoheadrightarrow Z$ is a surjective, monotone and open map such that for each proper subcontinuum W of X, $f(W) \neq Z$, then $\mathcal{T}_X$ is continuous for X.*

Proof. Let A be a closed subset of X. Since Z is locally connected, $\mathcal{T}_Z(f(A)) = f(A)$ (Theorem 3.1.31). Hence, $f\mathcal{T}_X f^{-1}f(A) = f(A)$, by Theorem 3.1.64 (c). Thus,

$$f^{-1}f\mathcal{T}_X f^{-1}f(A) = f^{-1}f(A)$$

and

$$\mathcal{T}_X f^{-1}f(A) \subset f^{-1}f(A).$$

Then $\mathcal{T}_X f^{-1}f(A) = f^{-1}f(A)$, by Remark 3.1.5. Since $A \subset f^{-1}f(A)$, $\mathcal{T}_X(A) \subset \mathcal{T}_X f^{-1}f(A) = f^{-1}f(A)$ (the inclusion holds by Proposition 3.1.7).

Suppose there exists $x \in f^{-1}f(A) \setminus \mathcal{T}_X(A)$. Then there exists a subcontinuum W of X such that $x \in Int(W) \subset W \subset X \setminus A$. Since f is open, $f(x) \in Int(f(W))$. For each $z \in Z \setminus f(Int(W))$, there exists a subcontinuum M_z of Z such that $z \in Int(M_z) \subset M_z \subset Z \setminus \{f(x)\}$ ($\mathcal{T}_Z(\{z\}) = \{z\}$). Since $Z \setminus f(Int(W))$ is compact, there exist $z_1, \ldots, z_n \in Z \setminus f(Int(W))$ such that $Z \setminus f(Int(W)) \subset \bigcup_{j=1}^{n} Int(M_{z_j}) \subset \bigcup_{j=1}^{n} M_{z_j}$. By choosing n as small as possible, we may assume that $M_{z_j} \cap f(W) \neq \emptyset$ for each $j \in \{1, \ldots, n\}$. Then

$$Z = f(W) \cup \left(\bigcup_{j=1}^{n} M_{z_j}\right), \text{ and } f^{-1}f(W) \cup \left(\bigcup_{j=1}^{n} f^{-1}(M_{z_j})\right) = X.$$

For each $j \in \{1, \ldots, n\}$, let $q_j \in f(W) \cap M_{z_j}$. Then there exists $p_j \in W$ such that $f(p_j) = q_j$. Since $p_j \in f^{-1}(M_{z_j})$, $W \cap f^{-1}(M_{z_j}) \neq \emptyset$, $j \in \{1, \ldots, n\}$. Let $Y = W \cup \left(\bigcup_{j=1}^{n} f^{-1}(M_{z_j})\right)$. Then Y is a subcontinuum of X. We assert that $Y \neq X$. To see this, recall that $x \in f^{-1}f(A)$. Thus, there exists $y \in A$ such that $f(y) = f(x)$. Then $y \in X \setminus W$ and $f(y) \in Z \setminus \left(\bigcup_{j=1}^{n} M_{z_j}\right)$ (recall the construction of the M_{z_j}'s). Hence, $y \in X \setminus \left(\bigcup_{j=1}^{n} f^{-1}(M_{z_j})\right)$. However, $f(Y) = Z$, contrary to our hypothesis. Therefore, $\mathcal{T}_X(A) = f^{-1}f(A)$, i.e., $\mathcal{T}_X = \Im(f) \circ 2^f$.

Since f is continuous and open, $\Im(f)$ and 2^f are continuous

(Theorems 1.8.24 and 1.8.22, respectively). Therefore, $\mathcal{T}_X$ is continuous.

Q.E.D.

3.2.3. Remark. As a consequence of Theorem 3.2.2, we have that the circle of pseudo–arcs [7] is a decomposable nonlocally connected continuum for which $\mathcal{T}$ is continuous. For other decomposable nonlocally connected continua for which $\mathcal{T}$ is continuous see [20].

3.2.4. Definition. Let $f\colon X \to Z$ be a map. We say that f *is* $\mathcal{T}_{XZ}$*-continuous* provided that always $f\mathcal{T}_X(A) \subset \mathcal{T}_Z f(A)$ for every $A \in \mathcal{P}(X)$, or equivalently, $f^{-1}\mathcal{T}_Z(B) \supset \mathcal{T}_X f^{-1}(B)$ for every $B \in \mathcal{P}(Z)$.

3.2.5. Theorem. *Let X be a continuum for which $\mathcal{T}_X$ is continuous, and let Z be a continuum. If $f\colon X \twoheadrightarrow Z$ is a $\mathcal{T}_{XZ}$-continuous surjective open map, then $\mathcal{T}_Z$ is continuous for Z.*

Proof. Since f is $\mathcal{T}_{XZ}$–continuous, by Theorem 3.1.64 (a),

$$f\mathcal{T}_X f^{-1}(B) = \mathcal{T}_Z(B)$$

for each $B \in \mathcal{P}(Z)$. Thus, $\mathcal{T}_Z = 2^f \circ \mathcal{T}_X \circ \Im(f)$. Since f is continuous and open, $\Im(f)$ and 2^f are continuous (Theorems 1.8.24 and 1.8.22, respectively). Hence, $\mathcal{T}_Z$ is a composition of three maps. Therefore, $\mathcal{T}_Z$ is continuous.

Q.E.D.

The next Lemma says that given a continuum X for which $\mathcal{T}$ is continuous, the family $\{\mathcal{T}(\{x\}) \mid x \in X\}$ resembles an upper semicontinuous decomposition.

3.2.6. Lemma. *Let X be a continuum, with metric d, for which $\mathcal{T}$ is continuous. If U is an open subset of X and K is a closed subset of X, then:*
(1) $A = \{x \in X \mid \mathcal{T}(\{x\}) \subset U\}$ is open in X.
(2) $B = \{x \in X \mid \mathcal{T}(\{x\}) \cap U \neq \emptyset\}$ is open in X.
(3) $C = \{x \in X \mid \mathcal{T}(\{x\}) \cap K \neq \emptyset\}$ is closed in X.
(4) $D = \{x \in X \mid \mathcal{T}(\{x\}) \subset K\}$ is closed in X.

Proof. Let $x \in A$. Then $\mathcal{T}(\{x\}) \subset U$, and there exists $\varepsilon > 0$ such that $\mathcal{V}_\varepsilon^d(\mathcal{T}(\{x\})) \subset U$. Let $\delta > 0$ be given by the uniform continuity of $\mathcal{T}$ for ε. Let $w \in \mathcal{V}_\delta^d(x)$. Then $\{w\} \in \mathcal{V}_\delta^\mathcal{H}(\{x\})$. Thus, $\mathcal{H}(\mathcal{T}(\{w\}), \mathcal{T}(\{x\})) < \varepsilon$. Hence, $\mathcal{T}(\{w\}) \subset \mathcal{V}_\varepsilon^d(\mathcal{T}(\{x\})) \subset U$. This implies that $w \in A$. Therefore, A is open.

Let $x \in B$. Then $\mathcal{T}(\{x\}) \cap U \neq \emptyset$. Let $y \in \mathcal{T}(\{x\}) \cap U$ and let $\varepsilon > 0$ such that $\mathcal{V}_\varepsilon^d(y) \subset U$. Let $\delta > 0$ be given by the uniform continuity of $\mathcal{T}$ for ε. Let $z \in \mathcal{V}_\delta^d(x)$. Then $\{z\} \in \mathcal{V}_\delta^\mathcal{H}(\{x\})$. Hence, $\mathcal{H}(\mathcal{T}(\{z\}), \mathcal{T}(\{x\})) < \varepsilon$. This implies that there exists $z' \in \mathcal{T}(\{z\})$ such that $d(z', y) < \varepsilon$. Thus, $\mathcal{T}(\{z\}) \cap U \neq \emptyset$. Hence, $z \in B$. Therefore, B is open.

Now, let Z be either C or D. Let $x \in Cl(Z)$. Then there exists a sequence $\{x_n\}_{n=1}^\infty$ of elements of Z converging to x. Since $\mathcal{T}$ is continuous, $\lim_{n\to\infty} \mathcal{T}(\{x_n\}) = \mathcal{T}(\{x\})$.

If Z is C, then for each $n \in \mathbb{N}$, there exists $k_n \in \mathcal{T}(\{x_n\}) \cap K$. Hence, without loss of generality, we assume that $\{k_n\}_{n=1}^\infty$ converges to a point k. Since $\lim_{n\to\infty} \mathcal{T}(\{x_n\}) = \mathcal{T}(\{x\})$ and K is closed, $k \in \mathcal{T}(\{x\}) \cap K$. Thus, $x \in C$. Therefore, C is closed.

If Z is D, then for each $n \in \mathbb{N}$, $\mathcal{T}(\{x_n\}) \subset K$. Since $\lim_{n\to\infty} \mathcal{T}(\{x_n\}) = \mathcal{T}(\{x\})$ and K is closed, $\mathcal{T}(\{x\}) \subset K$. Hence, $x \in D$. Therefore, D is closed.

$$\text{Q.E.D.}$$

3.2.7. Notation. Let X be a compactum. If A is any subset of X, then

$$CL(A) = \{B \in 2^X \mid B \subset A\}.$$

The following Theorem tells us that the idempotency of $\mathcal{T}$ is related to its continuity.

3.2.8. Theorem. *If X is a continuum for which $\mathcal{T}$ is continuous, then $\mathcal{T}$ is idempotent on X.*

Proof. Let W be a subcontinuum of X, and let $x \in Int(W)$. Since $X \setminus Int(W)$ is a closed subset of X, it is easy to see that

$CL(X \setminus Int(W))$ is closed in 2^X. Then $\mathcal{T}^{-1}(CL(X \setminus Int(W)))$ is a closed subset of 2^X by the continuity of $\mathcal{T}$. Note that

$$CL(X \setminus W) \subset \mathcal{T}^{-1}(CL(X \setminus Int(W)))$$

because $X \setminus W \subset X \setminus Int(W)$, and Remark 3.1.5.

Since, for $W \neq X$, $X \setminus Int(W)$ is a limit point of $CL(X \setminus W)$ (recall that $Cl(X \setminus W) = X \setminus Int(W)$), it follows that $\mathcal{T}(X \setminus Int(W)) \subset X \setminus Int(W)$. Then there exists a subcontinuum M of X such that $x \in Int(M) \subset M \subset Int(W)$. If $W = X$, then it suffices to choose $M = X$. Thus, in either case, the Theorem follows from Theorem 3.1.54.

Q.E.D.

3.2.9. Theorem. *If X is a point $\mathcal{T}$–symmetric continuum for which $\mathcal{T}$ is continuous, then $\mathcal{T}(\{p, q\}) = \mathcal{T}(\{p\}) \cup \mathcal{T}(\{q\})$ for every $p, q \in X$.*

Proof. Let $p \in X$. For each $x \in X$, define

$$A(x) = \{y \in X \mid T(\{x, y\}) \in \mathcal{C}(X)\}$$

and

$$B(x) = \{y \in X \mid \mathcal{T}(\{x, y\}) = \mathcal{T}(\{x\}) \cup \mathcal{T}(\{y\})\}.$$

We assert that $A(x)$ and $B(x)$ are both closed for every $x \in X$. Let $x \in X$, and let $Z(x) \in \{A(x), B(x)\}$. If $y \in Cl(Z(x))$, then there exists a sequence $\{y_n\}_{n=1}^{\infty}$ of elements of $Z(x)$ converging to y. Hence, the sequence $\{\{x, y_n\}\}_{n=1}^{\infty}$ converges to $\{x, y\}$ (note that for each $z, w \in X$, $\{z, w\} = \sigma(\{\{z\}, \{w\}\})$). Since σ is continuous (Lemma 1.8.11),

$$\lim_{n \to \infty} \{x, y_n\} = \lim_{n \to \infty} \sigma(\{\{x\}, \{y_n\}\}) = \sigma(\{\{x\}, \{y\}\}) = \{x, y\}.$$

Hence, by the continuity of $\mathcal{T}$, the sequence $\{\mathcal{T}(\{x, y_n\})\}_{n=1}^{\infty}$ converges to $\mathcal{T}(\{x, y\})$.

If $Z(x)$ is $A(x)$, then, since $\mathcal{C}(X)$ is compact (Theorem 1.8.5), $\mathcal{T}(\{x, y\}) \in \mathcal{C}(X)$. If $Z(x)$ is $B(x)$, then $\mathcal{T}(\{x, y\}) = \lim_{n \to \infty} \mathcal{T}(\{x, y_n\})$ $= \lim_{n \to \infty} [\mathcal{T}(\{x\}) \cup \mathcal{T}(\{y_n\})] = \mathcal{T}(\{x\}) \cup \lim_{n \to \infty} \mathcal{T}(\{y_n\}) = \mathcal{T}(\{x\}) \cup \mathcal{T}(\{y\})$ (the third equality follows from the continuity of the union map, σ). Hence, $A(x)$ and $B(x)$ are both closed subsets of X. Note that, by Corollary 3.1.63, $X = A(p) \cup B(p)$.

Let $x \in \mathcal{T}(\{p\})$. Then $\mathcal{T}(\{x\}) \subset \mathcal{T}^2(\{p\}) = \mathcal{T}(\{p\})$ (by Proposition 3.1.7 and the fact that $\mathcal{T}$ is idempotent, by Theorem 3.2.8). Since X is point $\mathcal{T}$–symmetric and $x \in \mathcal{T}(\{p\})$, $p \in \mathcal{T}(\{x\})$. Hence, $\mathcal{T}(\{p\}) \subset \mathcal{T}(\{x\})$. Therefore, $\mathcal{T}(\{p\}) = \mathcal{T}(\{x\})$. Also, since $\{p, x\} \subset \mathcal{T}(\{p\})$, $\mathcal{T}(\{p, x\}) \subset \mathcal{T}(\{p\})$. Thus, $\mathcal{T}(\{p, x\}) = \mathcal{T}(\{p\})$. Since $\mathcal{T}(\{p\})$ is a continuum (Theorem 3.1.21), $x \in A(p) \cap B(p)$. Now, let $x \in A(p) \cap B(p)$. Then $\mathcal{T}(\{p, x\}) = \mathcal{T}(\{p\}) \cup \mathcal{T}(\{x\})$, and this set is a continuum. Hence, $\mathcal{T}(\{p\}) \cap \mathcal{T}(\{x\}) \neq \emptyset$. Let $y \in \mathcal{T}(\{p\}) \cap \mathcal{T}(\{x\})$. Then $x \in \mathcal{T}(\{y\}) \subset \mathcal{T}^2(\{p\}) = \mathcal{T}(\{p\})$. Therefore, $A(p) \cap B(p) = \mathcal{T}(\{p\})$.

Now, suppose $B(p) \neq \mathcal{T}(\{p\})$. Let $y \in A(p)$, and let $x \in B(p) \setminus \mathcal{T}(\{p\})$ be arbitrary points. Then $\mathcal{T}(\{p, x\}) = \mathcal{T}(\{p\}) \cup \mathcal{T}(\{x\})$ and $\mathcal{T}(\{p\}) \cap \mathcal{T}(\{x\}) = \emptyset$. Hence, $\mathcal{T}(\{x\}) \cap A(p) = \emptyset$. (Suppose there exists $z \in \mathcal{T}(\{x\}) \cap A(p)$, then, as before, $\mathcal{T}(\{z\}) = \mathcal{T}(\{x\})$. Since $\{p, z\} \subset \mathcal{T}(\{p\}) \cup \mathcal{T}(\{z\}) = \mathcal{T}(\{p\}) \cup \mathcal{T}(\{x\}) \subset \mathcal{T}(\{p, x\})$, $\mathcal{T}(\{p, z\}) \subset \mathcal{T}^2(\{p, x\}) = \mathcal{T}(\{p, x\})$. Similarly, we have that $\mathcal{T}(\{p, x\}) \subset \mathcal{T}(\{p, z\})$. Hence, $\mathcal{T}(\{p, z\}) = \mathcal{T}(\{p, x\})$. Since $z \in A(p)$, $\mathcal{T}(\{p, z\})$ is connected. Thus, $\mathcal{T}(\{p, x\})$ is connected, a contradiction.) Let U be an open set such that $\mathcal{T}(\{x\}) \subset U$ and $Cl(U) \cap A(p) = \emptyset$. Now, let $q \in Bd(U)$. Then $q \in (X \setminus A(p)) \cap (X \setminus \mathcal{T}(\{x\}))$. Hence, $q \in X \setminus \mathcal{T}(\{p, x\})$. Since $\mathcal{T}$ is idempotent, $q \in X \setminus \mathcal{T}^2(\{p, x\})$. Thus, there exists a subcontinuum W of X such that $q \in Int(W) \subset W \subset X \setminus \mathcal{T}(\{p, x\})$. Then $W \subset B(p)$, since otherwise $(W \cap A(p)) \cup (W \cap B(p))$ is a separation of W (recall that $A(p) \cap B(p) = \mathcal{T}(\{p\})$ and $W \cap \mathcal{T}(\{p\}) = \emptyset$). Therefore, $y \in X \setminus W$, and $q \in X \setminus \mathcal{T}(\{x, y\})$ (if $q \in \mathcal{T}(\{x, y\})$, then $\{x, y\} \cap W \neq \emptyset$, a contradiction). Since q was an arbitrary point of $Bd(U)$,

$$\mathcal{T}(\{x, y\}) = (\mathcal{T}(\{x, y\}) \cap U) \cup (\mathcal{T}(\{x, y\}) \cap (X \setminus Cl(U))),$$

where $\mathcal{T}(\{x, y\}) \cap U$ and $\mathcal{T}(\{x, y\}) \cap (X \setminus Cl(U))$ are separated. Thus, $\mathcal{T}(\{x, y\})$ is not connected. Hence, $\mathcal{T}(\{x, y\}) = \mathcal{T}(\{x\}) \cup \mathcal{T}(\{y\})$, by Corollary 3.1.63. So, $x \in B(y)$. Thus, $B(p) \setminus \mathcal{T}(\{p\}) \subset B(y)$.

Note that $p \in Cl(B(p) \setminus \mathcal{T}(\{p\}))$. If not, $p \in Int(A(p))$, and since $A(p)$ is a continuum and $\mathcal{T}$ is idempotent (Theorem 3.2.8), there is a continuum M such that $p \in Int(M) \subset M \subset Int(A(p))$

(Theorem 3.1.54). Hence, M misses some point $q \in \mathcal{T}(\{p\})$, and $p \in X \setminus \mathcal{T}(\{q\})$, contradicting the point $\mathcal{T}$–symmetry of X. Thus, $p \in Cl(B(p) \setminus \mathcal{T}(\{p\}))$.

Since $B(y)$ is closed and $p \in Cl(B(p) \setminus \mathcal{T}(\{p\}))$, it follows that $p \in B(y)$, or that $y \in B(p)$. But $y \in A(p)$, so that $y \in \mathcal{T}(\{p\})$, and $A = \mathcal{T}(\{p\})$. By contrapositive, if $A(p) \neq \mathcal{T}(\{p\})$, then $B(p) = \mathcal{T}(\{p\})$. So, for each $p \in X$, either $A(p) = X$ or $B(p) = X$. Suppose there exists $p \in X$ such that $A(p) = X$. Let $q \in X$ be arbitrary. If $q \in \mathcal{T}(\{p\})$, then $A(q) = A(p) = X$. (As we did before, if $z \in A(p) \cup A(q)$, then, since $\mathcal{T}(\{p\}) = \mathcal{T}(\{q\})$, $\mathcal{T}(\{p, z\}) = \mathcal{T}(\{q, z\})$.) Suppose $q \in X \setminus \mathcal{T}(\{p\})$. Then $q \in A(p)$, so $p \in A(q)$. Since $p \in X \setminus \mathcal{T}(\{q\})$, $A(q) \neq \mathcal{T}(\{q\})$. Hence, $A(q) = X$. Thus, either $A(p) = X$ for every $p \in X$ or $B(p) = X$ for every $p \in X$. If $B(p) = X$ for every $p \in X$, the theorem is proved. So, suppose $A(p) = X$ for all $p \in X$. Then for each $p, q \in X$, $\mathcal{T}(\{p, q\})$ is a continuum. Hence, by Theorem 3.1.57, X is indecomposable. Thus, by Theorem 3.1.34, $B(p) = X$ for all $p \in X$ in this case also.

Q.E.D.

3.2.10. Theorem. *If X is a point $\mathcal{T}$–symmetric continuum for which $\mathcal{T}$ is continuous, then X is $\mathcal{T}$–additive.*

Proof. Let $\mathcal{F}(X) = \bigcup\limits_{n=1}^{\infty} \mathcal{F}_n(X)$. Then $\mathcal{F}(X)$ is the family of all finite subsets of X.

We show first that for each $A \in \mathcal{F}(X)$, $\mathcal{T}(A) = \bigcup\limits_{a \in A} \mathcal{T}(\{a\})$.

Suppose this is not true. Then there exists an $M \in \mathcal{F}(X)$ such that $\mathcal{T}(M) \neq \bigcup\limits_{p \in M} \mathcal{T}(\{p\})$. Take M with the smallest cardinality. By Theorem 3.2.9, M has at least three elements. We assert that $\mathcal{T}(M)$ is a continuum. If $\mathcal{T}(M)$ is not connected, then there exist two disjoint closed subsets of A and B of X such that $\mathcal{T}(M) = A \cup B$. Then, by Lemma 3.1.59, Corollary 3.1.14 and the minimality of M:

$$\mathcal{T}(M) = A \cup B = \mathcal{T}(M \cap A) \cup \mathcal{T}(M \cap B)$$

$$= \bigcup\limits_{p \in M \cap A} \mathcal{T}(\{p\}) \cup \bigcup\limits_{p \in M \cap B} \mathcal{T}(\{p\}) = \bigcup\limits_{p \in M} \mathcal{T}(\{p\}),$$

contrary to the choice of M. Further, if $p, q \in M$ are distinct points, $\mathcal{T}(\{p\}) \cap \mathcal{T}(\{q\}) = \emptyset$; otherwise, by point $\mathcal{T}$–symmetry and idempotency, $\mathcal{T}(\{p\}) = \mathcal{T}(\{q\})$ and then, since $M \subset \mathcal{T}(M \setminus \{p\})$:

$$\mathcal{T}(M) \subset \mathcal{T}^2(M \setminus \{p\}) = \mathcal{T}(M \setminus \{p\}) = \bigcup_{r \in M \setminus \{p\}} \mathcal{T}(\{r\}) \subset \mathcal{T}(M).$$

Thus, $\mathcal{T}(M) = \bigcup_{p \in M} \mathcal{T}(\{p\})$. This also contradicts the choice of M.

Now, let $p \in M$ be arbitrary, and let $N = M \setminus \{p\}$. Then N has at least two elements. Since N has cardinality smaller than the cardinality of M, $\mathcal{T}(N) = \bigcup_{r \in N} \mathcal{T}(\{r\})$, i.e., $\mathcal{T}(N)$ is a finite union of pairwise disjoint subcontinua. Hence, $\mathcal{T}(N)$ is not a continuum. Now, define

$$L = \left\{ x \in X \ \middle| \ \mathcal{T}(N \cup \{x\}) = \bigcup_{r \in N} \mathcal{T}(\{r\}) \cup \mathcal{T}(\{x\}) \right\}$$

and

$$K = \{ x \in X \mid \mathcal{T}(N \cup \{x\}) \in \mathcal{C}(X) \}.$$

Note that $N \subset L$ and $p \in K$. Thus, $L \neq \emptyset$ and $K \neq \emptyset$. We claim that L and K are closed subsets of X. To see this, let Z be either L or K. Let $x \in Cl(Z)$. Then there exists a sequence $\{x_n\}_{n=1}^{\infty}$ of elements of Z converging to x.

If Z is L, since $\mathcal{T}$ and the union map σ (Lemma 1.8.11) are continuous,

$$\mathcal{T}(N \cup \{x\}) = \lim_{n \to \infty} \mathcal{T}(N \cup \{x_n\}) = \lim_{n \to \infty} \left(\bigcup_{r \in N} \mathcal{T}(\{r\}) \cup \mathcal{T}(\{x_n\}) \right) =$$

$$\bigcup_{r \in N} \mathcal{T}(\{r\}) \cup \lim_{n \to \infty} \mathcal{T}(\{x_n\}) = \bigcup_{r \in N} \mathcal{T}(\{r\}) \cup \mathcal{T}(\{x\}).$$

Hence, $x \in L$. Therefore, L is closed.

If Z is K, since $\mathcal{T}$ and the union map σ are continuous, the sequence $\{\mathcal{T}(N \cup \{x_n\})\}_{n=1}^{\infty}$ is a sequence of continua converging to $\mathcal{T}(N \cup \{x\})$ and $\mathcal{C}(X)$ is closed in 2^X (Theorem 1.8.5), then $\mathcal{T}(N \cup \{x\}) \in \mathcal{C}(X)$. Hence, $x \in K$. Therefore, K is closed.

If $y \in K \cap L$, then $\mathcal{T}(N \cup \{y\}) = \bigcup_{r \in N} \mathcal{T}(\{r\}) \cup \mathcal{T}(\{y\})$ and this set is connected. Thus, $\mathcal{T}(\{y\}) \cap \mathcal{T}(\{r\}) \neq \emptyset$ for each $r \in N$. Then, by point $\mathcal{T}$–symmetry and idempotency, $\mathcal{T}(\{y\}) = \mathcal{T}(\{r\})$ for every $r \in N$, a contradiction to the fact that for $r_1, r_2 \in N$, if $r_1 \neq r_2$, then $\mathcal{T}(\{r_1\}) \neq \mathcal{T}(\{r_2\})$. Thus, $K \cap L = \emptyset$.

Let $x \in X \setminus L$. Then $\mathcal{T}(N \cup \{x\})$ is a continuum; otherwise, there exist two disjoint closed subsets A and B of X such that $\mathcal{T}(N \cup \{x\}) = A \cup B$. Hence, by Lemma 3.1.59 and Corollary 3.1.14:

$$\mathcal{T}(N \cup \{x\}) = A \cup B = \mathcal{T}((N \cup \{x\}) \cap A) \cup \mathcal{T}((N \cup \{x\}) \cap B)$$

$$= \bigcup_{r \in (N \cup \{x\}) \cap A} \mathcal{T}(\{r\}) \cup \bigcup_{r \in (N \cup \{x\}) \cap B} \mathcal{T}(\{r\}) = \bigcup_{r \in N} \mathcal{T}(\{r\}) \cup \mathcal{T}(\{x\}),$$

a contradiction to the fact that $x \in X \setminus L$. Therefore, $\mathcal{T}(N \cup \{x\})$ is a continuum. Thus, $x \in K$. Hence, $X = K \cup L$, and K and L are disjoint closed sets, a contradiction to the fact that X is connected.

Therefore, $\mathcal{T}(A) = \bigcup_{a \in A} \mathcal{T}(\{a\})$ for each $A \in \mathcal{F}(X)$.

Now, let $B \in 2^X$, and let $\varepsilon > 0$. Let $\delta > 0$ given by the uniform continuity of $\mathcal{T}$ for $\frac{\varepsilon}{2}$. Since $\mathcal{F}(X)$ is dense in 2^X (proof of Corollary 1.8.9), there exists $A \in \mathcal{F}(X)$ such that $\mathcal{H}(B, A) < \delta$.

Since $\mathcal{H}(B, A) < \delta$, $\mathcal{H}^2(\mathcal{F}_1(B), \mathcal{F}_1(A)) < \delta$. Hence, by Theorem 1.8.22, $\mathcal{H}^2(2^{\mathcal{T}}(\mathcal{F}_1(B)), 2^{\mathcal{T}}(\mathcal{F}_1(A))) < \frac{\varepsilon}{2}$. This implies, by Lemma 1.8.11, that

$$\mathcal{H}\left(\bigcup_{b \in B} \mathcal{T}(\{b\}), \bigcup_{a \in A} \mathcal{T}(\{a\})\right) < \frac{\varepsilon}{2}.$$

Therefore:

$$\mathcal{H}\left(\mathcal{T}(B), \bigcup_{b \in B} \mathcal{T}(\{b\})\right) \leq$$

$$\mathcal{H}(\mathcal{T}(B), \mathcal{T}(A)) + \mathcal{H}\left(\bigcup_{b \in B} \mathcal{T}(\{b\}), \bigcup_{a \in A} \mathcal{T}(\{a\})\right) < \frac{\varepsilon}{2} + \frac{\varepsilon}{2} = \varepsilon.$$

Since ε was arbitrary, $\mathcal{T}(B) = \bigcup_{b \in B} \mathcal{T}(\{b\})$. Now, the fact that X is $\mathcal{T}$–additive follows easily.

Q.E.D.

As a consequence of Theorems 3.1.49 and 3.2.10, we have the following:

3.2.11. Corollary. *If X is a continuum for which $\mathcal{T}$ is continuous, then X is point $\mathcal{T}$–symmetric if and only if X is $\mathcal{T}$–symmetric.*

Proof. Clearly, if X is $\mathcal{T}$–symmetric, then X is point $\mathcal{T}$–symmetric.

Suppose $\mathcal{T}$ is continuous for X and X is point $\mathcal{T}$–symmetric. Hence, X is $\mathcal{T}$–additive (Theorem 3.2.10). Thus, since X is point $\mathcal{T}$–symmetric and $\mathcal{T}$–additive, X is $\mathcal{T}$–symmetric (Theorem 3.1.49).

Q.E.D.

3.2.12. Theorem. *If X is a continuum for which $\mathcal{T}$ is continuous, W is a subcontinuum of X with nonempty interior, and U is an open subset of X such that $W \subset U$, then there is a point $p \in X$ such that $\mathcal{T}(\{p\}) \subset U$.*

Proof. If $X \setminus W$ is connected, let $p \in Int(W)$. Then $Cl(X \setminus W)$ is a continuum with every point outside W in its interior such that $Cl(X \setminus W) \subset X \setminus \{p\}$. Hence, $\mathcal{T}(\{p\}) \subset W \subset U$.

Suppose $X \setminus W$ is not connected. Thus, there exist two nonempty disjoint open sets M and N of X such that $X \setminus W = M \cup N$. Then if $x \in M$, $\mathcal{T}(\{x\}) \subset Cl(M)$, since $W \cup N$ is a continuum (Lemma 1.7.18) having every point outside $Cl(M)$ in its interior and such that $W \cup N \subset X \setminus \{x\}$. Similarly, if $x \in N$, $\mathcal{T}(\{x\}) \subset Cl(N)$. Now, let

$$A = \{x \in X \mid \mathcal{T}(\{x\}) \cap M \cap (X \setminus U) \neq \emptyset\}$$

and

$$B = \{x \in X \mid \mathcal{T}(\{x\}) \cap M \neq \emptyset\}.$$

Note that $A \subset B$, $B \neq \emptyset$ (since $M \subset B$), B is open (Lemma 3.2.6 (2)) while A is closed. (Note that $X \setminus A = \{x \in X \mid \mathcal{T}(\{x\}) \subset (X \setminus M) \cup U\}$. Since $Cl(M) \setminus M \subset U$, $(X \setminus M) \cup U$ is open. Thus, $X \setminus A$ is open, by Lemma 3.2.6 (1)). Since $N \cap B = \emptyset$, $B \neq X$. Thus,

$A \neq B$ since X is connected. Let $x \in B \setminus A$. Then $\mathcal{T}(\{x\}) \cap M \neq \emptyset$ and $\mathcal{T}(\{x\}) \cap M \cap (X \setminus U) = \emptyset$. Let $p \in \mathcal{T}(\{x\}) \cap M$. Then $\mathcal{T}(\{p\}) \subset Cl(M) \cap \mathcal{T}(\{x\})$ ($\mathcal{T}$ is idempotent by Theorem 3.2.8); in particular:

$$\mathcal{T}(\{p\}) \cap (X \setminus U) \subset Cl(M) \cap \mathcal{T}(\{x\}) \cap (X \setminus U) = \emptyset,$$

since $Cl(M) \setminus M \subset U$. Thus, $\mathcal{T}(\{p\}) \subset U$.

Q.E.D.

3.2.13. Theorem. *If X is a $\mathcal{T}$–additive continuum for which $\mathcal{T}$ is continuous and W is a continuum domain of X, then $\mathcal{T}(W) = W$.*

Proof. Let $L = \{p \in X \mid \mathcal{T}(\{p\}) \subset W\}$. Let $x \in W$, and let M be a subcontinuum of X such that $x \in Int(M)$. Then $Int(M) \cap Int(W) \neq \emptyset$. Let $y \in Int(M) \cap Int(W)$. Then, since $\mathcal{T}$ is idempotent (Theorem 3.2.8),

$$y \in X \setminus (\mathcal{T}(X \setminus Int(M)) \cup \mathcal{T}(X \setminus Int(W)))$$

by Theorem 3.1.54. By additivity:

$$y \in X \setminus \mathcal{T}((X \setminus Int(M)) \cup (X \setminus Int(W)))$$

and

$$y \in X \setminus \mathcal{T}(X \setminus (Int(M) \cap Int(W))).$$

Hence, there exists a subcontinuum N of X such that $y \in Int(N)$ and $N \subset Int(M) \cap Int(W)$. Then, by Theorems 3.2.12 and 3.1.54, there exists $p \in N$ such that $p \in Int(N)$ and $\mathcal{T}(\{p\}) \subset N$. Thus, $\mathcal{T}(\{p\}) \subset W$ so that $p \in L$. Hence, $M \cap L \neq \emptyset$ and $x \in \mathcal{T}(L)$. Since x was an arbitrary point of W, $W \subset \mathcal{T}(L)$. By definition of L and additivity, $\mathcal{T}(L) \subset W$. Thus, $\mathcal{T}(W) = \mathcal{T}^2(L) = \mathcal{T}(L) = \mathcal{T}(W)$.

Q.E.D.

The following Theorem says that for continua for which $\mathcal{T}$ is continuous $\mathcal{T}$–additivity is equivalent to $\mathcal{T}$–symmetry.

3.2.14. Theorem. *If X is a continuum for which $\mathcal{T}$ is continuous, then X is $\mathcal{T}$–additive if and only if X is $\mathcal{T}$–symmetric.*

Proof. If X is $\mathcal{T}$–symmetric, by Theorem 3.1.44, X is $\mathcal{T}$–additive.

Suppose X is $\mathcal{T}$–additive. Let A and B be two closed subsets of X such that $A \cap \mathcal{T}(B) = \emptyset$. Then, by definition of $\mathcal{T}$, compactness and Corollary 3.1.56, there exist a finite collection $\{W_j\}_{j=1}^n$ such that W_j is a continuum domain, for each $j \in \{1, \ldots, n\}$, $A \subset \bigcup_{j=1}^{n} Int(W_j)$ and $B \cap \left(\bigcup_{j=1}^{n} W_j \right) = \emptyset$. Then, by additivity and Theorem 3.2.13, $\mathcal{T} \left(\bigcup_{j=1}^{n} W_j \right) = \bigcup_{j=1}^{n} W_j$. Hence, $\mathcal{T}(A) \subset \bigcup_{j=1}^{n} W_j$. Therefore, $\mathcal{T}(A) \cap B = \emptyset$.

$$\textbf{Q.E.D.}$$

As a consequence of Corollary 3.2.11 and Theorem 3.2.14, we have the following:

3.2.15. Corollary. *If X is a continuum for which $\mathcal{T}$ is continuous, then the following are equivalent:*

(1) X is point $\mathcal{T}$–symmetric;

(2) X is $\mathcal{T}$–symmetric;

(3) X is $\mathcal{T}$–additive.

3.2.16. Corollary. *If X is an aposyndetic continuum for which $\mathcal{T}$ is continuous, then X is locally connected.*

Proof. Since X is aposyndetic, by Theorem 3.1.28, X is point $\mathcal{T}$–symmetric. Hence, by Corollary 3.2.15, X is $\mathcal{T}$–additive. Thus, by Theorem 3.1.47, X is locally connected.

$$\textbf{Q.E.D.}$$

3.2.17. Theorem. *If X is a continuum for which $\mathcal{T}$ is continuous and X is almost connected im kleinen at $p \in X$, then X is semi–locally connected at p.*

Proof. Let
$$\mathcal{L} = \{A \in 2^X \mid p \in Int(A)\}.$$

Note that $X \in \mathcal{L}$. Hence, $\mathcal{L} \neq \emptyset$. By the almost connected-ness im kleinen of X and Theorem 3.2.12, the set $B(A) = \{x \in X \mid \mathcal{T}(\{x\}) \subset A\}$ is nonempty for each $A \in \mathcal{L}$. By continuity of $\mathcal{T}$, $B(A)$ is closed (Lemma 3.2.6 (4)) for each $A \in \mathcal{L}$. Hence, $\{B(A) \mid A \in \mathcal{L}\}$ is a filterbase of closed subsets, and $\bigcap_{A \in \mathcal{L}} B(A) \neq \emptyset$.

But,

$$\bigcap_{A \in \mathcal{L}} B(A) \subset \bigcap_{A \in \mathcal{L}} A = \{p\}.$$

Thus, $\mathcal{T}(\{p\}) \subset \bigcap_{A \in \mathcal{L}} A = \{p\}$. Therefore, by Theorem 3.1.29, X is semi–locally connected at p.

Q.E.D.

3.2.18. Theorem. *Let X be a continuum. If $\mathcal{T}$ is both additive and continuous for X and $p \in X$, then the following are equivalent:*

(1) X is connected im kleinen at p.

(2) X is almost connected im kleinen at p.

(3) X is semi–locally connected at p.

Proof. Note that, by Corollary 3.2.15, X is $\mathcal{T}$–symmetric. Thus, by Theorem 3.1.38, X is connected im kleinen at p if and only if X is semi–locally connected at p.

Clearly, if X is connected im kleinen at p, then X is almost connected im kleinen at p.

By Theorem 3.2.17, if X is almost connected im kleinen at p, then X is semi–locally connected at p.

Q.E.D.

3.2.19. Theorem. *Let X be a continuum for which $\mathcal{T}$ is continuous. Then X is semi–locally connected if and only if X is locally connected.*

Proof. First note that if X is either semi–locally connected or locally connected, then $\mathcal{T}(\{p\}) = \{p\}$ for each $p \in X$ (Theorems 3.1.29 and 3.1.31, respectively). Hence, X is point $\mathcal{T}$–symmetric. Thus, by Corollary 3.2.15, X is $\mathcal{T}$–symmetric. Therefore, by Theorems 3.1.38 and 1.7.12, X is semi–locally connected if and only if X is locally connected.

Q.E.D.

As a consequence of Theorems 3.2.17 and 3.2.19, we have the following:

3.2.20. Corollary. *Let X be a continuum for which $\mathcal{T}$ is continuous. If X is almost connected im kleinen at each of its points, then X is locally connected.*

3.3 Applications

We present several applications of the set function $\mathcal{T}$; most of which are taken from [4], [6], [12] and [19].

We begin with a couple of definitions.

3.3.1. Definition. *A continuum X is m–aposyndetic provided that for each subset K of X with m points, $\mathcal{T}(K) = K$. We say X is countable closed aposyndetic if for each countable closed subset K of X, $\mathcal{T}(K) = K$.*

3.3.2. Definition. *Let X be a continuum, and let A and B be two nonempty disjoint subsets of X. A subset C of X is a separator of A and B (or separates X between A and B) if $X \setminus C = U \cup V$, where U and V are separated (i.e., $Cl(U) \cap V = \emptyset$ and $U \cap Cl(V) = \emptyset$), $A \subset U$ and $B \subset V$.*

3.3.3. Theorem. *Let X be a continuum. If $A \in \mathcal{P}(X)$, then $\mathcal{T}(A)$ intersects each closed separator of A and some point of $\mathcal{T}(A)$.*

Proof. Suppose the theorem is not true. Then there exist $p \in \mathcal{T}(A)$ and a closed subset B of X such that $X \setminus B = H \cup K$, $A \subset H$, $p \in K$, where H and K are disjoint open subsets of X and $B \cap \mathcal{T}(A) = \emptyset$. Hence, for each $b \in B$, there exists a subcontinuum W_b of X such that $b \in Int(W_b) \subset W_b \subset X \setminus A$. Since B is compact, there exist $b_1, \ldots, b_n \in B$ such that $B \subset \bigcup_{j=1}^{n} Int(W_{b_j})$.

Since $Cl(K) \setminus K \subset B$, by Theorem 1.7.22, $Y = K \cup \bigcup_{j=1}^{n} W_{b_j}$ has a finite number of components. Note that Y is closed in X since it is the union of a finite number of continua and $Cl(K) \setminus K \subset Y$. Since $p \in K$ and $K \subset Y$, $p \in Y$. Let C be the component of Y such that $p \in C$. By Lemma 1.6.2, $p \in Int(C)$. This implies that $p \in X \setminus \mathcal{T}(A)$, a contradiction. Therefore, $B \cap \mathcal{T}(A) \neq \emptyset$.

Q.E.D.

3.3.4. Corollary. *Let X be a continuum. If $A \in \mathcal{P}(X)$, then each component of $\mathcal{T}(A)$ intersects $Cl(A)$.*

Proof. Suppose the corollary is not true. Then there exists a component C of $\mathcal{T}(A)$ such that $C \cap Cl(A) = \emptyset$. Hence, by Theorem 1.6.8, there exist two disjoint closed subset H and K of $\mathcal{T}(A)$ such that $\mathcal{T}(A) = H \cup K$, $Cl(A) \subset H$ and $C \subset K$. Since X is a metric space, there exist two disjoint open subsets U and V of X such that $H \subset U$ and $K \subset V$. Let $B = X \setminus (U \cup V)$. Then B is a closed separator of A and C and $B \cap \mathcal{T}(A) = \emptyset$, a contradiction to Theorem 3.3.3.

Q.E.D.

3.3.5. Corollary. *Let X be a continuum, and let A and B be two closed subsets of X. If K is a component of $\mathcal{T}(A \cup B)$ such that $K \setminus (\mathcal{T}(A) \cup \mathcal{T}(B)) \neq \emptyset$, then $K \cap A \neq \emptyset$ and $K \cap B \neq \emptyset$.*

Proof. Since X is a continuum, by Corollary 3.1.14, $\mathcal{T}(\emptyset) = \emptyset$. By Corollary 3.3.4, $K \cap (A \cup B) \neq \emptyset$. Now, by Theorem 3.1.60, $\mathcal{T}((A \cup B) \cap K) = K$. Since $K \setminus (\mathcal{T}(A) \cup \mathcal{T}(B)) \neq \emptyset$, $(A \cup B) \cap K$ meets both A and B. Thus, $K \cap A \neq \emptyset$ and $K \cap B \neq \emptyset$.

Q.E.D.

3.3.6. Corollary. *Let X be a continuum, and let A and B be closed subsets of X. If $\mathcal{T}(A \cup B) \neq \mathcal{T}(A) \cup \mathcal{T}(B)$, then there exists a subcontinuum K of X such that $K \subset \mathcal{T}(A \cup B)$, $K \cap A \neq \emptyset$ and $K \cap B \neq \emptyset$.*

Proof. Let K be a component of $\mathcal{T}(A \cup B) \setminus \mathcal{T}(A) \cup \mathcal{T}(B)$ and apply Corollary 3.3.5.

Q.E.D.

3.3.7. Corollary. *If X is a continuum and A is a closed subset of X such that $\mathcal{T}(A)$ is totally disconnected then $\mathcal{T}(A) = A$.*

Proof. Let A be a closed subset of X such that $\mathcal{T}(A)$ is a totally disconnected subset of X. By Remark 3.1.5, $A \subset \mathcal{T}(A)$. By Corollary 3.3.4, each component of $\mathcal{T}(A)$ intersects $Cl(A) = A$. Since all the components of $\mathcal{T}(A)$ are singletons, $\mathcal{T}(A) \subset A$.

Q.E.D.

The following Theorem tells us that $\mathcal{T}$ of a product of continua behaves like the identity map at the product of two proper closed subsets.

3.3.8. Theorem. *Let X and Y be continua. If A and B are proper closed subsets of X and Y, respectively, then $\mathcal{T}(A \times B) = A \times B$.*

Proof. Let A and B be two proper closed subsets of X and Y, respectively. By Remark 3.1.5, $A \times B \subset \mathcal{T}(A \times B)$.

Let $(x, y) \in (X \times Y) \setminus (A \times B)$; without loss of generality, we assume that $x \in X \setminus A$. Then there exists an open subset U of X such that $x \in U \subset Cl(U) \subset X \setminus A$. Hence, $(Cl(U) \times Y) \cap (A \times B) = \emptyset$.

Let $z \in Y \setminus B$. Then $(X \times \{z\}) \cap (A \times B) = \emptyset$. Thus, $(x, y) \in (Cl(U) \times Y) \cup (X \times \{z\}) \subset (X \times Y) \setminus (A \times B)$. Since $(Cl(U) \times Y) \cup (X \times \{z\})$ is a continuum containing (x, y) in its interior, $(x, y) \in (X \times Y) \setminus \mathcal{T}(A \times B)$. Therefore, $\mathcal{T}(A \times B) = A \times B$.

Q.E.D.

As a consequence of Theorem 3.1.28 and Theorem 3.3.8, we have the following Corollary:

3.3.9. Corollary. *If X and Y are continua, then $X \times Y$ is an aposyndetic continuum.*

3.3.10. Theorem. *Let X and Y be continua. If A and B are two closed totally disconnected subsets of X and Y, respectively, then for each closed subset K of $A \times B$, $\mathcal{T}(K) = K$.*

Proof. Let K be a closed subset of $A \times B$. By Theorem 3.3.8, $\mathcal{T}(A \times B) = A \times B$. Since K is a closed subset of $A \times B$, by Proposition 3.1.7, $\mathcal{T}(K) \subset \mathcal{T}(A \times B) = A \times B$. Hence, $\mathcal{T}(K)$ is totally disconnected. By Corollary 3.3.7, $\mathcal{T}(K) = K$.

Q.E.D.

Now we are ready to prove our application of $\mathcal{T}$ to Cartesian products.

3.3.11. Theorem. *Let X and Y be continua. If K is a countable closed subset of $X \times Y$, then $\mathcal{T}(K) = K$. In particular, $X \times Y$ is m–aposyndetic for each $m \in \mathbb{N}$.*

Proof. Let K be a countable closed subset of $X \times Y$. Let $\pi_X : X \times Y \twoheadrightarrow X$ and $\pi_Y : X \times Y \twoheadrightarrow Y$ be the projection maps. Since K is countable and closed, $\pi_X(K)$ and $\pi_Y(K)$ are countable closed subsets of X and Y, respectively. Then $K \subset \pi_X(K) \times \pi_Y(K)$. Thus, by Theorem 3.3.10, $\mathcal{T}(K) = K$.

Q.E.D.

3.3.12. Corollary. *Let Y and Z be continua. If $X = Y \times Z$ is a continuum for which $\mathcal{T}$ is continuous, then X is locally connected. In particular, Y and Z are locally connected.*

Proof. Let A be a nonempty closed subset of X. By the proof of Corollary 1.8.9, there exists a sequence $\{K_m\}_{m=1}^{\infty}$ of finite subsets of X converging to A. Since $\mathcal{T}$ is continuous, $\{\mathcal{T}(K_m)\}_{m=1}^{\infty}$ converges to $\mathcal{T}(A)$. By Theorem 3.3.11, $\mathcal{T}(K_m) = K_m$ for each $m \in \mathbb{N}$. Hence, $\mathcal{T}(A) = A$. Since A was an arbitrary nonempty closed subset of X, by Theorem 3.1.31, X is locally connected. Note that Y and Z are locally connected by Theorem 3.1.65.

Q.E.D.

Our application of $\mathcal{T}$ to symmetric products is the following:

3.3.13. Theorem. *Let X be a continuum and let $n \geq 2$ be an integer. If K is a countable closed subset of $\mathcal{F}_n(X)$, then $\mathcal{T}_{\mathcal{F}_n(X)}(K) = K$. In particular, $\mathcal{F}_n(X)$ is m–aposyndetic for each $m \in \mathbb{N}$.*

Proof. Let K be a countable closed subset of $\mathcal{F}_n(X)$. Then the function $f_n \colon X^n \twoheadrightarrow \mathcal{F}_n(X)$ is a continuous surjection (Lemma 1.8.6). Since for each $w \in \mathcal{F}_n(X)$, $f_n^{-1}(w)$ is finite and K is countable and closed, $f_n^{-1}(K) = \bigcup_{k \in K} f_n^{-1}(k)$ is a countable closed subset of X^n. Hence, by Theorem 3.3.11 and induction, $\mathcal{T}_{X^n} f_n^{-1}(K) = f_n^{-1}(K)$. This implies that $f_n \mathcal{T}_{X^n} f_n^{-1}(K) = f_n f_n^{-1}(K) = K$. Thus, by Theorem 3.1.64 (a), $\mathcal{T}_{\mathcal{F}_n(X)}(K) \subset f_n \mathcal{T}_{X^n} f_n^{-1}(K)$. Hence, $\mathcal{T}_{\mathcal{F}_n(X)}(K) \subset K$. Since $K \subset \mathcal{T}_{\mathcal{F}_n(X)}(K)$, by Remark 3.1.5, $\mathcal{T}_{\mathcal{F}_n(X)}(K) = K$.

Q.E.D.

3.3.14. Corollary. *Let X be a continuum and let $n \geq 2$ be an integer. If $\mathcal{T}_{\mathcal{F}_n(X)}$ is continuous for $\mathcal{F}_n(X)$, then X is locally connected.*

Proof. As in the proof of Corollary 3.3.12, $\mathcal{T}_{\mathcal{F}_n(X)}(A) = A$ for each closed subset A of $\mathcal{F}_n(X)$. Hence, $\mathcal{F}_n(X)$ is locally connected, by Theorem 3.1.31. Therefore, X is locally connected by Lemma 2 of [19].

Q.E.D.

3.3.15. Remark. Let us observe that Corollaries 3.3.12 and 3.3.14 follow from Theorems 3.3.11 and 3.3.13 and Corollary 3.2.16. We decided to present different proofs.

Regarding the continuity of $\mathcal{T}$ on the other hyperspaces, we have the following Theorem:

3.3.16. Theorem. *Let X be a continuum. If*

$$\mathcal{L} \in \left\{ 2^X, \mathcal{C}_n(X) \ (n \in \mathbb{N}) \right\}$$

and $\mathcal{T}_{\mathcal{L}}$ is continuous for $\mathcal{L}$, then X is locally connected.

Proof. Suppose $\mathcal{T}_{\mathcal{L}}$ is continuous for $\mathcal{L}$. Since $\mathcal{L}$ is aposyndetic (Theorem 1 of [13] for 2^X, and Corollary 6.3.3 for $\mathcal{C}_n(X)$), by Corollary 3.2.16, $\mathcal{L}$ is locally connected. Hence, X is locally connected ((1.92) of [22] for 2^X, and Theorem 6.1.4 for $\mathcal{C}_n(X)$).

Q.E.D.

The set function $\mathcal{T}$ may be used to show that continua are not contractible. First, we show the following Lemma:

3.3.17. Lemma. *Suppose X is a continuum, A is a closed subset of X and $r \in \mathcal{T}_X(A)$. If $H: X \times [0,1] \twoheadrightarrow X$ is a homotopy such that $H((x,0)) = x$ for each $x \in X$ and such that $H((r,1)) \in X \setminus \mathcal{T}_X(A)$, then $H((r,t)) \in A$ for some $t \in [0,1]$.*

Proof. Let $f: X \to X$ be given by $f(x) = H((x,1))$. Suppose the result is not true. Then $H^{-1}(A) \cap (\{r\} \times [0,1]) = \emptyset$. Thus, there exists an open subset U of X such that $r \in U$ and $H^{-1}(A) \cap (Cl(U) \times [0,1]) = \emptyset$. Since $f(r) = H((r,1)) \in X \setminus \mathcal{T}_X(A)$, there exists a subcontinuum W of X such that $f(r) \in Int(W) \subset W \subset X \setminus A$. Let $V = U \cap f^{-1}(Int(W))$. Then $r \in V$, $H^{-1}(A) \cap (Cl(V) \times [0,1]) = \emptyset$ and $f(Cl(V)) \subset W$. Let $L = (X \times \{0\}) \cup (Cl(V) \times [0,1])$. Then L is a subcontinuum of $X \times [0,1]$. Consider the disjoint union $L \cup W$. Let K be the quotient space of $L \cup W$ obtained by identifying each $(v,1) \in Cl(V) \times \{1\}$ with $f(v)$, and let $q: L \cup W \twoheadrightarrow K$ be the quotient map.

Define $G: K \twoheadrightarrow X$ by

$$G(q(\omega)) = \begin{cases} \omega & \text{if } \omega \in W; \\ H(\omega) & \text{if } \omega \in L. \end{cases}$$

Note that G is, in fact, surjective since $G(q((x,0))) = x$ for each $x \in X$. By Theorem 3.1.64 (a), $\mathcal{T}_X(A) \subset G\mathcal{T}_K G^{-1}(A)$. In particular, $r \in G\mathcal{T}_K G^{-1}(A)$. Hence, $G^{-1}(r) \cap \mathcal{T}_K G^{-1}(A) \neq \emptyset$. However, since $A \cap W = \emptyset$ and $H^{-1}(A) \cap (Cl(V) \times [0,1]) = \emptyset$, it follows that $G^{-1}(A) = q(A \times \{0\})$. Thus, $G^{-1}(r) \cap \mathcal{T}_K(q(A \times \{0\})) \neq \emptyset$.

Let $J = q(W \cup (Cl(V) \times [0,1]))$. Then J is a subcontinuum of K, and

$$M = q(W \cup (Cl(V) \times (0,1]) \cup (V \times [0,1])) \subset Int_K(J).$$

Since $G^{-1}(r) \cap q(X \times \{0\}) = q(\{(r,0)\})$, $G^{-1}(r) \subset M \subset Int_K(J)$. Since $q(A \times \{0\}) \cap J = \emptyset$, we have that $G^{-1}(r) \cap \mathcal{T}_K(q(A \times \{0\})) = \emptyset$, a contradiction. Therefore, there exists $t \in [0,1]$ such that $H((r,t)) \in A$.

Q.E.D.

The next Theorem gives a sufficient condition for a continuum not to be contractible.

3.3.18. Theorem. *If X is a continuum and A and B are closed subsets of X such that $A \cap \mathcal{T}(B) = \emptyset$, $\mathcal{T}(A) \cap B = \emptyset$ and $\mathcal{T}(A) \cap \mathcal{T}(B) \neq \emptyset$, then X is not contractible.*

Proof. Suppose X is contractible. Then there exists a homotopy $H \colon X \times [0,1] \twoheadrightarrow X$ such that $H((x,0)) = x$ and $H((x,1)) = p$ for each $x \in X$ and some $p \in X$. Without loss of generality, we assume that $p \in A$. Let $r \in \mathcal{T}(A) \cap \mathcal{T}(B)$. Since $H((r,1)) \in A$ and $H((r,1)) \in X \setminus \mathcal{T}(B)$, by Lemma 3.3.17, there exists $t \in [0,1]$ such that $H((r,t)) \in B$. Let t_A and t_B be the smallest elements of $[0,1]$ such that $H((r,t_A)) \in A$ and $H((r,t_B)) \in B$. Thus, either $t_A < t_B$ or $t_A > t_B$, since $t_A = t_B$ is impossible. If $t_A < t_B$, applying Lemma 3.3.17 to $H|_{X \times [0,t_A]}$, we obtain an element $t' < t_A$ such that $H((r,t')) \in B$, a contradiction. Similarly, if $t_A > t_B$, applying Lemma 3.3.17 to $H|_{X \times [0,t_B]}$, we obtain an element $t'' < t_B$ such that $H((r,t'')) \in A$, a contradiction. Therefore, X is not contractible.

Q.E.D.

Now, we present some connectivity properties of $\mathcal{T}$.

3.3.19. Lemma. *Let X be a continuum. If S is a nonempty totally disconnected closed subset of X such that there is a point $p \in \mathcal{T}(S) \setminus S$ and such that for each proper subset S' of S, $p \in X \setminus \mathcal{T}(S')$, then $\mathcal{T}(S)$ is connected.*

Proof. Let S be a totally disconnected closed subset of X satisfying the properties stated. Let S_0 be a nonempty subset of S which is both open and closed in S. Since $p \in X \setminus \mathcal{T}(S \setminus S_0)$, there exists a subcontinuum W of X such that $p \in Int(W) \subset W \subset X \setminus (S \setminus S_0)$. Let $\{U_n\}_{n=1}^{\infty}$ and $\{V_n\}_{n=1}^{\infty}$ be decreasing sequences of open subsets of X such that for each $n \in \mathbb{N}$, $S \setminus S_0 \subset U_n$, $S_0 \subset V_n$, $U_1 \cap Cl(V_1) = U_1 \cap W = Cl(V_1) \cap \{p\} = \emptyset$, $S \setminus S_0 = \bigcap_{n=1}^{\infty} U_n$ and $S_0 = \bigcap_{n=1}^{\infty} V_n$.

For each $n \in \mathbb{N}$, let C_n be the component of $X \setminus U_n$ such that $W \subset C_n$. Since p does not belong to the interior of the component of $C_n \setminus V_n$ in which it lies, there is a sequence $\{D_n^j\}_{j=1}^{\infty}$ of distinct components of $C_n \setminus V_n$ such that $p \in D_n = \lim_{j \to \infty} D_n^j$. Since, by Theorem 1.7.22, $D_n^j \cap Cl(V_n) \neq \emptyset$, for each $n \in \mathbb{N}$, D_n is a continuum, $D_n \subset C_n \setminus V_n$, $p \in D_n$ and $D_n \cap Cl(V_n) \neq \emptyset$. Let $D = \lim_{n \to \infty} D_n$. Then D is a continuum containing p and $D \cap S_0 \neq \emptyset$.

We assert that $D \subset \mathcal{T}(S)$. Suppose this is not true. Then there exist $q \in D \setminus \mathcal{T}(S)$ and a subcontinuum W' of X such that $q \in Int(W') \subset W' \subset X \setminus S$. It follows that there exists $N_1 \in \mathbb{N}$ such that for $n > N_1$, $W' \subset X \setminus (U_n \cup V_n)$. Since $q \in Int(W') \cap D$, there exists $N_2 \in \mathbb{N}$ such that for $n > N_2$, $(Int(W')) \cap D_n \neq \emptyset$. Let $m > \max\{N_1, N_2\}$. Since $(Int(W')) \cap D_m \neq \emptyset$, there exists $N_3 \in \mathbb{N}$ such that for $j > N_3$, $(Int(W')) \cap D_m^j \neq \emptyset$. Let $j > \max\{N_1, N_2, N_3\}$. Then $D_m^j \subset C_m$ and $W' \cap C_m \neq \emptyset$. Thus, $W' \subset C_m \setminus V_m$, a contradiction, since $j > \max\{N_1, N_2, N_3\}$ and $(Int(W')) \cap D_m^j \neq \emptyset$. Therefore, $D \subset \mathcal{T}(S)$.

Now, let $s \in S$. Then there exists a sequence $\{S_n\}_{n=1}^{\infty}$ of subsets of S such that for each $n \in \mathbb{N}$, S_n is open and closed in S, and $\{s\} = \bigcap_{n=1}^{\infty} S_n$. By the previous construction, for each $n \in \mathbb{N}$, there is a subcontinuum A_n of $\mathcal{T}(S)$ such that $p \in A_n$ and $A_n \cap S_n \neq \emptyset$. Since $\mathcal{C}(X)$ is compact (Theorem 1.8.5), without loss of generality, we assume that the sequence $\{A_n\}_{n=1}^{\infty}$ converges. Let $A_s = \lim_{n \to \infty} A_n$. Then A_s is a subcontinuum of $\mathcal{T}(S)$ and $\{p, s\} \subset A_s$.

Let $A = \bigcup_{s \in S} A_s$. Then A is connected (for each $s \in S$, $p \in A_s$) and $S \subset A \subset \mathcal{T}(S)$. For each $x \in \mathcal{T}(S)$, let C_x be the component of $\mathcal{T}(S)$ such that $x \in C_x$. By Corollary 3.3.4, $C_x \cap S \neq \emptyset$. Thus, $C_x \cap A \neq \emptyset$ for each $x \in \mathcal{T}(S)$, and it follows that

$$\mathcal{T}(S) = A \cup \left(\bigcup_{x \in \mathcal{T}(S)} C_x \right)$$

is connected.

Q.E.D.

The following Theorem shows that we may remove the "totally disconnected" hypothesis in Lemma 3.3.19.

3.3.20. Theorem. *Let X be a continuum. Suppose A is a nonempty closed subset of X, and there exists a point $p \in \mathcal{T}_X(A) \setminus A$ such that for each proper closed subset A' of A, $p \in X \setminus \mathcal{T}_X(A')$. Then $\mathcal{T}_X(A)$ is connected.*

Proof. If $\mathcal{T}_X(A) = X$, there is nothing to do. Then suppose $\mathcal{T}_X(A)$ is a proper closed subset of X. Let $\{A_\lambda\}_{\lambda \in \Lambda}$ be the family of components of A. Then $\mathcal{G} = \{A_\lambda\}_{\lambda \in \Lambda} \cup \{\{x\} \mid x \in X \setminus A\}$ is an upper semicontinuous decomposition of X and $X/\mathcal{G}$ is a continuum (Theorem 1.7.3). Let $q \colon X \twoheadrightarrow X/\mathcal{G}$ be the quotient map. Note that q is monotone.

Since A is a closed subset of X, $q(A) = q\left(\bigcup_{\lambda \in \Lambda} A_\lambda\right) = \bigcup_{\lambda \in \Lambda} q(A_\lambda)$ is a closed totally disconnected subset of $X/\mathcal{G}$. Thus, $\mathcal{T}_{X/\mathcal{G}}(q(A))$ is connected, by Lemma 3.3.19. Since q is monotone, by Theorem 3.1.64 (b), we have that $\mathcal{T}_X(A) = \mathcal{T}_X(q^{-1}q(A)) \subset q^{-1}\mathcal{T}_{X/\mathcal{G}}(q(A))$. By Theorem 3.1.64 (c), we have $q^{-1}\mathcal{T}_{X/\mathcal{G}}(q(A)) = q^{-1}q\mathcal{T}_X(q^{-1}(q(A)))$ $= q^{-1}q\mathcal{T}_X(A)$. Since $A \subset \mathcal{T}(A)$, $q^{-1}q\mathcal{T}_X(A) = \mathcal{T}_X(A)$. Hence, $q^{-1}\mathcal{T}_{X/\mathcal{G}}(q(A)) = \mathcal{T}_X(A)$. Therefore, $\mathcal{T}_X(A)$ is connected.

Q.E.D.

3.3.21. Theorem. *Let X be a continuum. Suppose A is a nonempty closed subset of X and there exists $x \in \mathcal{T}(A) \setminus A$. Then there exists a closed subset D of X such that:*

(a) $D \subset A$,

(b) $x \in \mathcal{T}(D)$,

(c) if E is a nonempty closed subset of X and $E \subsetneq D$, then $x \in X \setminus \mathcal{T}(E)$, and

(d) $\mathcal{T}(D)$ is a continuum.

Proof. Let $\{B_n\}_{n=1}^{\infty}$ be a decreasing sequence of nonempty closed subsets of X such that for each $n \in \mathbb{N}$, $x \in \mathcal{T}(B_n)$. Let $B = \bigcap_{n=1}^{\infty} B_n$. Then B is a nonempty closed subset of X. If $x \in X \setminus \mathcal{T}(B)$, then there exists a subcontinuum W of X such that $x \in Int(W) \subset W \subset X \setminus B$. Since W is compact, there exists $N \in \mathbb{N}$ such that $W \subset \bigcup_{n=1}^{N}(X \setminus B_n)$, a contradiction to the fact that $x \in \mathcal{T}(B_N)$. Hence, $x \in \mathcal{T}(B)$. By the Brouwer Reduction Theorem (see (11.1) (p. 17) of [26]), there exists a minimal element D satisfying conditions (a), (b) and (c). Condition (d) follows from Theorem 3.3.20.

Q.E.D.

In 5.12 of [23] it is shown that if X is a continuum and x is a point at which X is not connected im kleinen, then there is a subcontinuum K of X such that $x \in K$ and X is not connected im kleinen at any of the points of K. In the following theorem, we show that something similar happens with the images of $\mathcal{T}$.

3.3.22. Theorem. *Let X be a continuum. Suppose A is a nonempty closed subset of X such that there exists $x \in \mathcal{T}(A) \setminus A$. Then there exists a nondegenerate subcontinuum K of X such that $x \in K \subset \mathcal{T}(A)$.*

Proof. Let A be a nonempty closed subset of X and let $x \in \mathcal{T}(A) \setminus A$. By Theorem 3.3.21, there exists a nonempty closed subset D of X satisfying conditions (a), (b), (c) and (d). Let V be an open subset of X such that $x \in V \subset Cl(V) \subset X \setminus A$. Let H be the component of $V \cap \mathcal{T}(D)$ containing x, and let $K = Cl(H)$. Since $V \cap \mathcal{T}(D)$ is open in $\mathcal{T}(D)$, by Theorem 1.7.22, K intersects $Bd(V)$. Thus, K is a nondegenerate subcontinuum of X contained in $\mathcal{T}(A)$.

Q.E.D.

The following Theorem characterizes local connectedness in unicoherent continua.

3.3.23. Theorem. *Let X be a unicoherent continuum. Then X is locally connected if and only if for each nonempty closed subset C of X which separates X between two points x and y, there exists a component E of C which separates X between x and y.*

Proof. First, suppose that X is unicoherent and locally connected. Let C be a nonempty closed subset of X which separates X between x and y. Let A be the component of $X \setminus C$ containing x, and let B be the component of $X \setminus Cl(A)$ containing y. Since X is locally connected, by Lemma 1.7.11, A and B are open subsets of X. Also, $X \setminus (Cl(A) \cap Cl(B)) \subset (X \setminus Cl(B)) \cup B$, $x \in X \setminus Cl(B)$, and $y \in B$. Thus, $Cl(A) \cap Cl(B)$ is a closed subset of X which separates X between x and y.

Let $\{S_\lambda\}_{\lambda \in \Lambda}$ be the family of components of $X \setminus Cl(A)$ such that $S_\lambda \cap B = \emptyset$ for each $\lambda \in \Lambda$. Then, by Theorem 1.7.22,
$Cl(A) \cup \left(\bigcup_{\lambda \in \Lambda} Cl(S_\lambda) \right)$ is connected. Note that $X = Cl(A) \cup$
$Cl \left(\bigcup_{\lambda \in \Lambda} Cl(S_\lambda) \right) \cup Cl(B)$. Since X is unicoherent, $Cl(A) \cap Cl(B) =$
$\left(Cl(A) \cup Cl \left(\bigcup_{\lambda \in \Lambda} Cl(S_\lambda) \right) \right) \cap Cl(B)$ is a continuum. Let E be the component of C that contains $Cl(A) \cap Cl(B)$. If x and y belong to the same component of $X \setminus E \subset X \setminus (Cl(A) \cap Cl(B))$, then x and y are in the same component of $X \setminus (Cl(A) \cap Cl(B))$. Since this is not true, E separates X between x and y.

Now, suppose X is a unicoherent continuum satisfying the conditions stated, and suppose X is not locally connected, hence not connected im kleinen at some point $p \in X$. Then, by Theorem 3.1.24, there exists a nonempty closed subset A of X such that $p \in \mathcal{T}(A) \setminus A$. By Theorem 3.3.21, there exists a nonempty closed subset B of X satisfying conditions (a), (b), (c) and (d). Let $x \in B$ and let U_0 be an open subset of X such that $p \in U_0 \subset Cl(U_0) \subset X \setminus B$. Then $Bd(U_0)$ is a closed set which separates X between x and p. Hence, there exists a component N of $Bd(U_0)$ which separates X between x and p.

Let U and V be disjoint open sets such that $X \setminus N = U \cup V$, $p \in U$ and $x \in V$. Then, by Lemma 1.7.18, $H = N \cup U$ and $K = N \cup V$ are continua, and $X = H \cup K$. Since $x \in B \cap K$, then $B \cap K \neq \emptyset$. If $B \cap H = \emptyset$, then $p \in X \setminus \mathcal{T}(B)$, which is contrary to (b) of Theorem 3.3.21. Thus, $B \cap H \neq \emptyset$. Since $B \cap K$ and $B \cap H$ are two nonempty closed proper subsets of B, it follows from (c) of Theorem 3.3.21 that there exist two subcontinua L_1

and L_2 of X such that $p \in Int(L_1) \subset L_1 \subset X \setminus (B \cap H)$ and $p \in Int(L_2) \subset L_2 \subset X \setminus (B \cap K)$. Again, if $L_1 \cap (B \cap K) = \emptyset$ or $L_2 \cap (B \cap H) = \emptyset$, it follows that $p \in X \setminus \mathcal{T}(B)$. Thus, assume that $L_1 \cap (B \cap K) \neq \emptyset$ and $L_2 \cap (B \cap H) \neq \emptyset$. Let $X_1 = L_1 \cup K$ and $X_2 = L_2 \cup H$. Then X_1 and X_2 are continua and $X = X_1 \cup X_2$. Now, since X is unicoherent, $X_1 \cap X_2$ is a continuum. Note that $p \in Int(X_1 \cap X_2) \subset (X_1 \cap X_2) \subset X \setminus B$. Hence, $p \in X \setminus \mathcal{T}(B)$, which is contrary to condition (b) of Theorem 3.3.21. Therefore, X is locally connected.

Q.E.D.

REFERENCES

[1] D. P. Bellamy, Continua for Which the Set Function $\mathcal{T}$ is Continuous, Trans. Amer. Math. Soc., 1511 (1970), 581–587.

[2] D. P. Bellamy, Set Functions and Continuous Maps, in *General Topology and Modern Analysis*, (L. F. McAuley and M. M. Rao, eds.), Academic Press, (1981), 31–38.

[3] D. P. Bellamy, Some Topics in Modern Continua Theory, in *Continua Decompositions Manifolds*, (R. H. Bing, W. T. Eaton and M. P. Starbird, eds.), University of Texas Press, (1983), 1–26.

[4] D. P. Bellamy and J. J. Charatonik, The Set Function $\mathcal{T}$ and Contractibility of Continua, Bull. Acad. Polon. Sci. Sér. Sci. Math. Astronom. Phys., 25 (1977), 47–49.

[5] D. P. Bellamy and H. S. Davis, Continuum Neighborhoods and Filterbases, Proc. Amer. Math. Soc., 27 (1971), 371–374.

[6] D. E. Bennet, A Characterization of Locally Connectedness By Means of the Set Function $\mathcal{T}$, Fund. Math. 86 (1974), 137–141.

[7] R. H. Bing and F. B. Jones, Another Homogeneous Plane Continuum, Trans. Amer. Math. Soc., 90 (1959), 171–192.

[8] K. Borsuk and S. Ulam, On Symmetric Products of Topological Spaces, Bull. Amer. Math. Soc., 37 (1931), 875-882.

[9] H. S. Davis, A Note on Connectedness im Kleinen, Proc. Amer. Math. Soc., 19 (1968), 1237–1241.

[10] H. S. Davis, Relationships Between Continuum Neighborhoods in Inverse Limit Spaces and Separations in Inverse Limit Sequences, Proc. Amer. Math. Soc., 64 (1977), 149–153.

[11] H. S. Davis and P. H. Doyle, Invertible Continua, Portugal. Math., 26 (1967), 487–491.

[12] R. W. FitzGerald, The Cartesian Product of Non–degenerate Compact Continua is n–point Aposyndetic, Topology Conference (Arizona State Univ., Tempe, Ariz., 1967), Arizona State University, Tempe, Ariz., (1968), pp. 324–326.

[13] J. T. Goodykoontz, Jr., Aposyndetic Properties of Hyperspaces, Pacific J. Math., 47 (1973), 91–98.

[14] J. G. Hocking and G. S. Young, *Topology*, Dover Publications, Inc., New York, 1988.

[15] F. B. Jones, Concerning the Boundary of a Complementary Domain of a Continuous Curve, Bull. Amer. Math. Soc., 45 (1939), 428–435.

[16] F. B. Jones, Aposyndetic Continua and Certain Boundary Problems, Amer. J. Math., 53 (1941), 545–553.

[17] F. B. Jones, Concerning Nonaposyndetic Continua, Amer. Math., 70 (1948), 403–413.

[18] K. Kuratowski, *Topology*, Vol. II, Academic Press, New York, N. Y., 1968.

[19] S. Macías, Aposyndetic Properties of Symmetric Products of Continua, Topology Proc., 22 (1997), 281–296.

[20] S. Macías, A Class of One–dimensional Nonlocally Connected Continua for Which the Set Function $\mathcal{T}$ is Continuous, to appear in Houston Journal of Mathematics.

[21] M. A. Molina, *Algunos Aspectos Sobre la Función $\mathcal{T}$ de Jones*, Tesis de Licenciatura, Facultad de Ciencias, U. N. A. M., 1998. (Spanish)

[22] S. B. Nadler, Jr., *Hyperspaces of Sets*, Monographs and Textbooks in Pure and Applied Math., Vol. 49, Marcel Dekker, New York, Basel, 1978.

[23] S. B. Nadler, Jr., *Continuum Theory: An Introduction,* Monographs and Textbooks in Pure and Applied Math., Vol. 158, Marcel Dekker, New York, Basel, Hong Kong, 1992.

[24] E. L. VandenBoss, *Set Functions and Local Connectivity*, Ph. D. Dissertation, Michigan State University (1970). University Microfilms # 71–11997.

[25] G. T. Whyburn, Semi–locally–connected Sets, Amer. J. Math., 61 (1939), 733–749.

[26] G. T. Whyburn, *Analytic Topology,* Amer. Math. Soc. Colloq. Publ., vol. 28, Amer. Math. Soc., Providence, R. I., 1942.

[27] S. Willard, *General Topology,* Addison–Wesley Publishing Co., 1970.

Chapter 4

A THEOREM OF E. G. EFFROS

We present a topological proof of a Theorem by E. G. Effros [4] and a consequence of it, due to C. L. Hagopian [5], which has been very useful in the theory of homogeneous continua. We present the proof of Effros's result given by Fredric G. Ancel [1].

Before showing Effros's Theorem, we present some background on topological groups and actions of topological groups on metric spaces.

4.1 Topological Groups

We introduce topological groups and give some of its elementary properties.

4.1.1. Definition. We say that a group G, which has a topology τ, is a *topological group* provided that the group operations are continuous, i.e., the functions given by:

$$\pi\colon G \times G \to G \quad \pi((g_1, g_2)) = g_1 \cdot g_2$$

and

$$\xi \colon G \to G \qquad \xi(g) = g^{-1}$$

are continuous.

4.1.2. Proposition. *Let G be a group with a topology τ. Then G is a topological group if and only if the function $\omega \colon G \times G \to G$ given by $\omega((g_1, g_2)) = g_1 \cdot g_2^{-1}$ is continuous.*

Proof. Suppose G is a topological group. Observe that $\omega = \pi \circ (1_G \times \xi)$, where 1_G denotes the identity map of G. Hence, ω is continuous.

Now, suppose that ω is continuous. Note that $\xi(g) = \omega((e_G, g))$, where e_G is the identity element of G. Thus, ξ is continuous. Since $\pi = \omega \circ (1_G \times \xi)$, π is continuous.

Q.E.D.

4.1.3. Notation. Let G be a group. If H and K are nonempty subsets of G, then define

$$H \cdot K = \{h \cdot k \mid h \in H \text{ and } k \in K\}$$

and

$$H^{-1} = \{h^{-1} \mid h \in H\}.$$

If $H = \{h\}$, then we write $h \cdot K$ instead of $\{h\} \cdot K$. Similarly, we write $H \cdot k$ instead of $H \cdot \{k\}$, when $K = \{k\}$. We define, inductively, $H^{n+1} = H \cdot H^n$ for each $n \in \mathbb{N}$.

4.1.4. Notation. If G is a topological group and $g \in G$, then $\mathcal{N}(g)$ denotes the family of all neighborhoods of g in G. Also, e_G denotes the identity element of G.

4.1.5. Remark. If G is a topological group, then we may describe the continuity of the group operations as follows: Let g_1 and g_2 be two elements of G. For the continuity of π: for each $U \in \mathcal{N}(g_1 \cdot g_2)$, there exist $V \in \mathcal{N}(g_1)$ and $W \in \mathcal{N}(g_2)$ such that $V \cdot W \subset U$. For the continuity of ξ: for each $U \in \mathcal{N}(g_1^{-1})$, there exists $V \in \mathcal{N}(g_1)$ such that $V^{-1} \subset U$.

4.1.6. Example. It is easy to see that $(\mathbb{R}^n, +)$, $(\mathbb{C}, +)$ and $(\mathbb{C} \setminus \{0\}, \cdot)$ are topological groups with the usual operations and the usual topologies.

4.1.7. Proposition. *Let G be a topological group. If g_0 is a fixed element of G, then the functions $\psi_{g_0}, \varphi_{g_0} \colon G \to G$ given by $\psi_{g_0}(g) = g_0 \cdot g$ and $\varphi_{g_0}(g) = g \cdot g_0$ are homeomorphisms.*

Proof. Note that ψ_{g_0} and φ_{g_0} are restrictions of the operation π to $\{g_0\} \times G$ and $G \times \{g_0\}$, respectively. Hence, both functions are continuous. Also note that $\psi_{g_0} \circ \psi_{g_0^{-1}} = \psi_{g_0^{-1}} \circ \psi_{g_0} = 1_G$ and that $\varphi_{g_0} \circ \varphi_{g_0^{-1}} = \varphi_{g_0^{-1}} \circ \varphi_{g_0} = 1_G$. Therefore, ψ_{g_0} y φ_{g_0} are homeomorphisms, since, clearly, $\psi_{g_0^{-1}}$ and $\varphi_{g_0^{-1}}$ are continuous functions.

Q.E.D.

4.1.8. Definition. Let G be a topological group. The maps ψ_{g_0} and φ_{g_0} of Proposition 4.1.7 are called *left translation by g_0* and *right translation by g_0*, respectively.

4.1.9. Corollary. *If G is a topological group and ψ_{g_0} and φ_{g_0} are the left and right translation by g_0, respectively, then $\psi_{g_0}^{-1} = \psi_{g_0^{-1}}$ and $\varphi_{g_0}^{-1} = \varphi_{g_0^{-1}}$.*

4.1.10. Proposition. *Let G be a topological group. Then the map $\xi \colon G \to G$ given by $\xi(g) = g^{-1}$ is a homeomorphism.*

Proof. By Definition 4.1.1, ξ is continuous. Since $\xi \circ \xi = 1_G$, ξ is a homeomorphism.

Q.E.D.

4.1.11. Corollary. *Each topological group is a homogeneous space.*

Proof. Let g_1 and g_2 be two elements of G. Let $\zeta \colon G \to G$ be given by $\zeta(g) = g \cdot g_1^{-1} \cdot g_2$. Note that $\zeta(g_1) = g_2$ and that $\zeta = \varphi_{g_1^{-1} \cdot g_2}$. Therefore, ζ is a homeomorphism.

Q.E.D.

4.1.12. Proposition. *Let G be a topological group, and let $\mathcal{N}^{\circ}(e_G)$ be a local basis of open sets of G containing e_G. Let $\mathcal{L}_g = \{g \cdot U \mid U \in \mathcal{N}^{\circ}(e_G)\}$ and let $\mathcal{R}_g = \{U \cdot g \mid U \in \mathcal{N}^{\circ}(e_G)\}$, for each $g \in G$. If $\mathcal{L} = \bigcup_{g \in G} \mathcal{L}_g$ and $\mathcal{R} = \bigcup_{g \in G} \mathcal{R}_g$, then $\mathcal{L}$ and $\mathcal{R}$ both form a basis for the topology of G.*

Proof. Let W be an open subset of G and let g_0 be an element of W. Since $\psi_{g_0^{-1}}$ is a homeomorphism (Proposition 4.1.7), $\psi_{g_0^{-1}}(W) = g_0^{-1} \cdot W$ is an open subset of G such that $e_G \in g_0^{-1} \cdot W$. Since $\mathcal{N}^{\circ}(e_G)$ is a local basis of e_G, there exists $U \in \mathcal{N}^{\circ}(e_G)$ such that $U \subset g_0^{-1} \cdot W$. Hence, $\psi_{g_0}(e_G) \in \psi_{g_0}(U) \subset \psi_{g_0}(g_0^{-1} \cdot W)$, i.e., $g_0 \in g_0 \cdot U \subset (g_0 \cdot g_0^{-1}) \cdot W = W$. Therefore, $\mathcal{L}$ is a basis for the topology of G.

The proof of the fact that $\mathcal{R}$ forms a basis for the topology of G is similar.

Q.E.D.

4.1.13. Definition. Let G be a topological group and let S be a nonempty subset of G. We say that S is *symmetric* if $S = S^{-1}$. The family of symmetric neighborhoods of e_G is denoted by $\mathcal{N}^{*}(e_G)$, i.e., $\mathcal{N}^{*}(e_G) = \{V \in \mathcal{N}(e_G) \mid V = V^{-1}\}$.

The next three Propositions say that each of the families of symmetric neighborhoods, powers of neighborhoods and closure of neighborhoods of the identity element of a topological group forms a local basis.

4.1.14. Proposition. *If G is a topological group and if $U \in \mathcal{N}(e_G)$, then there exists $V \in \mathcal{N}^{*}(e_G)$ such that $V \subset U$.*

Proof. Let $U \in \mathcal{N}(e_G)$. Since ξ is a homeomorphism (Proposition 4.1.10), $\xi(U) \in \mathcal{N}(e_G)$. Note that $\xi(U) = U^{-1}$. Let $V = U \cap U^{-1}$. Then, clearly, $V \in \mathcal{N}^{*}(e_G)$ and $V \subset U$.

Q.E.D.

4.1.15. Proposition. *If G is a topological group and $U \in \mathcal{N}(e_G)$, then for each $n \in \mathbb{N}$, there exists $V \in \mathcal{N}(e_G)$ such that $V^n \subset U$.*

Proof. Let $U \in \mathcal{N}(e_G)$. We do the proof by induction over n. For $n = 1$, let $V = U$.

Let $n \geq 2$ and suppose there exists $W \in \mathcal{N}(e_G)$ such that $W^n \subset U$. We show there exists $V \in \mathcal{N}(e_G)$ such that $V^{n+1} \subset U$.

Since the operation π is continuous and $e_G \cdot e_G = e_G$, there exist $V_1, V_2 \in \mathcal{N}(e_G)$ such that $V_1 \cdot V_2 \subset W$. Let $V = V_1 \cap V_2$. Then $V \in \mathcal{N}(e_G)$ and $V^2 \subset W$. Thus, $V^{n+1} = V^2 \cdot V^{n-1} \subset W \cdot W^{n-1} = W^n \subset U$.

Q.E.D.

4.1.16. Proposition. *If G is a topological group and $U \in \mathcal{N}(e_g)$, then there exists $V \in \mathcal{N}(e_G)$ such that $Cl(V) \subset U$.*

Proof. Let $U \in \mathcal{N}(e_G)$. By Proposition 4.1.15, there exists $W \in \mathcal{N}(e_G)$ such that $W^2 \subset U$. By Proposition 4.1.14, there exists $V \in \mathcal{N}^*(e_G)$ such that $V \subset W$. Hence, $V^2 \subset U$. Let $g \in Cl(V)$. Then $g \cdot V \in \mathcal{N}(g)$ (Proposition 4.1.12). Thus, $g \cdot V \cap V \neq \emptyset$. This implies that there exist $g_1, g_2 \in V$ such that $g \cdot g_1 = g_2$. Hence, $g = g_2 \cdot g_1^{-1} \in V \cdot V^{-1} = V^2 \subset U$. Therefore, $Cl(V) \subset U$.

Q.E.D.

4.1.17. Proposition. *Let G be a topological group. Let S, T, U and R be nonempty subsets of G, and let $g_0 \in G$. Then the following statements are true:*

(1) If U is an open subset of G, then the sets: $g_0 \cdot U$, $U \cdot g_0$, U^{-1}, $R \cdot U$ and $U \cdot R$ are open subsets of G.

(2) If S is a closed subset of G, then the sets: $g_0 \cdot S$, $S \cdot g_0$ and S^{-1} are closed subsets of G.

(3) If S and T are compact subsets of G, then the sets: $S \cdot T$ and S^{-1} are compact subsets of G.

(4) If S is a compact subset of G and T is a closed subset of G, then $S \cdot T$ and $T \cdot S$ are closed subsets of G.

(5) $Cl(S) = \bigcap_{W \in \mathcal{N}(e_G)} S \cdot W$ and $Cl(S) = \bigcap_{W \in \mathcal{N}(e_G)} W \cdot S$.

Proof. We show (1) is true. Note that, since the maps ψ_{g_0}, φ_{g_0} (Proposition 4.1.7) and ξ (Proposition 4.1.10) are homeomorphisms, $g_0 \cdot U$, $U \cdot g_0$ and U^{-1} are open subsets of G.

Since $R \cdot U = \bigcup_{r \in R} r \cdot U$ and $U \cdot R = \bigcup_{r \in R} U \cdot r$, $R \cdot U$ and $U \cdot R$ are open subsets of G.

The proof of (2) is similar to the one given in (1).

Note that (3) follows from the continuity of the operations π and ξ and the fact that S and T are compact sets.

We show (4) is true. To this end, we see that $T \cdot S$ is closed. The proof of the fact that $S \cdot T$ is closed is similar.

To see $T \cdot S$ is closed, we show $G \setminus (T \cdot S)$ is an open subset of G. Let $g \in G \setminus (T \cdot S)$. Then for each $s \in S$, $T \cdot s$ is a closed subset of G (part (2)). Thus, there exists $U_s \in \mathcal{N}^*(e_G)$ such that $g \cdot U_s \cap T \cdot s = \emptyset$. By Proposition 4.1.15, there exists $W_s \in \mathcal{N}(e_G)$ such that $W_s^2 \subset U_s$. By Proposition 4.1.14, there exists $V_s \in \mathcal{N}^*(e_G)$ such that $V_s \subset W_s$. Hence, for each $s \in S$, $V_s^2 \subset U_s$. Note that $g \cdot V_s \cap (T \cdot s) \cdot V_s = \emptyset$ for every $s \in S$.

Now, observe that the family of neighborhoods $\{s \cdot V_s \mid s \in S\}$ covers S. Since S is compact, there exist $s_1, \dots, s_n \in S$ such that $S \subset \bigcup_{j=1}^{n} s_j \cdot V_{s_j}$. Let $W' = \bigcap_{j=1}^{n} V_{s_j}$. Then $W' \in \mathcal{N}^*(e_G)$, $g \cdot W' \in \mathcal{N}(g)$ and $g \cdot W' \cap (T \cdot s_j) \cdot V_{s_j} = \emptyset$, for each $j \in \{1, \dots, n\}$. Thus, $g \cdot W' \cap T \cdot S = \emptyset$. Therefore, $G \setminus (T \cdot S)$ is open.

We prove (5) is satisfied. Let $W \in \mathcal{N}(e_G)$. Then there exists $V \in \mathcal{N}^*(e_G)$ such that $V \subset W$. Note that $S \subset S \cdot V \subset S \cdot W$. Let $g \in Cl(S)$. Since $V \in \mathcal{N}(e_G)$, $g \cdot V \cap S \neq \emptyset$. Thus, there exist $v \in V$ and $s \in S$ such that $g \cdot v = s$. Hence, $g = s \cdot v^{-1} \in S \cdot V^{-1} = S \cdot V \subset S \cdot W$. Therefore, $Cl(S) \subset \bigcap_{W \in \mathcal{N}(e_G)} S \cdot W$.

Next, let $g \in \bigcap_{W \in \mathcal{N}(e_G)} S \cdot W$, and let $V \in \mathcal{N}(g)$. We show that $V \cap S \neq \emptyset$, which implies that $g \in Cl(S)$. To this end, note that $V^{-1} \cdot g \in \mathcal{N}(e_G)$. Then $g \in S \cdot (V^{-1} \cdot g)$. Thus, there exist $s \in S$ and $v \in V$ such that $g = s \cdot v^{-1} \cdot g$. Hence, $s = v$. Therefore, $S \cap V \neq \emptyset$.

The proof of the other equality is similar.

Q.E.D.

4.2   Group Actions and a Theorem of Effros

We present elementary properties of action of a topological group on a topological space. We also give a topological proof of a Theorem of E. G. Effros (Theorem 4.2.25).

4.2.1. Definition. An *action of a topological group G on a metric space X* is a map $\theta\colon G \times X \to X$ such that:

(1) $\theta(g, \theta(g', x)) = \theta(g \cdot g', x)$ for each $g', g \in G$ and each $x \in X$; and

(2) $\theta(e_G, x) = x$ for each $x \in X$.

4.2.2. Notation. Instead of $\theta(g, x)$, we write $g \cdot x$. In this way, the properties (1) and (2) of Definition 4.2.1 become:

(1) $g \cdot (g' \cdot x) = (g \cdot g') \cdot x$; and

(2) $e_G \cdot x = x$.

4.2.3. Remark. Let G be a topological group acting on a metric space X. Note that for each $g \in G$, the function $\eta_g\colon X \twoheadrightarrow X$, given by $\eta_g(x) = g \cdot x$, is a homeomorphism, whose inverse is the map $\eta_{g^{-1}}$.

4.2.4. Example.

(1) If G is topological group, then G acts on G^n, for each $n \in \mathbb{N}$.

(2) If G is $\mathbb{R} \setminus \{0\}$ (or G is $\mathbb{C} \setminus \{0\}$), then G acts on $\mathbb{R}^n$ (or on $\mathbb{C}^n$, respectively).

(3) If G is $\mathcal{S}^1$, then G acts on $\mathbb{C}$.

4.2.5. Notation. Let G be a topological group acting on a metric space X. If K is a nonempty subset of G and $x \in X$, then $K \cdot x = \{g \cdot x \mid g \in K\}$.

4.2.6. Definition. Let G be a topological group acting on a metric space X. We say that G *acts transitively on* X, if $G \cdot x = X$ for each $x \in X$. We say that G *acts micro–transitively on* X if for each $x \in X$ and each $U \in \mathcal{N}(e_G)$, $U \cdot x$ is a neighborhood of x in X.

4.2.7. Remark. From now on, G denotes a topological group whose topology is given by a complete metric acting on a metric space X. We call such a group a *complete metric group.*

4.2.8. Definition. Let G be a complete metric group acting on a metric space X. For each $x \in X$, define the map $\gamma_x \colon G \twoheadrightarrow X$ by $\gamma_x(g) = g \cdot x$ for every $g \in G$.

4.2.9. Lemma. *If G is a complete metric group acting transitively on a metric space X, then the following are equivalent:*

(a) G acts micro–transitively on X;

(b) γ_x is an open map for each $x \in X$;

(c) γ_x is an open map for some $x \in X$.

Proof. First assume (a). We show (b). Let x be a point of X and let U be an open subset of G. Take $g \in U$. Note that $g^{-1} \cdot U$ is an open subset of G such that $e_G \in g^{-1} \cdot U$ (Proposition 4.1.7). Since G acts micro–transitively on X, $(g^{-1} \cdot U) \cdot x$ is a neighborhood of x in X. Hence, $U \cdot x = \gamma_x(U)$ is a neighborhood of $g \cdot x = \gamma_x(g)$. Therefore, γ_x is an open map.

Obviously, if (b) is true, then (c) is true. Assume (c). We prove
(b). Suppose γ_x is an open map for some $x \in X$. Let $z \in X$. We see
that γ_z is also an open map. Since G acts transitively on X, there
exists $g_0 \in G$ such that $z = g_0 \cdot x$. Since φ_{g_0} is a homeomorphism
(Proposition 4.1.7) and $\gamma_z = \gamma_x \circ \varphi_{g_0}$, γ_z is an open map.

Finally, assume (b). We see (a). Let $x \in X$ and let $U \in \mathcal{N}(e_g)$.
Since γ_x is an open map, $\gamma_x(Int_G(U)) = (Int_G(U)) \cdot x$ is an open
subset of X such that $x \in (Int_G(U)) \cdot x$. Hence, $U \cdot x$ is a neighborhood of x in X. Therefore, G acts micro–transitively on X.

Q.E.D.

4.2.10. Definition. Let G be a complete metric group acting on
a metric space X. We say that X is G–*countably covered* if for
each $x \in X$ and each $U \in \mathcal{N}(e_G)$, there is a sequence $\{h_n\}_{n=1}^\infty$ of
homeomorphisms of X such that the family $\{h_n(U \cdot x)\}_{n=1}^\infty$ covers
X.

4.2.11. Definition. Let G be a complete metric group acting on a
metric space X. We say that G *acts weakly micro–transitively on* X
if for each $x \in X$ and each $U \in \mathcal{N}(e_G)$, $Cl(U \cdot x)$ is a neighborhood
of x in X.

The following Lemma provides examples of metric spaces which
are G–countably covered.

4.2.12. Lemma. *Let G be a complete metric group acting transi-*
tively on a metric space X. If G is separable, then X is G–countably
covered.

Proof. Suppose G is separable. Let $x \in X$ and let $U \in \mathcal{N}(e_G)$.
Then the collection $\{g \cdot U \mid g \in G\}$ covers G. Since G is a separable
metric space, there is a sequence, $\{g_n\}_{n=1}^\infty$, of elements of G such
that the sequence of open sets $\{g_n \cdot U\}_{n=1}^\infty$ (Proposition 4.1.7) covers
G. Because G acts transitively on X, $\{(g_n \cdot U) \cdot x\}_{n=1}^\infty$ covers X.
Note that for each $n \in \mathbb{N}$, $(g_n \cdot U) \cdot x = \eta_{g_n}(U \cdot x)$ and each η_{g_n}
is a homeomorphism (Remark 4.2.3). Therefore, X is G–countably
covered.

Q.E.D.

4.2.13. Lemma. *Let G be a complete metric group acting on a metric space X. If X is of the second category and is G–countably covered, then G acts weakly micro–transitively on X.*

Proof. Assume X is of the second category and is G–countably covered. Let $x \in X$, and let $U \in \mathcal{N}(e_G)$. We show that $Cl(U \cdot x)$ is a neighborhood of x in X.

Since G is a topological group, there exists $V \in \mathcal{N}(e_G)$ such that $V \cdot V^{-1} \subset U$ (Proposition 4.1.2). By hypothesis, there is a sequence, $\{h_n\}_{n=1}^{\infty}$, of homeomorphisms of X such that the family $\{h_n(V \cdot x)\}_{n=1}^{\infty}$ covers X. Since X is of the second category, there exists $n \in \mathbb{N}$ such that $Cl(h_n(V \cdot x))$ has nonempty interior. Hence, $Cl(V \cdot x)$ has nonempty interior. Since any open nonempty subset of $Cl(V \cdot x)$ must intersect $V \cdot x$, it follows that there exists $g \in V$ such that $g \cdot x \in Int(Cl(V \cdot x))$. Thus, $x \in g^{-1} \cdot (Int(Cl(V \cdot x))) = Int(Cl((g^{-1} \cdot V) \cdot x)) \subset Int(Cl((V^{-1} \cdot V) \cdot x)) \subset Int(Cl(U \cdot x))$. Therefore, $Cl(U \cdot x)$ is a neighborhood of x in X.

Q.E.D.

Let G be a complete metric group acting transitively on a metric space X. If G also acts micro–transitively on X, then G acts weakly micro–transitively on X. The next Lemma shows that the converse is also true.

4.2.14. Lemma. *Let G be a complete metric group acting transitively on a metric space X. If G acts weakly micro–transitively on X, then G acts micro–transitively on X.*

Proof. Suppose G acts weakly micro–transitively on X. Let ρ_X be a metric on X and let ρ_G be a complete metric on G.

Let $x_0 \in X$ and let $U \in \mathcal{N}(e_G)$. We show that $U \cdot x_0$ is a neighborhood of x_0 in X. To this end, let $U_0 \in \mathcal{N}(e_G)$ such that $(Cl(U_0))^{-1} \cdot (Cl(U_0)) \subset U$ (Propositions 4.1.2 and 4.1.16). Since G acts weakly micro–transitively on X, there exists an open subset M_0 of X such that $x_0 \in M_0 \subset Cl(U_0 \cdot x_0)$. We prove that $M_0 \subset U \cdot x_0$.

Let $y_0 \in M_0$; we must find $g \in U$ such that $g \cdot x_0 = y_0$. At this point we follow a technique introduced by Homma [7]. What we do is to move from x_0 to y_0 and from y_0 to x_0, planning ahead each movement so that the next movement is possible and is close to e_G.

First, let $V_0 = U_0$. Invoke the weak micro–transitivity action of G on X to obtain an open subset N_0 such that $y_0 \in N_0 \subset Cl(V_0 \cdot y_0)$. We construct eight sequences, namely:

two sequences, $\{g_n\}_{n=1}^{\infty}$ and $\{h_n\}_{n=1}^{\infty}$, of elements of G,

two sequences, $\{x_n\}_{n=0}^{\infty}$ and $\{y_n\}_{n=0}^{\infty}$, of elements of X,

two sequences, $\{U_n\}_{n=0}^{\infty}$ and $\{V_n\}_{n=0}^{\infty}$, of neighborhoods of e_G in G and

two sequences, $\{M_n\}_{n=0}^{\infty}$ and $\{N_n\}_{n=0}^{\infty}$, of open subsets of X.

These eight sequences are constructed to satisfy the following thirteen properties:

(1_n) $g_n \in U_{n-1}$. $\qquad\qquad$ (2_n) $h_n \in V_{n-1}$.

(3_n) $x_n = g_n \cdot x_{n-1}$. $\qquad$ (4_n) $y_n = h_n \cdot y_{n-1}$.

(5_n) $x_n \in N_{n-1}$. $\qquad\qquad$ (6_n) $y_n \in M_n$.

(7_n) $U_n \cdot g_n \cdot \ldots \cdot g_1 \subset U_0$. $\qquad$ (8_n) $V_n \cdot h_n \cdot \ldots \cdot h_1 \subset V_0$.

(9_n) $\operatorname{diam}(U_n \cdot g_n \cdot \ldots \cdot g_1) < \dfrac{1}{2^n}$. $\quad$ (10_n) $\operatorname{diam}(V_n \cdot h_n \cdot \ldots \cdot h_1) < \dfrac{1}{2^n}$.

(11_n) $x_n \in M_n \subset Cl(U_n \cdot x_n)$. $\quad$ (12_n) $y_n \in N_n \subset Cl(V_n \cdot y_n)$.

(13_n) $\operatorname{diam}(M_n) < \dfrac{1}{n}$.

The construction of these eight sequences is done by induction. We already have chosen U_0, V_0, x_0, y_0, M_0 and N_0. Let $n \geq 1$ and inductively assume that for $k \in \{1, \ldots, n-1\}$, we have chosen g_k, h_k, x_k, y_k, U_k, V_k, M_k and N_k in such a way that $(1_k), \ldots, (13_k)$ are satisfied.

Note that from (6_{n-1}), (12_{n-1}) and (11_{n-1}), we have that

$$y_{n-1} \in M_{n-1} \cap N_{n-1} \subset Cl(U_{n-1} \cdot x_{n-1}).$$

Hence, there exists $g_n \in U_{n-1}$ such that $g_n \cdot x_{n-1} \in N_{n-1}$; so (1_n) holds. Let $x_n = g_n \cdot x_{n-1}$. Then (3_n) and (5_n) also hold. From (7_{n-1}) and (1_n) follows that $g_n \cdot g_{n-1} \cdot \ldots \cdot g_1 \in U_0$. Thus, we may choose $U_n \in \mathcal{N}(e_G)$ in such a way that (7_n) and (9_n) are satisfied. Since G acts weakly micro–transitively on X, there exists an open subset M_n of X such that satisfies (11_n) and (13_n). So far, we have chosen g_n, x_n, U_n and M_n in such a way that (1_n), (3_n), (5_n), (7_n), (9_n), (11_n) and (13_n) hold.

From (5_n), (11_n) and (12_{n-1}), we conclude that

$$x_n \in N_{n-1} \cap M_n \subset Cl(V_{n-1} \cdot y_{n-1}).$$

Hence, there exists $h_n \in V_{n-1}$ such that $h_n \cdot y_{n-1} \in M_n$, so (2_n) holds. Let $y_n = h_n \cdot y_{n-1}$. Then (4_n) and (6_n) are also satisfied. From (8_{n-1}) and (2_n), we obtain that $h_n \cdot h_{n-1} \cdot \ldots \cdot h_1 \in V_0$. Thus, we may choose $V_n \in \mathcal{N}(e_G)$ such that (8_n) and (10_n) hold. Since G acts weakly micro–transitively on X, we may find an open subset N_n of X satisfying (12_n). So, we have completed the inductive construction of the eight sequences.

For each $n \geq 1$, let $\tilde{g}_n = g_n \cdot \ldots \cdot g_1$ and let $\tilde{h}_n = h_n \cdot \ldots \cdot h_1$. From (1_n), (2_n), (9_n) and (10_n), we conclude that $\{\tilde{g}_n\}_{n=1}^{\infty}$ and $\{\tilde{h}_n\}_{n=1}^{\infty}$ are Cauchy sequences with respect to the metric ρ_G of G. Since ρ_G is a complete metric, it follows that $\{\tilde{g}_n\}_{n=1}^{\infty}$ and $\{\tilde{h}_n\}_{n=1}^{\infty}$ converge to elements g and h of G, respectively. It follows from (7_n) and (8_n) that $\{\tilde{g}_n\}_{n=1}^{\infty} \subset U_0$ and $\{\tilde{h}_n\}_{n=1}^{\infty} \subset V_0$. Therefore, $g \in Cl(U_0)$ and $h \in Cl(V_0) = Cl(U_0)$. By the election of U_0, we have that $h^{-1} \cdot g \in U$.

It follows from (3_n) and (4_n) that for each $n \geq 1$, $x_n = \tilde{g}_n \cdot x_0$ and $y_n = \tilde{h}_n \cdot y_0$. Hence, the sequence $\{x_n\}_{n=1}^{\infty}$ converges to $g \cdot x_0$, and the sequence $\{y_n\}_{n=1}^{\infty}$ converges to $h \cdot y_0$. Since for each $n \in \mathbb{N}$, it follows from (6_n) and (11_n) that both x_n and y_n belong to M_n, then, from (13_n), we have that $\rho_X(x_n, y_n) < \dfrac{1}{n}$. Thus, $g \cdot x_0 = h \cdot y_0$. Therefore, $h^{-1} \cdot g \in U$ and $(h^{-1} \cdot g) \cdot x_0 = y_0$.

Q.E.D.

4.2.15. Theorem. *Let G be a complete metric group acting transitively on a metric space X. If G is separable, then the following are equivalent:*

(a) G acts micro–transitively on X.

(b) X is topologically complete.

(c) X is of the second category.

Proof. First, suppose G acts micro–transitively on X. To see X has a complete metric, let $x \in X$ and consider the map $\gamma_x \colon G \twoheadrightarrow X$ (Definition 4.2.8). Since G acts transitively on X, γ_x is surjective. Since G acts micro–transitively on X, by Lemma 4.2.9, γ_x is an open map. Hence, by Theorem 1.5.13, X has a complete metric.

By the Baire Category Theorem (Theorem 1.5.12), if X has a complete metric, then X is of the second category.

Finally, suppose X is of the second category. Since G is separable, by Lemma 4.2.12, X is G–countably covered. Hence, since X is of the second category and is G–countably covered, by Lemma 4.2.13, G acts weakly micro–transitively on X. Therefore, by Lemma 4.2.14, G acts micro–transitively on X.

Q.E.D.

Now, we present some definitions needed for Effros's Theorem.

4.2.16. Definition. Let G be a complete metric group acting on a metric space X. Then $G \cdot x$ is called the *orbit of x under the action of G on X*. Note that distinct orbits are disjoint. The set

$$\{G \cdot x \mid x \in X\}$$

of orbits is called the *orbit space determined by the action of G on X*, and it is denoted by X/G. We give X/G the quotient topology. The function $q \colon X \twoheadrightarrow X/G$ given by $q(x) = G \cdot x$ is the quotient map.

4.2.17. Lemma. *If G is a complete metric group acting on a metric space X, then quotient map $q \colon X \twoheadrightarrow X/G$ is open.*

Proof. It is enough to observe that if U is an open subset of X, then $q^{-1}(q(U)) = G \cdot U$. Since $G \cdot U = \bigcup \{g \cdot U \mid g \in G\} = \bigcup \{\eta_g(U) \mid g \in G\}$ and each η_g is a homeomorphism (Remark 4.2.3), $q^{-1}(q(U))$ is open in X.

Q.E.D.

4.2.18. Definition. Let G be a complete metric group acting on a metric space X. If $x \in X$, then

$$G_x = \{g \in G \mid g \cdot x = x\}$$

is a subgroup of G called the *stabilizer subgroup of x*.

4.2.19. Remark. Note that G_x acts on G by multiplication on the right. The orbit space of this action, G/G_x, is the set

$$G/G_x = \{g \cdot G_x \mid g \in G\}$$

and consists of all the left cosets of G_x. We give G/G_x the quotient topology. The function $q_x \colon G \to G/G_x$ given by $q_x(g) = g \cdot G_x$ is the quotient map.

4.2.20. Lemma. *Let G be a complete metric group acting on a metric space X. If $x \in X$, then quotient map $q_x \colon G \twoheadrightarrow G/G_x$ is open.*

Proof. It is enough to observe that if U is an open subset of G, then $q_x^{-1}(q_x(U)) = U \cdot G_x$. By Proposition 4.1.17 (1), $U \cdot G_x$ is open in G.

Q.E.D.

4.2.21. Definition. Let G be a complete metric group acting on a metric space X. If $x \in X$, define the function $\psi_x \colon G/G_x \to G \cdot x$ by $\psi_x(g \cdot G_x) = g \cdot x$.

In the following Lemma we show that ψ_x is a well defined function which is a continuous bijection.

4.2.22. Lemma. *Let G be a complete metric group acting on a metric space X. If $x \in X$, then the function $\psi_x \colon G/G_x \to G \cdot x$ of Definition 4.2.21 is a well defined continuous bijection. Moreover, the following diagram*

$$\begin{array}{ccc} & G & \\ {\scriptstyle q_x}\swarrow & & \searrow{\scriptstyle \gamma_x} \\ G/G_x & \xrightarrow[\psi_x]{} & (G \cdot x) \end{array}$$

is commutative.

Proof. Let $x \in X$. For $g', g \in G$, $g \cdot G_x = g' \cdot G_x$ if and only if $g \cdot x = g' \cdot x$. Hence, ψ_x is well defined and one–to–one. To see ψ_x is surjective, let $g \cdot x \in G \cdot x$. Then $\psi_x(g \cdot G_x) = g \cdot x$. Therefore, ψ_x is a bijection.

Next, let $g \in G$. Then $\psi_x \circ q_x(g) = \psi_x(q_x(g)) = \psi_x(g \cdot G_x) = g \cdot x = \gamma_x(g)$. Therefore, the diagram is commutative.

Finally, since $\psi_x \circ q_x = \gamma_x$, q_x is open (Lemma 4.2.20) and γ_x is continuous, we have that ψ_x is continuous.

Q.E.D.

4.2.23. Lemma. *Let G be a complete metric group acting on a metric space X. If $x \in X$, then G acts micro–transitively on the orbit $G \cdot x$ if and only if the map $\psi_x \colon G/G_x \to G \cdot x$ of Definition 4.2.21 is a homeomorphism.*

Proof. By Lemma 4.2.22, ψ_x is continuous. Since $\psi_x \circ q_x = \gamma_x$ and q_x is continuous and open (Lemma 4.2.20), ψ_x is open if and only if γ_x is open. By Lemma 4.2.9, γ_x is open if and only if G acts micro–transitively on the orbit $G \cdot x$.

Q.E.D.

4.2.24. Lemma. *If G is a complete metric group acting on a separable complete metric space X, then each orbit is a G_δ subset of X if and only if X/G is a T_0 space.*

Proof. First, assume that each orbit is a G_δ subset of X. We show that X/G is a T_0 space. Let $x, y \in X$. Suppose that $q(x) \in Cl_{X/G}(\{q(y)\})$ and $q(y) \in Cl_{X/G}(\{q(x)\})$. We see that $q(x) = q(y)$.

We assert that $G \cdot x \subset Cl_X(G \cdot y)$. Indeed, suppose there exists $g \in G$ such that $g \cdot x \notin Cl_X(G \cdot y)$. Then there exists an open subset U of X such that $g \cdot x \in U$ and $U \cap G \cdot y = \emptyset$. Since the map $q \colon X \to X/G$ is open (Lemma 4.2.17), $q(U)$ is an open subset of X/G such that $q(g \cdot x) = q(x) \in q(U)$. Note that $q(G \cdot y) = \{q(y)\}$ and $q(y) \notin q(U)$. A contradiction to the fact that $q(x) \in Cl_{X/G}(\{q(y)\})$. Therefore, $G \cdot x \subset Cl_X(G \cdot y)$. Similarly, $G \cdot y \subset Cl_X(G \cdot x)$. We conclude that $G \cdot x \cup G \cdot y \subset Cl_X(G \cdot x) \cap Cl_X(G \cdot y)$.

Since $G \cdot x$ is dense in $Cl_X(G \cdot x)$ and $G \cdot y$ is dense in $Cl_X(G \cdot y)$, $G \cdot x$ and $G \cdot y$ both are dense in $Cl_X(G \cdot x) \cap Cl_X(G \cdot y)$. By hypothesis,

$G \cdot x$ and $G \cdot y$ are both G_δ subsets of $Cl_X(G \cdot x) \cap Cl_X(G \cdot y)$. Since $Cl_X(G \cdot x) \cap Cl_X(G \cdot y)$ is a closed subset of X and X is a complete metric space, $Cl_X(G \cdot x) \cap Cl_X(G \cdot y)$ has a complete metric (Lemma 1.5.3). By Lemma 1.5.9, $G \cdot x \cap G \cdot y$ is dense in $Cl_X(G \cdot x) \cap Cl_X(G \cdot y)$. In particular, $G \cdot x \cap G \cdot y \neq \emptyset$. Hence, $G \cdot x = G \cdot y$ and $q(x) = q(y)$. Therefore, X/G is a T_0 space.

Next, suppose X/G is a T_0 space. Since X is a separable metric space, X has a countable basis $\{U_n\}_{n=1}^\infty$. We assert that $\{q(U_n)\}_{n=1}^\infty$ is a countable basis for X/G. Clearly, this family is countable. Since the map $q \colon X \to X/G$ is open (Lemma 4.2.17), $q(U_n)$ is an open subset of X/G for each $n \in \mathbb{N}$. Furthermore, if $x \in X$ and V is an open subset of X/G such that $q(x) \in V$, then there exists $n \in \mathbb{N}$ such that $q(x) \in q(U_n) \subset V$ (by the continuity of q and the fact that $\{U_n\}_{n=1}^\infty$ is a basis for X). Thus, $\{q(U_n)\}_{n=1}^\infty$ is a countable basis for X/G.

Now, take $x \in X$. For each $n \in \mathbb{N}$, let

$$
V_n = \begin{cases} q(U_n) & \text{if } q(x) \in q(U_n), \\ X/G \setminus q(U_n) & \text{if } q(x) \notin q(U_n). \end{cases}
$$

Since X/G is a T_0 space, $\bigcap_{n=1}^\infty V_n = \{q(x)\}$. Hence, $\bigcap_{n=1}^\infty q^{-1}(V_n) = q^{-1}(q(x)) = G \cdot x$. Since each V_n is either open or closed in X/G, $q^{-1}(V_n)$ is either open or closed in X. In either case, $q^{-1}(V_n)$ is a G_δ subset of X (see Lemma 1.5.7 when $q^{-1}(V_n)$ is closed). Therefore, $G \cdot x$ is a G_δ subset of X.

Q.E.D.

We are ready to prove Effros's Theorem [4].

4.2.25. Theorem. *If G is a separable complete metric group acting on a separable complete metric space X, then the following are equivalent:*

(a) For each $x \in X$, the map $\psi_x \colon G/G_x \to G \cdot x$ of Definition 4.2.21 is a homeomorphism.

(b) G acts micro–transitively on each orbit.

(c) Each orbit is of the second category (in itself).

(d) Each orbit is a G_δ subset of X.

(e) X/G is a T_0 space.

Proof. By Lemma 4.2.23, (a) and (b) are equivalent. By Theorem 4.2.15, (b) and (c) are equivalent. By Theorem 1.5.8, (d) is equivalent to the fact that each orbit is topologically complete. This latter statement is equivalent to (b) by Theorem 4.2.15. Hence, (d) is equivalent to (b). Finally, by Lemma 4.2.24, (d) is equivalent to (e).

Q.E.D.

4.2.26. Definition. Let X and Y be continua. Define

$$\mathcal{C}(X, Y) = \{f \colon X \to Y \mid f \text{ is a map}\}.$$

We topologize $\mathcal{C}(X, Y)$ with the sup *metric, ρ,* given by

$$\rho((f, g)) = \sup\{d(f(x), g(x)) \mid x \in X\},$$

for every $g, f \in \mathcal{C}(X, Y)$.

4.2.27. Definition. If X is a continuum, then $\mathcal{H}(X)$ denotes the *group of homeomorphisms of X*. $\mathcal{H}(X)$ acts on X as follows: If $h \in \mathcal{H}(X)$ and $x \in X$, then $\theta((h, x)) = h(x)$.

4.2.28. Theorem. *If X is a continuum, then $\mathcal{C}(X, X)$ is a separable complete metric space and $\mathcal{H}(X)$ is a G_δ subset of $\mathcal{C}(X, X)$.*

Proof. By Theorem 1 (p. 244) of [8], $\mathcal{C}(X, X)$ is a separable space. By Theorem 3 (p. 90) of [9], $\mathcal{C}(X, X)$ is a complete metric space. By Theorem 1 (p. 91) of [9], $\mathcal{H}(X)$ is a G_δ subset of $\mathcal{C}(X, X)$.

Q.E.D.

4.2.29. Corollary. *If X is a continuum, then $\mathcal{H}(X)$ has a separable complete metric, ρ'.*

Proof. By Theorem 4.2.28, $\mathcal{H}(X)$ is a G_δ subset of $\mathcal{C}(X, X)$. Hence, by Theorem 1.5.8, $\mathcal{H}(X)$ has a complete metric. In fact, it is known that ρ' is given by $\rho'(f, g) = \rho(f, g) + \rho(f^{-1}, g^{-1})$ for every $f, g \in \mathcal{H}(X)$.

Q.E.D.

4.2.30. Definition. A metric space X has the *Property of Effros* provided that for each $\varepsilon > 0$, there exists $\delta > 0$ such that if $x, y \in X$ and $d(x, y) < \delta$, then there exists $h \in \mathcal{H}(X)$, such that $h(x) = y$ and $d(z, h(z)) < \varepsilon$ for every $z \in X$. The number δ is called an *Effros number for the given ε*. A homeomorphism $h \in \mathcal{H}(X)$ satisfying that $d(z, h(z)) < \varepsilon$ for each $z \in X$ is called an *ε-homeomorphism*.

The following Theorem is known as Effros's Theorem in the theory of homogeneous continua. A slightly different version of this result was first shown by C. L. Hagopian [5].

4.2.31. Theorem. *If X is a homogeneous continuum, with metric d, then X has the property of Effros.*

Proof. By Corollary 4.2.29, $\mathcal{H}(X)$ is a separable complete metric space. Since X is a homogeneous continuum, $\mathcal{H}(X)$ acts transitively on X. Let $x \in X$, by Theorem 4.2.25, the map

$$\psi_x \colon \mathcal{H}(X)/\mathcal{H}(X)_x \twoheadrightarrow \mathcal{H}(X) \cdot x$$

is a homeomorphism. By Lemma 4.2.23, $\mathcal{H}(X)$ acts micro–transitively on X. Hence, by Lemma 4.2.9, the map γ_x is open.

Let $\varepsilon > 0$ be given and let $U = \mathcal{V}_{\frac{\varepsilon}{2}}^{\rho'}(1_X)$, where 1_X is the identity map of X. Then for each $h \in U$, h is an $\frac{\varepsilon}{2}$–homeomorphism of X. Since γ_x is open, $\gamma_x(U)$ is an open subset of X such that $x \in \gamma_x(U)$. Let $\delta_x > 0$ such that $\mathcal{V}_{\delta_x}^d(x) \subset \gamma_x(U)$. Thus, if $y \in \mathcal{V}_{\delta_x}^d(x)$, then there exists $h \in U$ such that $\gamma_x(h) = y$, i.e., h is an $\frac{\varepsilon}{2}$–homeomorphism such that $h(x) = y$.

Note that $\{\mathcal{V}_{\delta_x}^d(x) \mid x \in X\}$ is an open cover of X. Let δ be a Lebesgue number for this cover (Theorem 1.6.6). Let $x, y \in X$ such that $d(x, y) < \delta$. Then there exists $z \in X$ such that $x, y \in \mathcal{V}_{\delta_z}^d(z)$. By the previous paragraph, there exist $h_1, h_2 \in \gamma_z(U)$ such that

$h_1(z) = x$ and $h_2(z) = y$. Let $h = h_2 \circ h_1^{-1}$. Then $h \colon X \twoheadrightarrow X$ is a homeomorphism such that $h(x) = y$ $(h(x) = h_2 \circ h_1^{-1}(x) = h_2\left(h_1^{-1}(x)\right) = h_2(z) = y)$.

Let $w \in X$. Since

$$
\begin{aligned}
d(w, h(w) &= d\left(w, h_2\left(h_1^{-1}(w)\right)\right)\\
&\leq d\left(w, h_1^{-1}(w)\right) + d\left(h_1^{-1}(w), h_2\left(h_1^{-1}(w)\right)\right)\\
&= d\left(h_1\left(h_1^{-1}(w)\right), h_1^{-1}(w)\right) + d\left(h_1^{-1}(w), h_2\left(h_1^{-1}(w)\right)\right)\\
&< \frac{\varepsilon}{2} + \frac{\varepsilon}{2} = \varepsilon,
\end{aligned}
$$

h is an ε–homeomorphism.

Q.E.D.

As our first application of Effros's Theorem, we show that Jones's set function $\mathcal{T}$ is idempotent on homogeneous continua.

4.2.32. Theorem. *If X is a homogeneous continuum, with metric d, then for each closed subset A of X, $\mathcal{T}^2(A) = \mathcal{T}(A)$.*

Proof. Let A be a closed subset of X. Note that, by Remark 3.1.5, $\mathcal{T}(A) \subset \mathcal{T}^2(A)$.

Next, let $x \in X \setminus \mathcal{T}(A)$. Then there exists a subcontinuum W of X such that $x \in Int(W) \subset W \subset X \setminus A$. Let $\varepsilon > 0$ such that $\varepsilon < d(W, A)$ and $\mathcal{V}_\varepsilon^d(x) \subset Int(W)$. Let $\delta > 0$ be an Effros number for this ε. We assume that $\delta < \varepsilon$. Since W is compact, there exist $w_1, \ldots, w_m \in W$ such that $W \subset \bigcup\limits_{j=1}^{m} \mathcal{V}_\delta^d(w_j)$.

Let $j \in \{1, \ldots, m\}$. Then for each $y \in \mathcal{V}_\delta^d(w_j)$, there exists an ε–homeomorphism $h_y \colon X \twoheadrightarrow X$ such that $h_y(w_j) = y$. When $y = w_j$, we let $h_y = 1_X$, the identity map of X. Let

$$
M_j = Cl\left(\bigcup_{y \in \mathcal{V}_\delta^d(w_j)} h_y(W)\right).
$$

Then M_j is a subcontinuum of X such that $w_j \in \mathcal{V}_\delta^d(w_j) \subset M_j \subset X \setminus A$. Let $M = \bigcup\limits_{j=1}^{m} M_j$. Then M is a subcontinuum of X and $W \subset$

$$\bigcup_{j=1}^{m} \mathcal{V}_{\delta}^{d}(w_j) \subset Int(M) \subset M \subset X \setminus A. \text{ Since } Int(M) \subset X \setminus \mathcal{T}(A),$$

$x \in Int(W) \subset W \subset X \setminus \mathcal{T}(A)$. Hence, $x \in X \setminus \mathcal{T}^2(A)$. Therefore, $\mathcal{T}$ is idempotent.

Q.E.D.

We finish this chapter showing that any connected metric space with the property of Effros is homogeneous.

4.2.33. Theorem. *If X is a connected metric space, with metric d, satisfying the property of Effros, then X is homogeneous.*

Proof. Let $x, y \in X$, and let $\varepsilon > 0$. Since X has the property of Effros, there exists an Effros number $\delta > 0$ for this ε. Since X is connected, there exist finitely many points $z_1 = x, z_2, \ldots, z_{n-1}, z_n = y$ in X such that $d(z_{j-1}, z_j) < \delta$ for each $j \in \{2, \ldots, n\}$. Hence, there exists a homeomorphism $h_j \colon X \twoheadrightarrow X$ such that $h_j(z_{j-1}) = z_j$, $j \in \{2, \ldots, n\}$. Then $h = h_n \circ \ldots \circ h_2$ is a homeomorphism of X onto itself such that $h(x) = y$. Therefore, X is homogeneous.

Q.E.D.

REFERENCES

[1] F. D. Ancel, An Alternative Proof and Applications of a Theorem of E. G. Effros, Michigan Math. J., 34 (1987), 39–55.

[2] G. E. Bredon, *Introduction to Compact Transformation Groups*, Monographs and Textbooks in Pure and Applied Mathematics, Vol. 46, Academic Press, Inc., New York, 1972.

[3] G. E. Bredon, *Topology and Geometry*, Graduate Texts in Mathematics, Vol. 139, Springer–Verlag, New York, Inc., 1993.

[4] E. G. Effros, Transformation Groups and C^*–Algebras, Ann. of Math., (2) 81 (1965), 38–55.

[5] C. L. Hagopian, Homogeneous Plane Continua, Houston J. Math., 1 (1975), 35–41.

[6] C. Hernández, O. J. Rendón, M. Tkačenko and L. M. Villegas, *Grupos Topológicos*, Libros de Texto, Manuales de Prácticas y Antologías, Universidad Autónoma Metropolitana, Unidad Iztapalpa, 1997. (Spanish)

[7] T. Homma, On the Embedding of Polyhedra in Manifolds, Yokohama Math. J., 10 (1962), 5–10.

[8] K. Kuratowski, *Topology*, Vol. I, Academic Press, New York, N. Y., 1966.

[9] K. Kuratowski, *Topology*, Vol. II, Academic Press, New York, N. Y., 1968.

[10] G. McCarty, *Topology: An Introduction with Applications to Topological Groups*, Dover Publications, Inc., New York, 1988.

Chapter 5

DECOMPOSITION THEOREMS

We present a proof of Jones's Aposyndetic Decomposition Theorem and Rogers's Terminal Decomposition Theorem. These theorems are proven using Jones's set function $\mathcal{T}$ (Chapter 3) and Effros's Theorem (Chapter 4). We also give a construction of the Case continuum and present a sketch of the construction of the Minc–Rogers continua. Finally, we study covering spaces of any solenoid, the Menger curve, the Case continuum and one of the Minc–Rogers examples.

5.1   Jones's Theorem

We give a proof of Jones's Aposyndetic Decomposition Theorem (Theorem 5.1.19).

We begin this section proving that the set function $\mathcal{T}$ commutes with homeomorphisms.

225

5.1.1. Lemma. *Let X be a compactum. If $h\colon X \twoheadrightarrow X$ is a homeomorphism, then $h(\mathcal{T}(A)) = \mathcal{T}(h(A))$ for each closed subset A of X.*

Proof. Let A be a closed subset of X. Let $x \in X \setminus h(\mathcal{T}(A))$. Then $h^{-1}(x) \in X \setminus \mathcal{T}(A)$. Hence, there exists a subcontinuum W of X such that $h^{-1}(x) \in Int(W) \subset W \subset X \setminus A$. This implies that $x \in Int(h(W)) \subset h(W) \subset X \setminus h(A)$. Thus, $x \in X \setminus \mathcal{T}(h(A))$.

Now, let $x \in X \setminus \mathcal{T}(h(A))$. Then there exists a subcontinuum K of X such that $x \in Int(K) \subset K \subset X \setminus h(A)$. This implies that $h^{-1}(x) \in Int(h^{-1}(K)) \subset h^{-1}(K) \subset X \setminus A$. Thus, $h^{-1}(x) \in X \setminus \mathcal{T}(A)$. Hence, $x \in X \setminus h(\mathcal{T}(A))$.

Therefore, $h(\mathcal{T}(A)) = \mathcal{T}(h(A))$.

Q.E.D.

The following Theorem says that, for a homogeneous continuum, the images of all singletons under $\mathcal{T}$ form a decomposition of the continuum.

5.1.2. Theorem. *Let X be a homogeneous continuum with metric d. If $\mathcal{G} = \{\mathcal{T}(\{x\}) \mid x \in X\}$, then $\mathcal{G}$ is a decomposition of X.*

Proof. Let x be a point of X. We show that if $y \in \mathcal{T}(\{x\})$, then $\mathcal{T}(\{x\}) = \mathcal{T}(\{y\})$.

Let $y \in \mathcal{T}(\{x\})$. By Proposition 3.1.7 and the fact that $\mathcal{T}$ is idempotent (Theorem 4.2.32), $\mathcal{T}(\{y\}) \subset \mathcal{T}^2(\{x\}) = \mathcal{T}(\{x\})$.

To show $\mathcal{T}(\{x\}) \subset \mathcal{T}(\{y\})$ we use an argument of Bellamy and Lum in Lemma 5 of [6]. We find a point x_0 of X such that $\mathcal{T}(\{x_0\}) \subset \mathcal{T}(\{z\})$ for each $z \in \mathcal{T}(\{x_0\})$. Then, by Lemma 5.1.1, the homogeneity of X ensures that the same is true for every point of X.

We give a partial order to $\mathcal{G}$ as follows. $\mathcal{T}(\{z\}) < \mathcal{T}(\{w\})$ if $\mathcal{T}(\{w\}) \subset \mathcal{T}(\{z\})$. Let $\mathcal{K} = \{\mathcal{T}(\{x_\lambda\})\}_{\lambda \in \Lambda}$ be a (set theoretic) chain of elements of $\mathcal{G}$. We show that $\mathcal{K}$ has an upper bound. Since each $\mathcal{T}(\{x_\lambda\})$ is a continuum (Theorem 3.1.21), and $\mathcal{K}$ is a chain, $\bigcap_{\ell \in \Lambda} \mathcal{T}(\{x_\lambda\}) \neq \emptyset$. Let $x' \in \bigcap_{\lambda \in \Lambda} \mathcal{T}(\{x_\lambda\})$. Then, by the previous argument, for each $\lambda \in \Lambda$, $\mathcal{T}(\{x'\}) \subset \mathcal{T}(\{x_\lambda\})$. Hence, $\mathcal{T}(\{x'\}) \subset \bigcap_{\lambda \in \Lambda} \mathcal{T}(\{x_\lambda\})$. Thus, $\mathcal{K}$ has an upper bound. By Kuratowski–Zorn

Lemma, $\mathcal{G}$ has a maximal element, i.e., there exists $x_0 \in X$ such that for each $z \in \mathcal{T}(\{x_0\})$, $\mathcal{T}(\{x_0\}) \subset \mathcal{T}(\{z\})$.

Since X is homogeneous, there is a homeomorphism $h\colon X \twoheadrightarrow X$ such that $h(x_0) = x$. Since $y \in \mathcal{T}(\{x\})$, by Lemma 5.1.1, $h^{-1}(y) \in \mathcal{T}(\{x_0\})$. Hence, by the previous paragraph, we have that $\mathcal{T}(\{x_0\}) \subset \mathcal{T}(\{h^{-1}(y)\})$. Thus, $\mathcal{T}(\{x\}) = h(\mathcal{T}(\{x_0\})) \subset h(\mathcal{T}(\{h^{-1}(y)\})) = \mathcal{T}(\{y\})$ (Lemma 5.1.1). Consequently, $\mathcal{T}(\{x\}) \subset \mathcal{T}(\{y\})$.

Therefore, $\mathcal{G}$ is a decomposition of X.

Q.E.D.

5.1.3. Definition. Let $\mathcal{G}$ be a decomposition of the compactum X. Let $\mathcal{H}$ be a family of homeomorphisms of X. We say $\mathcal{H}$ *respects* $\mathcal{G}$ provided that for each pair $G_1, G_2 \in \mathcal{G}$ and each $h \in \mathcal{H}$, either $h(G_1) = G_2$ or $h(G_1) \cap G_2 = \emptyset$.

5.1.4. Theorem. *Let X be a homogeneous continuum, with metric d, and let $\mathcal{G}$ be a decomposition of X such that the elements of $\mathcal{G}$ are continua. If the homeomorphism group, $\mathcal{H}(X)$, of X respects $\mathcal{G}$, then the following hold:*

(1) $\mathcal{G}$ is a continuous decomposition of X.

(2) The elements of $\mathcal{G}$ are homogeneous mutually homeomorphic continua.

(3) The quotient space $X/\mathcal{G}$ is a homogeneous continuum.

Proof. First, we show that $\mathcal{G}$ is continuous. To this end, we prove first that $\mathcal{G}$ is upper semicontinuous.

Let $G \in \mathcal{G}$, and let U be an open subset of X such that $G \subset U$. Let $\varepsilon = d(G, X \setminus U)$. Since G and $X \setminus U$ are disjoint compacta, $\varepsilon > 0$. Let $\delta > 0$ be an Effros number for this ε (Theorem 4.2.31); without loss of generality, we assume that $\delta < \varepsilon$. Let $V = \mathcal{V}_\delta^d(G)$. Then V is an open subset of X contained in U. Let $G' \in \mathcal{G}$ such that $G' \cap V \neq \emptyset$. Let $y \in G' \cap V$, and let $x \in G$ such that $d(x,y) < \delta$. Since δ is an Effros number, there exists an ε–homeomorphism $h\colon X \twoheadrightarrow X$ such that $h(x) = y$. Since $\mathcal{H}(X)$ respects $\mathcal{G}$ and $h(G) \cap G' \neq \emptyset$, $h(G) = G'$. Hence, $G' \subset \mathcal{V}_\varepsilon^d(G)$ (h is an ε–homeomorphism), i.e., $G' \subset U$. Thus, $\mathcal{G}$ is upper semicontinuous.

Next, we show that $\mathcal{G}$ is lower semicontinuous. Let $G \in \mathcal{G}$, let $p, q \in G$ and let U be an open subset of X such that $p \in U$. Let $\varepsilon > 0$ such that $\mathcal{V}_\varepsilon^d(p) \subset U$. Let δ be an Effros number for this ε, and let $V = \mathcal{V}_\delta^d(q)$. Take $G' \in \mathcal{G}$ such that $G' \cap V \neq \emptyset$, and take $z \in G' \cap V$. Hence, $d(q, z) < \delta$. Since δ is an Effros number, there exists an ε–homeomorphism $k \colon X \twoheadrightarrow X$ such that $k(q) = z$. In particular, $d(p, k(p)) < \varepsilon$. Thus, $k(p) \in U$. Since $\mathcal{H}(X)$ respects $\mathcal{G}$, and $k(q) \in G'$, $k(G) = G'$. Hence, $G' \cap U \neq \emptyset$. Thus, $\mathcal{G}$ is lower semicontinuous.

Therefore, $\mathcal{G}$ is continuous.

Observe that since $\mathcal{H}(X)$ respects $\mathcal{G}$, all the elements of $\mathcal{G}$ are homeomorphic. Let $G \in \mathcal{G}$, and let $x, y \in G$. Since X is homogeneous, there exists a homeomorphism $\ell \colon X \twoheadrightarrow X$ such that $\ell(x) = y$. Note that, since $\mathcal{H}(X)$ respects $\mathcal{G}$, $\ell(G) = G$. Hence, $\ell|_G \colon G \twoheadrightarrow G$ is a homeomorphism sending x to y. Therefore, G is homogeneous.

Finally, we prove that the quotient space $X/\mathcal{G}$ is a homogeneous continuum. By Theorem 1.7.3, $X/\mathcal{G}$ is a continuum. Let $q \colon X \twoheadrightarrow X/\mathcal{G}$ be the quotient map.

Let $\chi_1, \chi_2 \in X/\mathcal{G}$ be two points. Let $x_1, x_2 \in X$ be such that $q(x_1) = \chi_1$ and $q(x_2) = \chi_2$. Since X is homogeneous, there exists a homeomorphism $h \colon X \twoheadrightarrow X$ such that $h(x_1) = x_2$. Since $\mathcal{H}(X)$ respects $\mathcal{G}$, $x_1 \in q^{-1}(\chi_1)$, $x_2 \in q^{-1}(\chi_2)$, and $h(x_1) = x_2$, we have that $h(q^{-1}(\chi_1)) = q^{-1}(\chi_2)$.

Define $f \colon X/\mathcal{G} \twoheadrightarrow X/\mathcal{G}$ by

$$f(\chi) = q \circ h\left(q^{-1}(\chi)\right).$$

Then f is well defined (because $\mathcal{H}(X)$ respects $\mathcal{G}$) and $f(\chi_1) = \chi_2$. Since $\mathcal{G}$ is a continuous decomposition, q is an open map (Theorem 1.2.23). Hence, since q is open and h is continuous, f is continuous. Note that $f^{-1} \colon X/\mathcal{G} \twoheadrightarrow X/\mathcal{G}$ is given by $f^{-1}(\chi) = q \circ h^{-1}\left(q^{-1}(\chi)\right)$ and it is continuous also. Thus, f is a homeomorphism. Therefore, $X/\mathcal{G}$ is homogeneous.

Q.E.D.

5.1.5. Definition. A surjective map $f \colon X \twoheadrightarrow Y$ between continua is *completely regular* if for each $\varepsilon > 0$ and each point $y \in Y$, there exists an open set V in Y containing y such that if $y' \in V$, then there exists a homeomorphism $h \colon f^{-1}(y) \twoheadrightarrow f^{-1}(y')$ such that for each $x \in f^{-1}(y)$, $d(x, h(x)) < \varepsilon$.

5.1.6. Remark. Note that the fibres of completely regular maps are all homeomorphic.

5.1.7. Proposition. *Let X and Y be continua. If $g\colon X \twoheadrightarrow Y$ is a completely regular map, then g is open.*

Proof. Let U be an open subset of X. Let $y \in g(U)$. Then there exists $x \in U$ such that $g(x) = y$. Let $\varepsilon > 0$ such that $\mathcal{V}_\varepsilon^d(x) \subset U$. Since g is completely regular, there is an open subset V of Y such that $y \in V$ and if $y' \in V$, then there exists a homeomorphism $h\colon g^{-1}(y) \twoheadrightarrow g^{-1}(y')$ such that $d(z, h(z)) < \varepsilon$ for all $z \in g^{-1}(y)$. In particular, $d(x, h(x)) < \varepsilon$. Thus, $h(x) \in U$. Thus, $V \subset g(U)$. Since y was an arbitrary point of $g(U)$, $g(U)$ is open. Therefore, g is an open map.

Q.E.D.

5.1.8. Theorem. *Let X be a homogeneous continuum, with metric d, and let $\mathcal{G}$ be a decomposition of X whose elements are proper nondegenerate subcontinua of X. If the homeomorphism group, $\mathcal{H}(X)$, of X respects $\mathcal{G}$, then the elements of $\mathcal{G}$ are nowhere dense, and the quotient map $q\colon X \twoheadrightarrow X/\mathcal{G}$ is completely regular.*

Proof. Let $G \in \mathcal{G}$ and suppose $Int(G) \neq \emptyset$. Let $g' \in Int(G)$, and let $\varepsilon > 0$ such that $\mathcal{V}_\varepsilon^d(g') \subset G$. Let $\delta > 0$ be an Effros number for this ε (Theorem 4.2.31). We assume that $\delta < \varepsilon$. Let $x \in X \setminus G$ such that $d(x, G) < \delta$. Since G is compact, there exists $g \in G$ such that $d(x, g) = d(x, G)$. Now, since $d(x, g) < \delta$, there exists an ε–homeomorphism $h\colon X \twoheadrightarrow X$ such that $h(g) = x$. Hence, since $h(g') \in \mathcal{V}_\varepsilon^d(g') \subset G$ and $\mathcal{H}(X)$ respects $\mathcal{G}$, $h(G) = G$. Thus, $x \in h(G) \setminus G$, a contradiction. Therefore, $Int(G) = \emptyset$.

To show the quotient map $q\colon X \twoheadrightarrow X/\mathcal{G}$ is completely regular, let $\varepsilon > 0$, and let $\delta > 0$ be an Effros number for this ε (Theorem 4.2.31). Since $\mathcal{H}(X)$ respects $\mathcal{G}$, by Theorem 5.1.4 (1), $\mathcal{G}$ is a continuous decomposition. Thus, by (Theorem 1.2.23), q is open.

Let $\chi \in X/\mathcal{G}$, and let $V = q\left(\mathcal{V}_\delta^d(x)\right)$, where $x \in q^{-1}(\chi)$. Then V is an open subset of $X/\mathcal{G}$ containing χ. Let $\chi' \in V$, and let $x' \in q^{-1}(\chi') \cap \mathcal{V}_\delta^d(x)$. Then since $d(x, x') < \delta$, there exists an ε–homeomorphism $h\colon X \twoheadrightarrow X$ such that $h(x) = x'$. Since

$\mathcal{H}(X)$ respects $\mathcal{G}$ and $h(q^{-1}(\chi)) \cap q^{-1}(\chi') \neq \emptyset$, $h(q^{-1}(\chi)) = q^{-1}(\chi')$. Thus, $h|_{q^{-1}(\chi)}\colon q^{-1}(\chi) \twoheadrightarrow q^{-1}(\chi')$ is a homeomorphism such that $d(z, h(z)) < \varepsilon$ for each $z \in q^{-1}(\chi)$. Therefore, q is completely regular.

Q.E.D.

5.1.9. Definition. Let X be a continuum. A subcontinuum Z of X is said to be *terminal* if each subcontinuum Y of X that intersects Z satisfies either $Y \subset Z$ or $Z \subset Y$. A decomposition $\mathcal{G}$ of X such that the elements of $\mathcal{G}$ are continua is said to be *terminal* if each element of $\mathcal{G}$ is a terminal subcontinuum of X.

5.1.10. Remark. It is easy to see that in the topologist sine curve X (Example 2.4.5), $\{0\} \times [-1, 1]$ is a terminal subcontinuum of X.

5.1.11. Lemma. *Let X and Y be continua. If $f\colon X \twoheadrightarrow Y$ is a monotone surjective map and W is a terminal subcontinuum of X, then $f(W)$ is a terminal subcontinuum of Y.*

Proof. Let W be a terminal subcontinuum of X. Let K be a subcontinuum of Y such that $K \cap f(W) \neq \emptyset$. Since f is a monotone map, $f^{-1}(K)$ is a subcontinuum of X (Lemma 2.1.12) such that $f^{-1}(K) \cap W \neq \emptyset$. Hence, since W is a terminal subcontinuum of X, either $W \subset f^{-1}(K)$ or $f^{-1}(K) \subset W$. This implies that either $f(W) \subset K$ or $K \subset f(W)$ (f is surjective). Therefore, $f(W)$ is a terminal subcontinuum of Y.

Q.E.D.

5.1.12. Corollary. *Let X be a continuum. If Z is a terminal subcontinuum of X and $h\colon X \twoheadrightarrow X$ is a homeomorphism, then $h(Z)$ is a terminal subcontinuum of X.*

5.1.13. Lemma. *Let X be a continuum. If X is aposyndetic, then X does not contain nondegenerate proper terminal subcontinua.*

Proof. Suppose Y is a nondegenerate proper terminal subcontinuum of X. Let $y \in Y$. Let $y' \in Y \setminus \{y\}$. Then since X is aposyndetic, there exists a subcontinuum W of X such that $y \in Int(W) \subset W \subset X \setminus \{y'\}$. Since Y is terminal and $Y \setminus W \neq \emptyset$, $W \subset Y$. Thus, y is an interior point of Y. Since y was an arbitrary point of Y, all the points of Y are interior points. Hence, Y is a nonempty open and closed proper subset of X. This contradicts the fact that X is connected. Therefore, X does not contain nondegenerate proper terminal subcontinua.

Q.E.D.

5.1.14. Theorem. *Let X be a continuum. If A is a terminal subcontinuum of X, if B is a subcontinuum of X disjoint from A, and if $f \colon A \to Y$ is a map from A into the absolute neighborhood retract Y, then there exists a map $F \colon X \to Y$ such that $F|_A = f$ and $F|_B$ is homotopic to a constant map.*

Proof. Since Y is an absolute neighborhood retract, there exist an open subset U of X containing A and a map $g \colon U \to Y$ such that $g|_A = f$. Without loss of generality, we assume that $U \cap B = \emptyset$. By Theorem 7.1 (p. 96) of [13], Y is locally contractible. Hence, there exists an open neighborhood V of a point a of A such that $V \subset U$ and $g|_V$ is homotopic to a constant map.

Note that $A \setminus V$ and $X \setminus U$ are two closed subsets of $X \setminus V$ such that no connected subset of $X \setminus V$ intersects both $A \setminus V$ and $X \setminus U$. (If K is a connected subset of $X \setminus V$ such that $K \cap (A \setminus V) \neq \emptyset$ and $K \cap (X \setminus U) \neq \emptyset$, then $Cl(K)$ is a continuum in $X \setminus V$ intersecting A and $X \setminus A$. Since A is terminal, $A \subset Cl(K)$. This implies that $V \cap Cl(K) \neq \emptyset$, a contradiction.) By Theorem 1.6.8, there exist two disjoint closed subsets X_1 and X_2 of X such that $X \setminus V = X_1 \cup X_2$, $A \setminus V \subset X_1$ and $X \setminus U \subset X_2$.

Note that $X_1 \cup A$ and X_2 are two disjoint closed subsets of X. Then, by Urysohn's Lemma, there exists a map $h \colon X \twoheadrightarrow [0, 1]$ such that $h(X_1 \cup A) = \{0\}$ and $h(X_2) = \{1\}$. Let $M = h^{-1}\left(\left[0, \frac{1}{2}\right]\right)$ and let $N = h^{-1}\left(\left[\frac{1}{2}, 1\right]\right)$. Then $X = M \cup N$, $A \subset M$, $X \setminus U \subset N$ and $M \cap N \subset V \setminus A$. Since $g|_{M \cap N}$ is homotopic to a constant map (because $M \cap N \subset V$ and $g|_V$ is homotopic to a constant map), by (15.A.1) of [8], there exists a map $k \colon N \to Y$ such that k is homotopic to a constant map and $k|_{M \cap N} = g|_{M \cap N}$.

Let $F \colon X \to Y$ be given by

$$F(x) = \begin{cases} g(x) & \text{if } x \in M, \\ k(x) & \text{if } x \in N. \end{cases}$$

Then F is well defined and continuous. Note that $F|_A = g|_A = f$ and $F|_B = k|_B$. Since $k|_B$ is homotopic to a constant map (because $B \subset X \setminus U \subset N$ and k is homotopic to a constant map), F is the desired extension of f.

Q.E.D.

5.1.15. Definition. A continuum X is *cell–like* if each map of X into a compact absolute neighborhood retract is homotopic to a constant map.

5.1.16. Definition. Let X and Y be a continua. We say that a surjective map $f \colon X \twoheadrightarrow Y$ is *cell–like* if for each $y \in Y$, $f^{-1}(y)$ is a cell–like continuum.

5.1.17. Theorem. *Let X and Z be nondegenerate continua. Suppose that $g \colon X \twoheadrightarrow Z$ is a monotone and completely regular map. If z_1 is a point of Z such that $g^{-1}(z_1)$ is terminal subcontinuum of X, then g is a cell–like map.*

Proof. Let Y be a compact absolute neighborhood retract, and let

$$f \colon g^{-1}(z_1) \to Y$$

be a map. Let $z_2 \in Z \setminus \{z_1\}$, and let $F \colon X \to Y$ be an extension of f such that $F|_{g^{-1}(z_2)}$ is homotopic to a constant map (Theorem 5.1.14).

Since Y is a compact absolute neighborhood retract, by Theorem 1.1 (p. 111) of [13], there exists $\varepsilon > 0$ such that for any metric space W and for any two maps $k, k' \colon W \to Y$ such that $d(k(w), k'(w)) < \varepsilon$ for every $w \in W$, k and k' are homotopic.

Since F is uniformly continuous, for this ε, there exists $\delta > 0$ such that if $d(x, x') < \delta$, then $d(F(x), F(x')) < \varepsilon$.

Let $Z' = \{z \in Z \mid F|_{g^{-1}(z)}$ is homotopic to a constant map$\}$. We show that Z' is open in Z. Let $z \in Z'$. Since g is completely regular, there exists an open subset V of Z such that $z \in V$ and if $z' \in V$, then there is a homeomorphism $h \colon g^{-1}(z) \twoheadrightarrow g^{-1}(z')$ with $d(x, h(x)) < \delta$ for every $x \in g^{-1}(z)$. Let $z' \in V$, and let h be the homeomorphism guaranteed by the complete regularity of g. Hence, $d\big(F|_{g^{-1}(z')}(x), (F|_{g^{-1}(z)}) \circ h^{-1}(x)\big) < \varepsilon$ for every $x \in g^{-1}(z')$. Thus, $F|_{g^{-1}(z')}$ is homotopic to $(F|_{g^{-1}(z)}) \circ h^{-1}$. Since $z \in Z'$, $(F|_{g^{-1}(z)}) \circ h^{-1}$ is homotopic to a constant map. Thus, $F|_{g^{-1}(z')}$ is homotopic to a constant map. Therefore, $z' \in Z'$. A similar argument shows that Z' is closed in Z.

Since Z' is a nonempty open and closed subset of Z and Z is connected, $Z' = Z$. This implies that $F|_{g^{-1}(z_1)} = f$ is homotopic to a constant map. Thus, each map from $g^{-1}(z_1)$ into a compact absolute neighborhood retract is homotopic to a constant map. Hence, $g^{-1}(z_1)$ is cell–like. Since fibres of completely regular maps defined on continua are homeomorphic, we have that g is a cell–like map.

Q.E.D.

A proof of the following result may be found in 2.1 of [15]. The Theorem was originally proved by E. Dyer.

5.1.18. Theorem. *Let X and Y be nondegenerate continua. If $f \colon X \twoheadrightarrow Y$ is a surjective, monotone and open map, then there exists a dense G_δ subset W of Y having the following property: for each $y \in W$, for each subcontinuum B of $f^{-1}(y)$, for each $x \in Int_{f^{-1}(y)}(B)$ and for each neighborhood U of B in X, there exist a subcontinuum Z of X containing B and a neighborhood V of y in Y such that $x \in Int_X(Z)$, $(f|_Z)^{-1}(V) \subset U$ and $f|Z \colon Z \to Y$ is a monotone surjective map.*

Now, we are ready to prove Jones's Aposyndetic Decomposition Theorem.

5.1.19. Theorem. *Let X be a decomposable homogeneous continuum, with metric d, which is not aposyndetic. If $\mathcal{G} = \{\mathcal{T}(\{x\}) \mid x \in X\}$, then the following hold:*

(1) $\mathcal{G}$ is a continuous, monotone and terminal decomposition of X.

(2) The elements of $\mathcal{G}$ are indecomposable, cell–like, homogeneous and mutually homeomorphic continua of the same dimension as X.

(3) The quotient map $q\colon X \twoheadrightarrow X/\mathcal{G}$ is completely regular.

(4) The quotient space $X/\mathcal{G}$ is a one–dimensional aposyndetic homogeneous continuum, which does not contain nondegenerate proper terminal subcontinua.

Proof. Since X is decomposable, there exist two points of X such that X is aposyndetic at one of them with respect to the other. Hence, the elements of $\mathcal{G}$ are nondegenerate proper subcontinua (Theorem 3.1.21) of X. By Theorem 5.1.2, $\mathcal{G}$ is a decomposition of X. By Lemma 5.1.1, the homeomorphism group of X respects $\mathcal{G}$. Hence, by Theorem 5.1.4, $\mathcal{G}$ is a continuous decomposition, the elements of $\mathcal{G}$ are mutually homeomorphic homogeneous continua and the quotient space $X/\mathcal{G}$ is a homogeneous continuum. A proof of the fact that the elements of $\mathcal{G}$ have the same dimension as X may be found in Corollary 9 of [32].

Now, we show that all the elements of $\mathcal{G}$ are terminal subcontinua. To this end, suppose there exists a point $x \in X$ such that $\mathcal{T}(\{x\})$ is not terminal. Hence, there exists a subcontinuum Y of X such that $Y \cap \mathcal{T}(\{x\}) \neq \emptyset$, $Y \setminus \mathcal{T}(\{x\}) \neq \emptyset$ and $\mathcal{T}(\{x\}) \setminus Y \neq \emptyset$. Without loss of generality, we assume that $x \in \mathcal{T}(\{x\}) \setminus Y$. Let $p \in Y \setminus \mathcal{T}(\{x\})$, and let $y \in Y \cap \mathcal{T}(\{x\})$.

Since $p \in Y \setminus \mathcal{T}(\{x\})$, there exists a subcontinuum W of X such that $p \in Int(W) \subset W \subset X \setminus \{x\}$. Let $K = Y \cup W$. Then K is a subcontinuum of X, $p \in Int(K)$ and $K \subset X \setminus \{x\}$. Let $\varepsilon > 0$ such that $\mathcal{V}_\varepsilon^d(x) \subset K$, $\mathcal{V}_{2\varepsilon}^d(K) \subset X \setminus \{x\}$ and $\varepsilon < d(x,y)$. Let $\delta > 0$ be an Effros number for this ε (Theorem 4.2.31). Hence, for each $y' \in \mathcal{V}_\delta^d(y)$, there exists an ε–homeomorphism $h_{y'}\colon X \twoheadrightarrow X$ such that $h_{y'}(y) = y'$ (for $y' = y$, we take $h_y = 1_X$). Then

$$M = Cl \left(\bigcup_{y' \in \mathcal{V}_\delta(y)} h_{y'}(K) \right)$$

is a subcontinuum of X such that $\mathcal{V}_\delta^d(y) \subset M \subset X \setminus \{x\}$. This contradicts the choice of y. Therefore, all the elements of $\mathcal{G}$ are terminal. Note that, this implies that all the elements of $\mathcal{G}$ are cell–like (Theorem 5.1.17) since the quotient map is completely regular (Theorem 5.1.8).

Next, we prove that all the elements of $\mathcal{G}$ are indecomposable. Suppose this is not true. Then there exists a point $x' \in X$ such that $\mathcal{T}(\{x'\})$ is decomposable. Hence, since X is homogeneous, $\mathcal{T}(\{x\})$ is decomposable for each $x \in X$ (Lemma 5.1.1). Let Υ be the G_δ dense subset of $X/\mathcal{G}$ guaranteed by Theorem 5.1.18. Let $x_0 \in X$ such that $q(x_0) \in \Upsilon$. Let B be a subcontinuum of $\mathcal{T}(\{x_0\})$ such that $Int_{\mathcal{T}(\{x_0\})}(B) \neq \emptyset$. Let $x \in Int_{\mathcal{T}(\{x_0\})}(B)$, and let $n \in \mathbb{N}$ such that $\mathcal{T}(\{x_0\}) \setminus \mathcal{V}_{\frac{1}{n}}^d(x) \neq \emptyset$. By Theorem 5.1.18, there exist a subcontinuum Z of X and a neighborhood Ω of $q(x_0)$ in $X/\mathcal{G}$ such that $x \in Int_X(Z)$, $B \subset Z$, $(q|_Z)^{-1}(\Omega) \subset \mathcal{V}_{\frac{1}{n}}^d(x)$ and $q|_Z \colon Z \twoheadrightarrow X/\mathcal{G}$ is a monotone surjective map. Since $\mathcal{T}(\{x_0\}) = q^{-1}(q(x_0))$ and $\mathcal{T}(\{x_0\}) \setminus \mathcal{V}_{\frac{1}{n}}^d(x) \neq \emptyset$, we obtain a contradiction. Hence, all proper subcontinua of $\mathcal{T}(\{x_0\})$ have empty interior. Thus, $\mathcal{T}(\{x_0\})$ is indecomposable (Corollary 1.7.21). Therefore, $\mathcal{T}(\{x\})$ is an indecomposable continuum for each $x \in X$.

To finish the proof, we show that $X/\mathcal{G}$ is aposyndetic. Let $\chi_1, \chi_2 \in X/\mathcal{G}$. We see that $X/\mathcal{G}$ is aposyndetic at χ_1 with respect to χ_2. Let $x_1 \in q^{-1}(\chi_1)$ and let $x_2 \in q^{-1}(\chi_2)$. Note that X is aposyndetic at x_1 with respect to x_2. Then there exists a subcontinuum W of X such that $x_1 \subset Int(W) \subset W \subset X \setminus \{x_2\}$. Since $\mathcal{G}$ is a terminal decomposition, $\mathcal{T}(\{x_1\}) \subset W$ and $W \cap \mathcal{T}(\{x_2\}) = \emptyset$ ($\mathcal{T}(\{x\})$ is nowhere dense for every $x \in X$ (Theorem 5.1.8)). Since $\mathcal{G}$ is a continuous decomposition, by Theorem 1.2.23, q is an open map. Then

$$\chi_1 = q(x_1) \in Int(q(W)) \subset q(W) \subset X/\mathcal{G} \setminus \{\chi_2\}.$$

Therefore, $X/\mathcal{G}$ is aposyndetic. The fact that $X/\mathcal{G}$ is one–dimensional may be found in Theorem 3 of [35].

The fact that $X/\mathcal{G}$ does not contain nondegenerate proper terminal subcontinua follows from Lemma 5.1.13.

Q.E.D.

As a consequence of Jones's Theorem we have the following three Corollaries:

5.1.20. Corollary. *Let X be a decomposable homogeneous continuum. If $x \in X$, then $\mathcal{T}(\{x\})$ is the maximal terminal proper subcontinuum of X containing x.*

Proof. If X is aposyndetic, then $\mathcal{T}(\{x\}) = \{x\}$ (Theorem 3.1.28). Since aposyndetic continua do not contain nondegenerate proper terminal subcontinua (Lemma 5.1.13), $\mathcal{T}(\{x\})$ is the maximal terminal proper subcontinuum of X containing x.

Suppose X is not aposyndetic, and let $x \in X$. Then, by Theorem 5.1.19 (1), $\mathcal{T}(\{x\})$ is a terminal subcontinuum of X. Suppose K is a terminal proper subcontinuum of X such that $\mathcal{T}(\{x\}) \subsetneq K$. By Lemma 5.1.11, $q(K)$ is a terminal subcontinuum of $X/\mathcal{G}$. Since $X/\mathcal{G}$ does not contain proper nondegenerate terminal subcontinua (Theorem 5.1.19 (4)), $q(K) = \{q(x)\}$. Hence, $K = q^{-1}(q(x)) = \mathcal{T}(\{x\})$, a contradiction. Therefore, $\mathcal{T}(\{x\})$ is the maximal terminal proper subcontinuum of X containing x.

Q.E.D.

5.1.21. Corollary. *If X is a hereditarily decomposable and homogeneous continuum, then X is aposyndetic.*

5.1.22. Corollary. *If X is an arcwise connected homogeneous continuum, then X is aposyndetic.*

5.1.23. Theorem. *Let X be a decomposable nonaposyndetic homogeneous continuum. If $\mathcal{G} = \{\mathcal{T}_X(\{x\}) \mid x \in X\}$, then $\mathcal{T}_X(Z) = q^{-1}\mathcal{T}_{X/\mathcal{G}}q(Z)$ for any nonempty closed subset Z of X, where $q \colon X \twoheadrightarrow X/\mathcal{G}$ is the quotient map.*

Proof. Let Z be a nonempty closed subset of X. We divide the proof in six steps.

Step 1. $q^{-1}q(Z) \subset \mathcal{T}_X(Z)$.

Let $x \in q^{-1}q(Z)$. Then $q(x) \in q(Z)$. Thus, there exists $z \in Z$ such that $q(z) = q(x)$. This implies that $\mathcal{T}_X(\{z\}) = \mathcal{T}_X(\{x\})$. Hence, since $\mathcal{T}_X$ is idempotent (Theorem 4.2.32), $\mathcal{T}_X(\{x\}) = \mathcal{T}_X(\{z\}) \subset \mathcal{T}_X(Z)$. Therefore, $x \in \mathcal{T}_X(Z)$, and $q^{-1}q(Z) \subset \mathcal{T}_X(Z)$.

Step 2. $\mathcal{T}_X(Z) = q^{-1}q\mathcal{T}_X(Z)$.

Clearly, $\mathcal{T}_X(Z) \subset q^{-1}q\mathcal{T}_X(Z)$. Let $x \in q^{-1}q\mathcal{T}_X(Z)$. Then $q(x) \in q\mathcal{T}_X(Z)$. Hence, there exists $y \in \mathcal{T}_X(Z)$ such that $q(y) = q(x)$. This implies that $\mathcal{T}_X(\{y\}) = \mathcal{T}_X(\{x\})$. Thus, since $\mathcal{T}_X$ is idempotent (Theorem 4.2.32), we have that $\mathcal{T}_X(\{y\}) \subset \mathcal{T}_X(Z)$. Hence, $x \in \mathcal{T}_X(Z)$, and $q^{-1}q\mathcal{T}_X(Z) \subset \mathcal{T}_X(Z)$. Therefore, $\mathcal{T}_X(Z) = q^{-1}q\mathcal{T}_X(Z)$.

Step 3. *If Z is connected and $Int_X(Z) \neq \emptyset$, then $Z = q^{-1}q(Z)$.*

Clearly, $Z \subset q^{-1}q(Z)$. Let $x \in q^{-1}q(Z)$. Then $q(x) \in q(Z)$. Thus, there exists $z \in Z$ such that $q(z) = q(x)$. This implies that $\mathcal{T}_X(\{z\}) = \mathcal{T}_X(\{x\})$. Since $\mathcal{T}_X(\{x\})$ is a nowhere dense terminal subcontinuum of X (Lemma 5.1.1, Theorem 5.1.8 and Theorem 3.1.21), $\mathcal{T}_X(\{x\}) \cap Z \neq \emptyset$ and $Int_X(Z) \neq \emptyset$, we have that $\mathcal{T}_X(\{x\}) \subset Z$. In particular, $x \in Z$. Therefore, $Z = q^{-1}q(Z)$.

Step 4. $q\mathcal{T}_X(Z) \subset \mathcal{T}_{X/\mathcal{G}}q(Z)$.

Let $\chi \in X/\mathcal{G} \setminus \mathcal{T}_{X/\mathcal{G}}q(Z)$. Then there exists a subcontinuum $\mathcal{W}$ of $X/\mathcal{G}$ such that $\chi \in Int_{X/\mathcal{G}}(\mathcal{W}) \subset \mathcal{W} \subset X/\mathcal{G} \setminus q(Z)$. From these inclusions we obtain that $q^{-1}(\chi) \subset Int_X(q^{-1}(\mathcal{W})) \subset q^{-1}(\mathcal{W}) \subset X \setminus q^{-1}q(Z) \subset X \setminus Z$. Hence, since q is monotone (Theorem 5.1.19), $q^{-1}(\chi) \cap \mathcal{T}_X(Z) = \emptyset$. Thus, $qq^{-1}(\chi) \cap q\mathcal{T}_X(Z) = \emptyset$. Therefore, $\chi \in X/\mathcal{G} \setminus q\mathcal{T}_X(Z)$, and $q\mathcal{T}_X(Z) \subset \mathcal{T}_{X/\mathcal{G}}q(Z)$.

Step 5. $\mathcal{T}_{X/\mathcal{G}}q(Z) \subset q\mathcal{T}_X(Z)$.

Let $\chi \in X/\mathcal{G} \setminus q\mathcal{T}_X(Z)$. Then $\{\chi\} \cap q\mathcal{T}_X(Z) = \emptyset$. This implies that $q^{-1}(\chi) \cap q^{-1}q\mathcal{T}_X(Z) = \emptyset$. Hence, by Step 2, $q^{-1}(\chi) \cap \mathcal{T}_X(Z) = \emptyset$. Since $\mathcal{T}_X$ is idempotent (Theorem 4.2.32), $q^{-1}(\chi) \cap \mathcal{T}_X^2(Z) = \emptyset$. Thus, there exists a subcontinuum W of X such that $q^{-1}(\chi) \subset Int_X(W) \subset W \subset X \setminus \mathcal{T}_X(Z) \subset X \setminus Z$. From these inclusions, since q is an open map (Theorem 5.1.19 and Proposition 5.1.7) we obtain that $\{\chi\} = qq^{-1}(\chi) \subset Int_{X/\mathcal{G}}(q(W)) \subset q(W) \subset q(X \setminus Z)$. To finish, we need to show that $q(W) \cap q(Z) = \emptyset$. Suppose there exists $\chi' \in q(W) \cap q(Z)$. Then, by Steps 3 and 1, $q^{-1}(\chi') \subset q^{-1}q(W) \cap$

$q^{-1}q(Z) = W \cap q^{-1}q(Z) \subset W \cap \mathcal{T}_X(Z)$, a contradiction to the election of W. Hence, $q(W) \cap q(Z) = \emptyset$, and $\chi \in X/\mathcal{G} \setminus \mathcal{T}_{X/\mathcal{G}}q(Z)$. Therefore, $\mathcal{T}_{X/\mathcal{G}}q(Z) \subset q\mathcal{T}_X(Z)$.

Step 6. $q\mathcal{T}_X(Z) = \mathcal{T}_{X/\mathcal{G}}q(Z)$.

The equality follows from Steps 4 and 5.

From Steps 2 and 6, we have that

$$\mathcal{T}_X(Z) = q^{-1}q\mathcal{T}_X(Z) = q^{-1}\mathcal{T}_{X/\mathcal{G}}q(Z).$$

Therefore, $\mathcal{T}_X(Z) = q^{-1}\mathcal{T}_{X/\mathcal{G}}q(Z)$.

Q.E.D.

5.1.24. Corollary. *Let X be a decomposable nonaposyndetic homogeneous continuum and let $\mathcal{G} = \{\mathcal{T}_X(\{x\}) \mid x \in X\}$. Then $\mathcal{T}_X$ is continuous if and only if $X/\mathcal{G}$ is locally connected.*

Proof. Let $q\colon X \twoheadrightarrow X/\mathcal{G}$ be the quotient map. Suppose $\mathcal{T}_X$ is continuous for X. Note that, by Step 4 of Theorem 5.1.23, q is $\mathcal{T}_{XX/\mathcal{G}}-$continuous (Definition 3.2.4). Thus, $\mathcal{T}_{X/\mathcal{G}}$ is continuous for $X/\mathcal{G}$ (Theorem 3.2.5). Since $X/\mathcal{G}$ is an aposyndetic continuum (Theorem 5.1.19) for which $\mathcal{T}_{X/\mathcal{G}}$ is continuous, $X/\mathcal{G}$ is locally connected (Corollary 3.2.16).

Next, suppose $X/\mathcal{G}$ is locally connected. Recall that $q\colon X \twoheadrightarrow X/\mathcal{G}$ is a monotone open map (Theorem 5.1.19). Let W be a proper subcontinuum of X, and let $x \in X \setminus W$. Since $\mathcal{T}_X(\{x\})$ is a terminal subcontinuum of X (Theorem 5.1.19), either $W \subset \mathcal{T}_X(\{x\})$ or $\mathcal{T}_X(\{x\}) \cap W = \emptyset$. In either case, $q(W)$ is a proper subcontinuum of $X/\mathcal{G}$. Therefore, by Theorem 3.2.2, $\mathcal{T}_X$ is continuous.

Q.E.D.

5.1.25. Remark. Note that Corollary 5.1.24 gives a partial answer to Question 7.2.2.

As a consequence of Theorem 3.1.34, Corollary 3.2.16 and Corollary 5.1.24, we have the following characterization of homogeneous continua for which the set function $\mathcal{T}$ is continuous:

5.1.26. Theorem. *Let X be a homogeneous continuum. Then $\mathcal{T}$ is continuous for X if and only if one of the following conditions holds:*

(1) X is indecomposable.

(2) X is not aposyndetic and $X/\mathcal{G}$ is locally connected, where $\mathcal{G} = \{\mathcal{T}(\{x\}) \mid x \in X\}$.

(3) X is locally connected.

Next, we prove an alternate version of Theorem 5.1.19. The awkward collection of hypotheses in this Theorem is precisely the situation encountered in Theorem 5.3.23.

5.1.27. Theorem. *Let Y be a continuum such that it is the union of two disjoint, nonempty sets K and E with the following properties:*

(1) Y is aposyndetic at each point of E,

(2) Y is not aposyndetic at any point of K,

(3) Y is aposyndetic at each point of K with respect to each point of E,

(4) K has a metric d such that (K, d) has the property of Effros,

(5) Each homeomorphism h of (K, d) can be extended to a homeomorphism $\hat{h}$ of Y, by defining $\hat{h}(z) = z$ for each $z \in E$,

(6) K is connected and open, and

(7) $\dim(E) = 0$ (i.e., E is totally disconnected).

If $\mathcal{G} = \{\mathcal{T}(\{x\}) \mid x \in Y\}$, then $\mathcal{G}$ is a continuous, terminal decomposition of Y with the following properties:

(8) The degenerate elements of $\mathcal{G}$ are precisely the points of E,

(9) The nondegenerate elements of $\mathcal{G}$ are mutually homeomorphic, cell–like, homogeneous, indecomposable continua of the same dimension as Y,

(10) The quotient space $Y/\mathcal{G}$ is aposyndetic, and

(11) The quotient space $K/\mathcal{G}'$ is homogeneous, where $\mathcal{G}' = \{\mathcal{T}(\{x\}) \mid x \in K\}$.

Proof. Since Y is aposyndetic at every point with respect to at least one other point (by (1) and (3)), the elements of $\mathcal{G}$ are proper subcontinua of Y (Theorem 3.1.21). Moreover, if $x \in K$, then $\mathcal{T}(\{x\}) \subset K$ since Y is aposyndetic at each point of E, by (1). By (2), the elements of $\mathcal{G}$ that are contained in K are nondegenerate. Furthermore, if $x \in E$, then $\mathcal{T}(\{x\}) = \{x\}$, since Y is aposyndetic at every point of K with respect to x, by (3).

Note that by (4) and by Theorem 4.2.33, K is a homogeneous space. By (5) and Lemma 5.1.1, the elements of $\mathcal{G}$ contained in K are mutually homeomorphic and homogeneous.

Thus, the elements of $\mathcal{G}$ are the points of E and some nondegenerate subcontinua of K, i.e., $\mathcal{G} = \{\{x\} \mid x \in E\} \cup \{\mathcal{T}(\{x\}) \mid x \in K\}$.

The proof of the fact that $\mathcal{G}$ is a decomposition of Y is similar to the one given in Theorem 5.1.2 (using (5) to extend the homeomorphisms defined on K to homeomorphisms defined on Y).

Now, we see that $\mathcal{G}$ is continuous. First, we show that $\mathcal{G}$ is upper semicontinuous.

Let $x \in E$. Then $\mathcal{T}(\{x\}) = \{x\}$. Suppose $\mathcal{G}$ is not upper semicontinuous at $\{x\}$. Then there exists an open subset U of Y such that $\{x\} \subset U$ and for each $n \in \mathbb{N}$, there exists $x_n \in \mathcal{V}^d_{\frac{1}{n}}(x)$ such that $\mathcal{T}(\{x_n\}) \setminus U \neq \emptyset$. Note that $\lim_{n \to \infty} x_n = x$. Since the hyperspace of subcontinua of Y, $\mathcal{C}(Y)$, is compact (Theorem 1.8.5), without loss of generality, we assume that the sequence $\{\mathcal{T}(\{x_n\})\}_{n=1}^{\infty}$ converges to a subcontinuum Z of Y. Note that $Z \setminus U \neq \emptyset$. Let $z \in Z \setminus U$. Since $z \in Y \setminus \mathcal{T}(\{x\})$, there exists a subcontinuum W of Y such that $z \in Int(W) \subset W \subset Y \setminus \{x\}$. Since $\lim_{n \to \infty} \mathcal{T}(\{x_n\}) = Z$, for each $n \in \mathbb{N}$, there exists $z_n \in \mathcal{T}(\{x_n\})$ such that $\lim_{n \to \infty} z_n = z$. Hence, there exists $N' \in \mathbb{N}$ such that $z_n \in Int(W)$ for every $n \geq N'$. Since $x \in Y \setminus W$, and $\lim_{n \to \infty} x_n = x$, there exists $N'' \in \mathbb{N}$ such that $x_n \in Y \setminus W$ for every $n \geq N''$. Let $n = \max\{N', N''\}$. Then $z_n \in Int(W)$ and $x_n \in Y \setminus W$. This implies that $z_n \in Y \setminus \mathcal{T}(\{x_n\})$, a contradiction to the choice of z_n. Hence, $\mathcal{G}$ is upper semicontinuous at $\{x\}$.

Let $x \in K$, and let U be an open subset of Y such that $\mathcal{T}(\{x\}) \subset U$. Without loss of generality, we assume that $U \subset K$. By a similar argument to the one given in Theorem 5.1.4, we can find an open subset V of Y such that $\mathcal{T}(\{x\}) \subset V \subset U$ and if $x' \in Y$ such that $\mathcal{T}(\{x'\}) \cap V \neq \emptyset$, then $\mathcal{T}(\{x'\}) \subset U$. Hence, $\mathcal{G}$ is upper semicontinuous at $\mathcal{T}(\{x\})$.

Therefore, $\mathcal{G}$ is upper semicontinuous.

Next, we prove that $\mathcal{G}$ is lower semicontinuous. Let $x \in E$. Then, since $\mathcal{T}(\{x\}) = \{x\}$, clearly, $\mathcal{G}$ is lower semicontinuous at $\{x\}$.

Let $x \in K$. Take $p, q \in \mathcal{T}(\{x\})$, and let U be an open subset of Y such that $p \in U$. Without loss of generality, we assume that $U \subset K$. By a similar argument to the one given in Theorem 5.1.4, we can find an open subset V of Y such that $q \in V$ and if $x' \in Y$ such that $\mathcal{T}(\{x'\}) \cap V \neq \emptyset$, then $\mathcal{T}(\{x'\}) \cap U \neq \emptyset$. Hence, $\mathcal{G}$ is lower semicontinuous at $\mathcal{T}(\{x\})$.

Therefore, $\mathcal{G}$ is a continuous decomposition.

Each degenerate element of $\mathcal{G}$ is obviously a terminal subcontinuum of Y. A proof of the fact that the nondegenerate elements of $\mathcal{G}$ are terminal subcontinua is similar to the one given in Theorem 5.1.19 (only the property of Effros is used).

The proof of the fact that the nondegenerate elements of $\mathcal{G}$ are indecomposable is similar to the proof given in Theorem 5.1.19.

By Theorem 1.7.3, $Y/\mathcal{G}$ is a continuum. Let $q \colon Y \twoheadrightarrow Y/\mathcal{G}$ be the quotient map. Then q is a monotone map. Since K has the property of Effros, $q|_K \colon K \twoheadrightarrow K/\mathcal{G}'$ is completely regular. Since the nondegenerate elements of $\mathcal{G}$ are terminal, by Theorem 5.1.17, each such element of $\mathcal{G}$ is cell–like. A proof of the fact that the elements of $\mathcal{G}$ have the same dimension as Y may be found in Corollary 9 of [32].

Since the elements of $\mathcal{G}$ are terminal subcontinua, the proof of the fact that $Y/\mathcal{G}$ is aposyndetic is similar to the one given in Theorem 5.1.19.

Since $\mathcal{G}$ is a continuous decomposition, the proof of the fact that $K/\mathcal{G}'$ is homogeneous is similar to the one given in Theorem 5.1.4.

Q.E.D.

5.2 Detour to Covering Spaces

We present the necessary definitions and results of the theory of covering spaces to state and prove Rogers's Terminal Decomposition Theorem (Theorem 5.3.28).

5.2.1. Definition. Let X and Y be metric spaces. A map $f\colon X \to Y$ is called a *local homeomorphism* provided that for each point $x \in X$, there exists an open subset U of X such that $x \in U$, $f(U)$ is an open subset of Y and $f|_U\colon U \twoheadrightarrow f(U)$ is a homeomorphism.

5.2.2. Proposition. *Let X and Y be metric spaces. If $f\colon X \twoheadrightarrow Y$ is a surjective local homeomorphism, then $f^{-1}(y)$ is a discrete subset of X for every $y \in Y$.*

Proof. Let $y \in Y$. Then, by definition, for each point $x \in f^{-1}(y)$ there exists an open subset U of X such that $x \in U$ and $f|_U\colon U \twoheadrightarrow f(U)$ is a homeomorphism. Then $U \cap f^{-1}(y) = \{x\}$. Hence, each point of $f^{-1}(y)$ is an isolated point in $f^{-1}(y)$. Therefore, $f^{-1}(y)$ is a discrete subset of X.

Q.E.D.

5.2.3. Corollary. *Let X and Y be metric spaces. If X is compact and $f\colon X \twoheadrightarrow Y$ is a surjective local homeomorphism, then $f^{-1}(y)$ is finite for every $y \in Y$.*

5.2.4. Definition. Let X, Y and Z be metric spaces. If $f\colon X \to Y$ and $g\colon Z \to Y$ are maps, then a *lifting of g relative to f* is a map $\tilde{g}\colon Z \to X$ such that $f \circ \tilde{g} = g$.

5.2.5. Proposition. *Let X, Y and Z be metric spaces. Let $f\colon X \to Y$ be a local homeomorphism. If Z is connected and $g\colon Z \to Y$ is a map, then two liftings $\tilde{g}, \hat{g}\colon Z \to X$ of g relative to f such that $\tilde{g}(z_0) = \hat{g}(z_0)$, for some point $z_0 \in Z$, are equal.*

Proof. Let $A = \{z \in Z \mid \tilde{g}(z) = \hat{g}(z)\}$. Then, since $\tilde{g}(z_0) = \hat{g}(z_0)$, $A \neq \emptyset$. Since X is a metric space, A is a closed subset of Z. To conclude that $\tilde{g} = \hat{g}$, it is enough to show that A is an open subset of Z (because Z is connected). Let $z \in A$. Then $\tilde{g}(z) = \hat{g}(z)$. Since f is a local homeomorphism, there exists an open subset U of X such that $\tilde{g}(z) = \hat{g}(z) \in U$ and $f|_U \colon U \twoheadrightarrow f(U)$ is a homeomorphism. Since $\tilde{g}$ and $\hat{g}$ are continuous, there exists an open subset V of Z such that $z \in V$ and $\tilde{g}(V) \cup \hat{g}(V) \subset U$. Then for each $z' \in V$, $(f \circ \tilde{g})(z') = g(z') = (f \circ \hat{g})(z')$. Hence, for each $z' \in V$ $\tilde{g}(z') = \hat{g}(z')$, since $f|_U$ is a homeomorphism. Thus, $V \subset A$. Therefore, $\tilde{g} = \hat{g}$.

Q.E.D.

5.2.6. Definition. Let X be a metric space. A *covering space* of X is a pair consisting of a metric space $\widetilde{X}$ and a map $\sigma \colon \widetilde{X} \twoheadrightarrow X$ such that the following condition holds: for each point $x \in X$, there exists an open subset U of X such that $x \in U$, $\sigma^{-1}(U) = \bigcup_{\lambda \in \Lambda} V_\lambda$,

where $\{V_\lambda\}_{\lambda \in \Lambda}$ is a family of pairwise disjoint open subsets of $\widetilde{X}$, and $\sigma|_{V_\lambda} \colon V_\lambda \twoheadrightarrow U$ is a homeomorphism for every $\lambda \in \Lambda$. Any such open subset U of X is called an *evenly covered* subset of X. The map σ is called a *covering map*.

5.2.7. Remark. Observe that, by definition, every covering map is a local homeomorphism. The converse is not necessarily true as can be easily seen using the interval $(0, 1)$ and the map $f \colon (0, 1) \twoheadrightarrow \mathcal{S}^1$ given by $f(t) = \exp(4\pi t)$.

The next Lemma says that the Cartesian product of covering spaces is a covering space.

5.2.8. Lemma. *Let X_1 and X_2 be metric spaces. If $\sigma_1 \colon \widetilde{X}_1 \twoheadrightarrow X_1$ and $\sigma_2 \colon \widetilde{X}_2 \twoheadrightarrow X_2$ are covering maps, then $(\sigma_1 \times \sigma_2) \colon \widetilde{X}_1 \times \widetilde{X}_2 \twoheadrightarrow X_1 \times X_2$ is a covering map.*

Proof. Let $(x_1, x_2) \in X_1 \times X_2$. Let U_j be an open subset of X_j such that $x_j \in U_j$ and U_j is evenly covered by σ_j, $j \in \{1, 2\}$. Suppose

$$\sigma_1^{-1}(U_1) = \bigcup_{\lambda \in \Lambda} V_\lambda^1 \text{ and } \sigma_2^{-1}(U_2) = \bigcup_{\gamma \in \Gamma} V_\gamma^2. \text{ Then}$$

$$(\sigma_1 \times \sigma_2)^{-1}(U_1 \times U_2) = \left(\bigcup_{\lambda \in \Lambda} V_\lambda^1 \right) \times \left(\bigcup_{\gamma \in \Gamma} V_\gamma^2 \right).$$

Note that $(\sigma_1 \times \sigma_2)|_{V_\lambda^1 \times V_\gamma^2} \colon V_\lambda^1 \times V_\gamma^2 \twoheadrightarrow U_1 \times U_2$ is a homeomorphism for every $\lambda \in \Lambda$ and $\gamma \in \Gamma$. Therefore, $\sigma_1 \times \sigma_2$ is a covering map.

Q.E.D.

5.2.9. Definition. Let X be a metric space. A covering space $\widetilde{X}$ of X is said to be *universal* if $\widetilde{X}$ is arcwise connected and its fundamental group is trivial.

5.2.10. Definition. Let $\sigma \colon \widetilde{X} \twoheadrightarrow X$ be a covering map. A homeomorphism

$$\varphi \colon \widetilde{X} \twoheadrightarrow \widetilde{X}$$

is *covering homeomorphism* of X if $\sigma \circ \varphi = \sigma$.

5.2.11. Definition. Let X and Y be metric spaces, with metrics d and d', respectively. A map $f \colon X \to Y$ is a *local isometry* provided that for each point $x \in X$, there exists an open subset U of X such that $x \in U$ and $d(x', x'') = d'(f(x'), f(x''))$ for every $x', x'' \in U$. The map f is an *isometry* if $d(x', x'') = d'(f(x'), f(x''))$ for every $x', x'' \in X$.

The following Proposition tells us that a connected covering space of a locally connected continuum has a metric such that the covering map is a local isometry.

5.2.12. Proposition. *Let X be a locally connected continuum with metric d. If $\sigma \colon \widetilde{X} \twoheadrightarrow X$ is a covering map, where $\widetilde{X}$ is connected, then there exists a metric $\tilde{d}$ for $\widetilde{X}$ such that σ is a local isometry and every covering homeomorphism of X is an isometry.*

Proof. First, we show the existence of a metric $\tilde{d}$ for $\widetilde{X}$ such that σ is a local isometry.

Since $\widetilde{X}$ is a covering space, for each $x \in X$, there exists an evenly covered open subset U_x of X. Thus, each $\sigma^{-1}(U_x)$ may be written as

$$\sigma^{-1}(U_x) = \bigcup_{\lambda \in \Lambda_x} V_\lambda(x),$$

where each $V_\lambda(x)$ is an open subset of $\widetilde{X}$ and $\sigma|_{V_\lambda(x)} \colon V_\lambda(x) \twoheadrightarrow U_x$ is a homeomorphism. Since X is locally connected, without loss of generality, we assume that each U_x is connected.

Let $\widetilde{U}_x = \bigcup_{\lambda \in \Lambda_x} V_\lambda(x)$. Then $\widetilde{U}_x$ is an open subset of $\widetilde{X}$. Note that $\mathcal{U} = \{\widetilde{U}_x \mid x \in X\}$ is an open cover of $\widetilde{X}$. Hence, since $\widetilde{X}$ is connected, by Theorem 3–4 of [12] for any two points $\tilde{x}, \tilde{y} \in \widetilde{X}$ there exists a finite subfamily $\{U_1, \dots, U_n\}$ of $\mathcal{U}$ such that $\tilde{x} \in U_1$, $\tilde{y} \in U_n$ and $U_j \cap U_k \neq \emptyset$ if and only if $|j - k| \leq 1$, $j, k \in \{1, \dots, n\}$.

Let $\tilde{z}_1 = \tilde{x}$, $\tilde{z}_n = \tilde{y}$ and let $\tilde{z}_j \in U_{j-1} \cap U_j$ for each $j \in \{2, \dots, n\}$. Then $\tilde{z}_1, \dots, \tilde{z}_n$ is a *chain of points from $\tilde{x}$ to $\tilde{y}$*. We define the length of this chain of points as

$$\sum_{j=2}^{n} d(z_{j-1}, z_j), \text{ where } z_j = \sigma(\tilde{z}_j), \ j \in \{1, \dots, n\}.$$

Define $\tilde{d}(\tilde{x}, \tilde{y})$ to be the infimum of the lengths of chains of points from $\tilde{x}$ to $\tilde{y}$. We assert that $\tilde{d}$ is a metric. Clearly, $\tilde{d}(\tilde{x}, \tilde{y}) \geq 0$ and $\tilde{d}(\tilde{x}, \tilde{y}) = \tilde{d}(\tilde{y}, \tilde{x})$ for every $\tilde{x}, \tilde{y} \in \widetilde{X}$.

We show the triangle inequality. Let $\tilde{x}, \tilde{y}, \tilde{z} \in \widetilde{X}$. Note that if $\tilde{a}_1, \dots, \tilde{a}_n$ is a chain of points from $\tilde{x}$ to $\tilde{y}$ and $\tilde{c}_1 \dots, \tilde{c}_m$ is a chain of points from $\tilde{y}$ to $\tilde{z}$, then $\tilde{a}_1, \dots, \tilde{a}_n, \tilde{c}_1 \dots, \tilde{c}_m$ is a chain of points from $\tilde{x}$ to $\tilde{z}$. Hence,

$$\sum_{j=2}^{n} d(a_{j-1}, a_j) + \sum_{j=2}^{m} d(c_{j-1}, c_j) \in$$

$$\left\{ \sum_{j=2}^{k} d(z_{j-1}, z_j) \ \middle| \ \tilde{z}_1, \dots, \tilde{z}_k \text{ is chain of points from } \tilde{x} \text{ to } \tilde{z} \right\}.$$

Since

$$\inf\left\{\sum_{j=2}^{n} d(a_{j-1}, a_j) + \sum_{j=2}^{m} d(c_{j-1}, c_j) \ \Big|\ \tilde{a}_1, \ldots, \tilde{a}_n, \tilde{c}_1 \ldots, \tilde{c}_m \text{ is a}\right.$$

$$\left.\text{chain of points from } \tilde{x} \text{ to } \tilde{z}\right\} \leq$$

$$\inf\left\{\sum_{j=2}^{n} d(a_{j-1}, a_j) \ \Big|\ \tilde{a}_1, \ldots, \tilde{a}_n \text{ is a chain of points from } \tilde{x} \text{ to } \tilde{y}\right\} +$$

$$\inf\left\{\sum_{j=2}^{m} d(c_{j-1}, c_j) \Big|\ \tilde{c}_1 \ldots, \tilde{c}_m \text{ is a chain of points from } \tilde{y} \text{ to } \tilde{z}\right\},$$

$\tilde{d}(\tilde{x}, \tilde{z}) \leq \tilde{d}(\tilde{x}, \tilde{y}) + \tilde{d}(\tilde{y}, \tilde{z})$. Hence, $\tilde{d}$ satisfies the triangle inequality. Therefore, $\tilde{d}$ is a metric for $\widetilde{X}$.

Next, we see that σ is a local isometry. Let $\tilde{x} \in \widetilde{X}$, and let $x = \sigma(\tilde{x})$. Let U_x be an open subset of X such that $x \in U_x$ and U_x is evenly covered by σ. Let $V_{\tilde{x}}$ be an open subset of $\sigma^{-1}(U_x)$ such that $\tilde{x} \in V_{\tilde{x}}$ and $\sigma|_{V_{\tilde{x}}} \colon V_{\tilde{x}} \twoheadrightarrow U_x$ is a homeomorphism.

Take $\tilde{w}, \tilde{z} \in V_{\tilde{x}}$, let $w = \sigma(\tilde{w})$ and let $z = \sigma(\tilde{z})$. Observe that $\tilde{w}, \tilde{z}$ is a chain of points from $\tilde{w}$ to $\tilde{z}$. Hence, $d(w, z)$ is the smallest length of chains of points from $\tilde{w}$ to $\tilde{z}$. Thus, $\tilde{d}(\tilde{w}, \tilde{z}) = d(w, z)$. Therefore, σ is a local isometry.

To finish the proof, we show that each covering homeomorphism is an isometry. Let $\varphi \colon \widetilde{X} \twoheadrightarrow \widetilde{X}$ be a covering homeomorphism. Let $\tilde{x}, \tilde{y} \in \widetilde{X}$. Let $\tilde{z}_1, \ldots, \tilde{z}_n$ be a chain of points from $\tilde{x}$ to $\tilde{y}$. Then $\varphi(\tilde{z}_1), \ldots, \varphi(\tilde{z}_n)$ is a chain of points from $\varphi(\tilde{x})$ to $\varphi(\tilde{y})$. Since $\sigma \circ \varphi = \sigma$,

$$d(\sigma(\tilde{z}_{j-1}), \sigma(\tilde{z}_j)) = d(\sigma \circ \varphi(\tilde{z}_{j-1}), \sigma \circ \varphi(\tilde{z}_j)).$$

Hence,

$$\left\{\sum_{j=2}^{n} d(\sigma(\tilde{z}_{j-1}), \sigma(\tilde{z}_j)) \ \Big|\ \tilde{z}_1, \ldots, \tilde{z}_n \text{ is a chain of points from } \tilde{x} \text{ to } \tilde{y}\right\}$$

$$\subset \left\{ \sum_{j=2}^{n} d(\sigma(\tilde{w}_{j-1}), \sigma(\tilde{w}_j)) \mid \tilde{w}_1, \ldots, \tilde{w}_n \text{ is a chain of points from}\right.$$

$$\left. \varphi(\tilde{x}) \text{ to } \varphi(\tilde{y}) \right\}.$$

Since φ is a covering homeomorphism, it is easy to see that φ^{-1} is also a covering homeomorphism. Thus, the reverse inclusion is also true. Consequently, $\tilde{d}(\tilde{x}, \tilde{y}) = \tilde{d}(\varphi(\tilde{x}), \varphi(\tilde{y}))$. Therefore, φ is an isometry.

Q.E.D.

The next Theorem says that, in certain cases, the property of Effros can be lifted to a covering space.

5.2.13. Theorem. *Let X be a compact and connected absolute neighborhood retract, and let $\sigma \colon \widetilde{X} \twoheadrightarrow X$ be a covering map, where $\widetilde{X}$ is connected. If M is a homogeneous subcontinuum of X, then $\widetilde{M} = \sigma^{-1}(M)$ has the property of Effros.*

Proof. Since X is an absolute neighborhood retract, X is locally connected (by (iii) (p. 339) of [16]). Hence, by Proposition 5.2.12, we assume that $\widetilde{X}$ has a metric $\tilde{d}$ such that σ is a local isometry and every covering homeomorphism is an isometry. Let $\mathcal{U}$ be a finite open cover of X by connected and evenly covered subsets of X. Let $\varepsilon > 0$ such that 2ε is a Lebesgue number for $\mathcal{U}$ (Theorem 1.6.6).

Since X is a compact absolute neighborhood retract, by Theorem 1.1 (p. 111) of [13], there exists $\beta > 0$ such that for any two maps $k, k' \colon \widetilde{X} \to X$ such that $d(k(\tilde{x}), k'(\tilde{x})) < \beta$ for every $\tilde{x} \in \widetilde{X}$, there exists a homotopy $F \colon \widetilde{X} \times [0,1] \to X$ such that $F((\tilde{x}, 0)) = k(\tilde{x})$, $F((\tilde{x}, 1)) = k'(\tilde{x})$, and $d(F((\tilde{x}, t)), F((\tilde{x}, s))) < \varepsilon$ for every $\tilde{x} \in \widetilde{X}$ and every $s, t \in [0,1]$ (such a homotopy is called an *ε-homotopy*). Without loss of generality, we assume that $\beta \leq \varepsilon$.

Let M be a homogeneous subcontinuum of X. Then, by Theorem 4.2.31, there exists an Effros number $\delta > 0$ for β. We assume that $\delta \leq \beta$. Let $\tilde{p}, \tilde{q} \in \widetilde{M}$ such that $\tilde{d}(\tilde{p}, \tilde{q}) < \delta$. We show there exists an ε-homeomorphism $\tilde{h} \colon \widetilde{M} \twoheadrightarrow \widetilde{M}$ such that $\tilde{h}(\tilde{p}) = \tilde{q}$.

Let $p = \sigma(\tilde{p})$ and let $q = \sigma(\tilde{q})$. Since σ is a local isometry, $\tilde{d}(\tilde{p}, \tilde{q}) = d(p, q)$ ($\delta < \varepsilon$ and 2ε is a Lebesgue number for $\mathcal{U}$). Hence, there exists a β–homeomorphism $h \colon M \twoheadrightarrow M$ such that $h(p) = q$. Since h is a β–homeomorphism, there exists an ε–homotopy F' between h and 1_M. Since 1_M can be extended to 1_X, there exist a map $f \colon X \to X$ such that $f|_M = h$ and an ε–homotopy F between f and 1_X such that $F|_{M \times [0,1]} = F'$ (Theorem 2.2 (p. 117) of [13]).

Let $G \colon \tilde{X} \times [0,1] \twoheadrightarrow X$ be given by $G = F \circ (\sigma \times 1_{[0,1]})$. Then G is a homotopy between $f \circ \sigma$ and σ because:

$$G((\tilde{x}, 0)) = F((\sigma(\tilde{x}), 0)) = f(\sigma(\tilde{x})) = f \circ \sigma(\tilde{x})$$

and

$$G((\tilde{x}, 1)) = F((\sigma(\tilde{x}), 1)) = \sigma(\tilde{x}).$$

Note that $1_{\tilde{X}}$ is a lifting of σ relative to σ. By Proposição 9 (p. 132) of [17], there exists a homotopy $\tilde{G} \colon \tilde{X} \times [0,1] \to \tilde{X}$ between $1_{\tilde{X}}$ and a map $\tilde{f} \colon \tilde{X} \to \tilde{X}$ with $\sigma \circ \tilde{f} = f \circ \sigma$ such that $\sigma \circ \tilde{G} = G$.

Let $\tilde{h} = \tilde{f}|_{\widetilde{M}}$. We show that $\tilde{h}$ is desired homeomorphism. Note that $\sigma \circ \tilde{h}(\widetilde{M}) = \sigma \circ \tilde{f}(\widetilde{M}) = f \circ \sigma(\widetilde{M}) = f \circ \sigma(\sigma^{-1}(M)) = f(M) = h(M)$. Hence, $\tilde{h}(\widetilde{M}) \subset \widetilde{M}$.

Now, we see that $\tilde{h}$ moves no point of $\widetilde{M}$ more than ε. Let $\tilde{x} \in \widetilde{M}$. Observe that $G|_{\widetilde{M} \times [0,1]}$ is an ε–homotopy. Hence, we have that $d(G((\tilde{x}, t)), G((\tilde{x}, s))) < \varepsilon$ for every $s, t \in [0,1]$. Thus, $G(\{\tilde{x}\} \times [0,1])$ is contained in an element of $\mathcal{U}$. Since $\sigma(\tilde{G}(\{\tilde{x}\} \times [0,1])) = G(\{\tilde{x}\} \times [0,1])$ and σ is a local isometry, $\tilde{d}(\tilde{G}((\tilde{x}, t)), \tilde{G}((\tilde{x}, s))) < \varepsilon$ for every $s, t \in [0,1]$. Since $\tilde{G}((\tilde{x}, 0)) = \tilde{x} = 1_{\widetilde{M}}(\tilde{x})$ and $\tilde{G}((\tilde{x}, 1)) = \tilde{f}(\tilde{x}) = \tilde{h}(\tilde{x})$, $\tilde{d}(\tilde{x}, \tilde{h}(\tilde{x})) < \varepsilon$. Therefore, $\tilde{h}$ moves no point of $\widetilde{M}$ more than ε.

To see that $\tilde{h}$ is a homeomorphism, follow the procedure of the above paragraphs to construct a lift $k \colon \widetilde{M} \to \widetilde{M}$ of the homeomorphism $h^{-1} \colon M \twoheadrightarrow M$ satisfying $\tilde{d}(\tilde{z}, k(\tilde{z})) < \varepsilon$ for all $\tilde{z} \in \widetilde{M}$. Let $\tilde{x} \in \widetilde{M}$. Then $\sigma \circ k \circ \tilde{h}(\tilde{x}) = h^{-1} \circ \sigma \circ \tilde{h}(\tilde{x}) = h^{-1} \circ h \circ \sigma(\tilde{x}) = \sigma(\tilde{x})$, and $\sigma \circ \tilde{h} \circ k(\tilde{x}) = h \circ \sigma \circ k(\tilde{s}) = h \circ h^{-1} \circ \sigma(\tilde{x}) = \sigma(\tilde{x})$. Hence, $k \circ \tilde{h}(\tilde{x}), \tilde{h} \circ k(\tilde{x}) \in \sigma^{-1}(\sigma(\tilde{x}))$. Since neither $k \circ \tilde{h}$ nor $\tilde{h} \circ k$ moves a point more than 2ε, it follows that $k \circ \tilde{h}(\tilde{x}) = \tilde{x}$ and $\tilde{h} \circ k(\tilde{x}) = \tilde{x}$. Therefore, $\tilde{h}$ is an ε–homeomorphism.

Since $h(p) = q$, f is an extension of h and $\tilde{f}$ is a lift of $f \circ \sigma$, it follows that $\tilde{h}(p) \in \sigma^{-1}(q)$. Since $\tilde{d}(\tilde{p}, \tilde{h}(\tilde{p})) < \varepsilon$,

$$\tilde{d}(\tilde{q}, \tilde{h}(\tilde{p})) \leq \tilde{d}(\tilde{q}, \tilde{p}) + \tilde{d}(\tilde{p}, \tilde{h}(\tilde{p})) < \delta + \varepsilon \leq 2\varepsilon.$$

Hence, $\tilde{h}(\tilde{p}) = \tilde{q}$. Therefore, $\widetilde{M}$ has the property of Effros.

Q.E.D.

5.3 Rogers's Theorem

We use covering spaces techniques due to James T. Rogers, Jr. to study homogeneous continua.

We begin with a discussion of Poincaré model of the hyperbolic plane $\mathbb{H}$ ([3], [9] and [36]).

Let $\mathbb{H}$ be the interior of the closed unit disk D in $\mathbb{R}^2$, and let $\mathcal{S}^1$ be its boundary.

5.3.1. Definition. A *geodesic* in $\mathbb{H}$ is the intersection of $\mathbb{H}$ and a circle C in $\mathbb{R}^2$ that intersects $\mathcal{S}^1$ orthogonally (straight lines through the origin are considered circles centered at ∞).

5.3.2. Definition. A *reflection* in a geodesic $C \cap \mathbb{H}$ is a Euclidean inversion in the circle. A *hyperbolic isometry* of $\mathbb{H}$ is a composition of reflections in geodesics.

5.3.3. Definition. The set $\mathbb{H}$ with the family of isometries is the *Poincaré model of the hyperbolic plane*. The boundary $\mathcal{S}^1$ of $\mathbb{H}$, which is not in $\mathbb{H}$, is called the *circle at ∞*.

Let $\mathbb{F}$ be a double torus. The universal covering space of $\mathbb{F}$ may be chosen to be $\mathbb{H}$ ([3] and [36]), and the group of covering homeomorphisms to be a subgroup of the orientation–preserving isometries of $\mathbb{H}$. Each covering homeomorphism (except the identity map $1_{\mathbb{H}}$) is a hyperbolic isometry. The pertinent property for us is that a hyperbolic isometry, when extended to the circle at ∞, has exactly two fixed points in $\mathcal{S}^1$ and none in $\mathbb{H}$ (Theorem 9–3, (p. 132) of [3] and p. 410 of [36]). The universal covering map $\sigma\colon \mathbb{H} \twoheadrightarrow \mathbb{F}$ may be chosen to be a local isometry.

5.3.4. Definition. A *geodesic in* $\mathbb{F}$ is the image under σ of a geodesic in $\mathbb{H}$. A geodesic is *simple* if it has no transverse intersection.

5.3.5. Definition. A *simple closed geodesic* is a geodesic in $\mathbb{F}$ that is a simple closed curve. A simple closed curve in $\mathbb{F}$ is *essential* if it does not bound a disk.

Each essential simple closed curve in $\mathbb{F}$ is isotopic to a unique simple closed geodesic (p. 339 of [33]).

Assume the universal covering space $(\mathbb{H}, \sigma)$ of $\mathbb{F}$ is constructed with the following properties. The geodesic $\mathbb{H} \cap x$–axis maps to a simple closed geodesic C_1 under σ. The geodesic $\mathbb{H} \cap y$–axis maps to a simple closed geodesic C_2 under σ. Furthermore, $C_1 \cup C_2$ is a figure eight W, and $C_1 \cap C_2 = \{v\}$.

A proof of the following Theorem may be found in Theorem 5 of [28].

5.3.6. Theorem. *Let* $\sigma\colon \mathbb{H} \twoheadrightarrow \mathbb{F}$ *be the universal covering map of* $\mathbb{F}$. *Let* $\mathcal{U}$ *be a finite cover of* $\mathbb{F}$ *by evenly covered sets, and let* 2ε *be a Lebesgue number for* $\mathcal{U}$. *If* $\mathbb{K}$ *is a compact subset of* $\mathbb{H}$ *and if* $\varphi\colon \mathbb{H} \twoheadrightarrow \mathbb{H}$ *is a covering homeomorphism different from the identity map* $1_{\mathbb{H}}$, *then there exists* $\tilde{x} \in \mathbb{K}$ *such that* $\tilde{d}(\tilde{x}, \varphi(\mathbb{K})) > \varepsilon$.

5.3.7. Theorem. *Let* $\sigma \colon \mathbb{H} \twoheadrightarrow \mathbb{F}$ *be the universal covering map of* $\mathbb{F}$. *Let* M *be a subspace of* $\mathbb{F}$, *and let* $\widetilde{M} = \sigma^{-1}(M)$. *Suppose* $\mathbb{L}$ *is a component of* $\widetilde{M}$. *If* $\mathbb{L}$ *is compact, then* $\sigma|_{\mathbb{L}}$ *is one–to–one.*

Proof. Suppose there exist $x \in M$, and two points $\tilde{x}_1, \tilde{x}_2 \in \mathbb{L} \cap \sigma^{-1}(x)$. Let $\varphi \colon \mathbb{H} \twoheadrightarrow \mathbb{H}$ be a covering homeomorphism such that $\varphi(\tilde{x}_1) = \tilde{x}_2$ (Proposiçao 8 (p. 162) of [17]). Since $\varphi \neq 1_{\mathbb{H}}$, by Theorem 5.3.6, there exists $\tilde{z} \in \varphi(\mathbb{L}) \setminus \mathbb{L}$. Hence, $\mathbb{L} \cup \varphi(\mathbb{L})$ is a subcontinuum of $\widetilde{M}$ which properly contains $\mathbb{L}$, a contradiction to the fact that $\mathbb{L}$ is a component of $\widetilde{M}$. Therefore, $\sigma|_{\mathbb{L}}$ is one–to–one.

$$\textbf{Q.E.D.}$$

5.3.8. Theorem. *Let* $\sigma \colon \mathbb{H} \twoheadrightarrow \mathbb{F}$ *be the universal covering map of* $\mathbb{F}$. *Let* M *be a subspace of* $\mathbb{F}$, *and let* $\widetilde{M} = \sigma^{-1}(M)$. *Suppose* $\mathbb{L}$ *is a component of* $\widetilde{M}$. *If* M *is arcwise connected, then* $\sigma(\mathbb{L}) = M$.

Proof. Let $z \in \sigma(\mathbb{L})$, and let $z' \in M$. Since M is arcwise connected, there exists map $\alpha \colon [0,1] \to M$ such that $\alpha(0) = z$ and $\alpha(1) = z'$. Let $\tilde{z} \in \mathbb{L}$ such that $\sigma(\tilde{z}) = z$. By a proof similar to the one given in Theorem 1.3.31, it is shown that there exists a map $\tilde{\alpha} \colon [0,1] \to \widetilde{M}$ such that $\tilde{\alpha}(0) = \tilde{z}$ and $\sigma \circ \tilde{\alpha} = \alpha$. Since $\tilde{\alpha}([0,1]) \cap \mathbb{L} \neq \emptyset$ and $\mathbb{L}$ is a component of $\widetilde{M}$, $\tilde{\alpha}([0,1]) \subset \mathbb{L}$. Hence, $\sigma(\tilde{\alpha}(1)) = z'$. Therefore, $\sigma(\mathbb{L}) = M$.

$$\textbf{Q.E.D.}$$

5.3.9. Definition. Let $\mathcal{Q}$ be the Hilbert cube, and let $\sigma \times 1_{\mathcal{Q}} \colon \mathbb{H} \times \mathcal{Q} \twoheadrightarrow \mathbb{F} \times \mathcal{Q}$ be the universal covering map of $\mathbb{F} \times \mathcal{Q}$. Let X be a continuum essentially embedded in $W \times \mathcal{Q}$ (this means that the embedding is not homotopic to a constant map; note that this eliminates some continua from consideration). Let $f \colon X \twoheadrightarrow W$ be the projection map. Let $\widetilde{X} = (\sigma \times 1_{\mathcal{Q}})^{-1}(X)$, and let $\tilde{f} \colon \widetilde{X} \twoheadrightarrow \widetilde{W}$ be the projection map. If $\mathbb{K}$ is a component of $\widetilde{X}$, then the set $\mathbb{E}(\mathbb{K}) = \{z \in \mathcal{S}^1 \mid z \text{ is a (Euclidean) limit point of } \tilde{f}(\mathbb{K})\}$ is called the *set of ends of* $\mathbb{K}$.

Let $\widetilde{W} = \sigma^{-1}(W)$. Then $\widetilde{W}$ is the universal covering space of W, the "infinite snowflake" pictured on top of page 253. The set $Cl(\widetilde{W}) \cap \mathcal{S}^1$ is a Cantor set; call it $\mathfrak{Z}$ (p. 341 of [33]).

5.3.10. Lemma. *Let X be a continuum. If $\ell\colon X \to W$ is a map which is not homotopic to a constant map, then X can be essentially embedded in $W \times Q$.*

Proof. By Theorem 1.1.16, there exists an embedding $g\colon X \to Q$. Let $h\colon X \to W \times Q$ be given by $h(x) = (\ell(x), g(x))$. Then h is an embedding of X into $W \times Q$ which is not homotopic to a constant map.

Q.E.D.

A proof of the following Theorem may be modeled from the proof of Theorem 8 of [28], using the results in [1] to obtain X as the remainder of a compactification of $[0, 1)$.

5.3.11. Theorem. *Let $\sigma \times 1_Q\colon \mathbb{H} \times Q \twoheadrightarrow \mathbb{F} \times Q$ be the universal covering map of $\mathbb{F} \times Q$. Let X be a continuum essentially embedded in $W \times Q$. If $\widetilde{X} = (\sigma \times 1_Q)^{-1}(X)$, then no component of $\widetilde{X}$ is compact.*

5.3.12. Theorem. *Let $\sigma \times 1_Q\colon \mathbb{H} \times Q \twoheadrightarrow \mathbb{F} \times Q$ be the universal covering map of $\mathbb{F} \times Q$. Let X be a continuum essentially embedded in $W \times Q$, and let $\widetilde{X} = (\sigma \times 1_Q)^{-1}(X)$. Then there exists $\delta > 0$ such that if $\tilde{x} \in \mathbb{K}$, $\tilde{x}' \in \mathbb{K}'$ and $\tilde{d}(\tilde{x}, \tilde{x}') < \delta$ (where $\mathbb{K}$ and $\mathbb{K}'$ are components of $\widetilde{X}$), then $\mathbb{E}(\mathbb{K}) = \mathbb{E}(\mathbb{K}')$.*

Proof. This is an immediate consequence of Theorem 5.2.13 and the fact that each bounded homeomorphism of $\widetilde{X}$ preserves ends.

Q.E.D.

5.3.13. Notation. Let $\sigma \times 1_Q\colon \mathbb{H} \times Q \twoheadrightarrow \mathbb{F} \times Q$ be the universal covering map of $\mathbb{F} \times Q$. Let X be a continuum essentially embedded in $W \times Q$. Let $\widetilde{X} = (\sigma \times 1_Q)^{-1}(X)$, and let $\mathbb{K}$ be a component of $\widetilde{X}$. Assume each of the geodesics C_1 and C_2 has length greater than one and that 2δ is an Effros number for $\varepsilon = \dfrac{1}{2}$. Cover $\{v\} \times Q$ with a finite collection $\mathcal{B}$ of open δ–balls. Since $\mathbb{K}$ is unbounded (Theorem 5.3.11) and locally compact, there exist a ball $B \in \mathcal{B}$, two liftings $\widetilde{B}_0$ and $\widetilde{B}_m$ of B and a subcontinuum M of $\widetilde{W}$ meeting both $\widetilde{B}_0$ and $\widetilde{B}_m$. Let $\varphi \times 1_Q\colon \mathbb{H} \times Q \twoheadrightarrow \mathbb{H} \times Q$ be a covering homeomorphism such that $(\varphi \times 1_Q)(\widetilde{B}_0) = \widetilde{B}_m$.

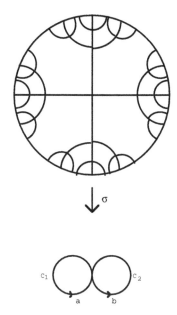

5.3.14. Theorem. *If* $\mathbb{K}$ *and* $\varphi \times 1_{\mathcal{Q}}$ *are as in Notation 5.3.13, then* $\mathbb{E}((\varphi \times 1_{\mathcal{Q}})(\mathbb{K})) = \mathbb{E}(\mathbb{K})$.

Proof. Let $\tilde{x}_1 \in \widetilde{B}_0 \cap M$ and let $\tilde{x}_2 \in \widetilde{B}_m \cap M$. Hence,

$$\tilde{d}(\tilde{x}_2, (\varphi \times 1_{\mathcal{Q}})(\tilde{x}_1)) < 2\delta.$$

By Theorem 5.3.12, $\mathbb{E}(\mathbb{K}) = \mathbb{E}((\varphi \times 1_{\mathcal{Q}})(\mathbb{K}))$.

Q.E.D.

Compactify $\mathbb{H} \times \mathcal{Q}$ with $\mathcal{S}^1 \times \mathcal{Q}$. Shrink each set of the form $\{z\} \times \mathcal{Q}$, where $z \in \mathcal{S}^1$, to a point to obtain another compactification of $\mathbb{H} \times \mathcal{Q}$. This time the remainder is $\mathcal{S}^1$. Let $\pi \colon (\mathbb{H} \times \mathcal{Q}) \cup \mathcal{S}^1 \twoheadrightarrow \mathbb{H} \cup \mathcal{S}^1$ be the map of this latter compactification onto the disk D obtained by naturally extending the projection map. Note that the restriction of π to $\widetilde{X}$ is just $\tilde{f}$. (If $\tilde{x} \in \widetilde{X}$, then $\sigma \circ \tilde{f}(\tilde{x}) = f \circ (\sigma \times 1_{\mathcal{Q}})(\tilde{x}) \in W$. Hence, $\tilde{f}(\tilde{x}) \in \widetilde{W}$.)

5.3.15. Definition. Let X be a continuum, with metric d, and let $f \colon X \twoheadrightarrow X$ be a surjective map. A point $x \in X$ is a *fixed attracting* *point* provided that $f(x) = x$ and there exists a neighborhood U of x in X such that $\lim\limits_{n\to\infty} d(x, f^n(y)) = 0$ for every $y \in U$.

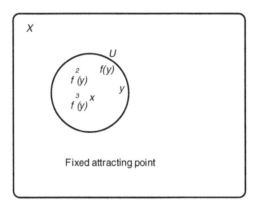

Fixed attracting point

5.3.16. Definition. Let X be a continuum, with metric d, and let $f\colon X \twoheadrightarrow X$ be a surjective map. A point $x \in X$ is a *fixed repelling point* provided that $f(x) = x$ and there exists a neighborhood U of x in X for each $y \in U \setminus \{x\}$, there exists $n \in \mathbb{N}$ such that $f^n(y) \in X \setminus U$.

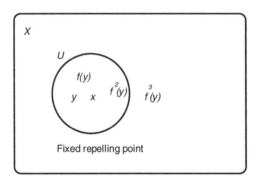

Fixed repelling point

5.3.17. Remark. It is known that a hyperbolic isometry of $\mathbb{H}$ has an attractive fixed point and a repelling fixed point on $\mathcal{S}^1$.

5.3.18. Theorem. *If $\mathbb{K}$ and $\varphi \times 1_\varrho$ are as in Notation 5.3.13, then the attracting point of φ belongs to $\mathbb{E}(\mathbb{K})$.*

Proof. Let $\mathbb{K}_n = Cl_{(\mathbb{H} \times \mathcal{Q}) \cup \mathcal{S}^1}((\varphi^n \times 1_\mathcal{Q})(\mathbb{K}))$. Since the hyperspace of subcontinua of $(\mathbb{H} \times \mathcal{Q}) \cup \mathcal{S}^1$, $\mathcal{C}((\mathbb{H} \times \mathcal{Q}) \cup \mathcal{S}^1)$ is compact (Theorem 1.8.5), without loss of generality, we assume that the sequence $\{\mathbb{K}_n\}_{n=1}^\infty$ converges to $\mathbb{P} \in \mathcal{C}((\mathbb{H} \times \mathcal{Q}) \cup \mathcal{S}^1)$. This implies that $\{\pi(\mathbb{K}_n)\}_{n=1}^\infty$ converges to $\pi(\mathbb{P})$.

Since $\pi(\mathbb{K}_n) \cap \mathcal{S}^1 = \mathbb{E}((\varphi^n \times 1_\mathcal{Q})(\mathbb{K})) = \mathbb{E}(\mathbb{K})$ for all $n \in \mathbb{N}$, by Theorem 5.3.14, and since $\{\pi(\mathbb{K}_n) \cap \mathcal{S}^1\}_{n=1}^\infty$ converges to $\pi(\mathbb{P}) \cap \mathcal{S}^1$, it follows that $\pi(\mathbb{P}) \cap \mathcal{S}^1 = \mathbb{E}(\mathbb{K})$. If $\tilde{y} \in \pi(\mathbb{K})$ and $\tilde{z}$ is the attracting point of φ, then $\{\varphi^n(\tilde{y})\}_{n=1}^\infty$ converges to $\tilde{z}$ in $\mathbb{H} \cup \mathcal{S}^1$. Since $\tilde{y} \in \pi(\mathbb{K})$, it follows that $\varphi^n(\tilde{y}) \in \pi((\varphi^n \times 1_\mathcal{Q})(\mathbb{K}))$, and so $\tilde{z} \in \pi(\mathbb{P})$. Hence, $\tilde{z} \in \mathbb{E}(\mathbb{K})$.

$$\textbf{Q.E.D.}$$

5.3.19. Theorem. *If $\mathbb{K}$ and $\varphi \times 1_\mathcal{Q}$ are as in Notation 5.3.13, then the repelling point of φ belongs to $\mathbb{E}(\mathbb{K})$.*

Proof. The repelling point of φ is the attracting point of φ^{-1}. Hence, the Theorem follows from Theorem 5.3.18.

$$\textbf{Q.E.D.}$$

5.3.20. Theorem. *Let $\sigma \times 1_\mathcal{Q} \colon \mathbb{H} \times \mathcal{Q} \twoheadrightarrow \mathbb{F} \times \mathcal{Q}$ be the universal covering map of $\mathbb{F} \times \mathcal{Q}$. Let X be a homogeneous continuum essentially embedded in $W \times \mathcal{Q}$. If $\widetilde{X} = (\sigma \times 1_\mathcal{Q})^{-1}(X)$ and $\mathbb{K}$ is a component of $\widetilde{X}$, then $\mathbb{E}(\mathbb{K})$ is either a two–point set or a Cantor set. Furthermore, $\mathbb{E}(\mathbb{K})$ contains a dense subset each point of which is a fixed point of a hyperbolic isometry φ such that $\varphi \times 1_\mathcal{Q}$ is a covering homeomorphism.*

Proof. Suppose $\mathbb{E}(\mathbb{K})$ has more than two points. Since $\mathbb{E}((\varphi^n \times 1_\mathcal{Q})(\mathbb{K})) = \mathbb{E}(\mathbb{K})$ for each $n \in \mathbb{N}$ (Theorem 5.3.14), and φ only fixes two points of $\mathcal{S}^1$ (p. 419 of [36]), it follows that $\mathbb{E}(\mathbb{K})$ is infinite. In fact, there exists a sequence of points $\{\tilde{z}_m\}_{m=1}^\infty$ in $\mathbb{E}(\mathbb{K})$ that converges to the attracting point, $\tilde{z}$, of φ.

Furthermore, if $\tilde{z}' \in \mathbb{E}(\mathbb{K})$, then arbitrarily close to $\tilde{z}'$ there is a point $\tilde{z}'' \in \mathbb{E}(\mathbb{K})$ such that $\tilde{z}''$ is the attracting fixed point of a hyperbolic isometry $\varphi'' \colon \mathbb{H} \twoheadrightarrow \mathbb{H}$ such that $\varphi'' \times 1_\mathcal{Q}$ is a covering homeomorphism. Such a hyperbolic isometry is constructed in the same way as φ was (Notation 5.3.13). In particular, φ'' has the same properties as φ. It follows that $\mathbb{E}(\mathbb{K})$ is perfect. Therefore, $\mathbb{E}(\mathbb{K})$ is a Cantor set.

Q.E.D.

A proof of the following Theorem may be found in Theorem 4.1 of [34].

5.3.21. Theorem. *Let $\sigma \times 1_{\mathcal{Q}} \colon \mathbb{H} \times \mathcal{Q} \twoheadrightarrow \mathbb{F} \times \mathcal{Q}$ be the universal covering map of $\mathbb{F} \times \mathcal{Q}$. Let X be a nonaposyndetic homogeneous continuum essentially embedded in $W \times \mathcal{Q}$. Let $\widetilde{X} = (\sigma \times 1_{\mathcal{Q}})^{-1}(X)$ and let $\mathbb{K}$ be a component of $\widetilde{X}$. Let $\mathbb{Y} = Cl_{(\mathbb{H} \times \mathcal{Q}) \cup \mathcal{S}^1}(\mathbb{K})$. Then $\mathbb{Y} \setminus \mathbb{K} = \mathbb{E}(\mathbb{K})$, and $\mathbb{Y}$ is connected im kleinen at each point of $\mathbb{E}(\mathbb{K})$.*

5.3.22. Corollary. *Let $\sigma \times 1_{\mathcal{Q}} \colon \mathbb{H} \times \mathcal{Q} \twoheadrightarrow \mathbb{F} \times \mathcal{Q}$ be the universal covering map of $\mathbb{F} \times \mathcal{Q}$. Let X be a nonaposyndetic homogeneous continuum essentially embedded in $W \times \mathcal{Q}$. Let $\widetilde{X} = (\sigma \times 1_{\mathcal{Q}})^{-1}(X)$ and let $\mathbb{K}$ be a component of $\widetilde{X}$. Let $\mathbb{Y} = Cl_{(\mathbb{H} \times \mathcal{Q}) \cup \mathcal{S}^1}(\mathbb{K})$. Then $\mathbb{Y}$ is aposyndetic at each point of $\mathbb{E}(\mathbb{K})$.*

5.3.23. Theorem. *Let $\sigma \times 1_{\mathcal{Q}} \colon \mathbb{H} \times \mathcal{Q} \twoheadrightarrow \mathbb{F} \times \mathcal{Q}$ be the universal covering map of $\mathbb{F} \times \mathcal{Q}$. Let X be a nonaposyndetic homogeneous continuum essentially embedded in $W \times \mathcal{Q}$. Let $\widetilde{X} = (\sigma \times 1_{\mathcal{Q}})^{-1}(X)$ and let $\mathbb{K}$ be a component of $\widetilde{X}$. Let $\mathbb{Y} = Cl_{(\mathbb{H} \times \mathcal{Q}) \cup \mathcal{S}^1}(\mathbb{K})$. Then the collection $\mathcal{G} = \{\mathcal{T}_{\mathbb{Y}}(\{\tilde{x}\}) \mid \tilde{x} \in \mathbb{Y}\}$ is a continuous, terminal decomposition of $\mathbb{Y}$ with the following properties:*

(1) The quotient space $\mathbb{Y}/\mathcal{G}$ is aposyndetic.

(2) $\mathcal{T}_{\mathbb{Y}}(\{\tilde{x}\})$ is degenerate if $\tilde{x} \in \mathbb{Y} \setminus \mathbb{K}$.

(3) The nondegenerate elements of $\mathcal{G}$ are mutually homeomorphic, cell–like, indecomposable homogeneous continua of the same dimension as $\mathbb{Y}$.

(4) $\mathbb{K}/\mathcal{G}'$ is homogeneous, where $\mathcal{G}' = \{\mathcal{T}_{\mathbb{Y}}(\{\tilde{x}\}) \mid \tilde{x} \in \mathbb{K}\}$.

Proof. Note that $\mathbb{Y} = \mathbb{K} \cup \mathbb{E}(\mathbb{K})$, where $\mathbb{K}$ is open and connected. By Theorem 5.3.20, $\dim(\mathbb{E}(\mathbb{K})) = 0$ and each homeomorphism h

of $\mathbb{K}$ can be extended to a homeomorphism $\hat{h}\colon \mathbb{Y} \twoheadrightarrow \mathbb{Y}$ such that $\hat{h}(\tilde{z}) = \tilde{z}$ for each $\tilde{z} \in \mathbb{E}(\mathbb{K})$. By Corollary 5.3.22, $\mathbb{Y}$ is aposyndetic at each point of $\mathbb{E}(\mathbb{K})$.

By Proposition 5.2.12, $(\mathbb{H} \times \mathcal{Q}) \cup \mathcal{S}^1$ has a metric $\tilde{d}$ such that $\sigma \times 1_{\mathcal{Q}}$ is a local isometry. Thus, by Theorem 5.2.13, $\widetilde{X}$ has a metric $\tilde{d}$ such that $\widetilde{X}$ has the property of Effros. Hence, $\mathbb{K}$ has a metric $\tilde{d}$ such that $\mathbb{K}$ has the property of Effros.

It can be shown that each point of $\mathbb{E}(\mathbb{K})$ has a local basis of open sets of $\mathbb{Y}$ whose complements are connected. Hence, $\mathbb{Y}$ is aposyndetic at each point of $\mathbb{K}$ with respect to each point of $\mathbb{E}(\mathbb{K})$.

Since X is not aposyndetic and homogeneous, X is not aposyndetic at any point with respect to any other. Let $\tilde{x} \in \mathbb{K}$ and suppose $\mathbb{Y}$ is aposyndetic at $\tilde{x}$ with respect to $\tilde{x}' \in \mathbb{K}$. Thus, there exists a subcontinuum $\mathbb{W}$ of $\mathbb{Y}$ such that $\tilde{x} \in Int_{\mathbb{Y}}(\mathbb{W}) \subset \mathbb{W} \subset \mathbb{Y} \setminus \{\tilde{x}'\}$; without loss of generality, we assume that $\mathbb{W} \subset \mathbb{K}$ and $(\sigma \times 1_{\mathcal{Q}})(\tilde{x}) \neq (\sigma \times 1_{\mathcal{Q}})(\tilde{x}')$. Hence, since $\sigma \times 1_{\mathcal{Q}}$ is a local isometry, $(\sigma \times 1_{\mathcal{Q}})(\mathbb{W})$ satisfies that $(\sigma \times 1_{\mathcal{Q}})(\tilde{x}) \in Int_X((\sigma \times 1_{\mathcal{Q}})(\mathbb{W})) \subset (\sigma \times 1_{\mathcal{Q}})(\mathbb{W}) \subset X \setminus \{(\sigma \times 1_{\mathcal{Q}})(\tilde{x}')\}$, a contradiction. Therefore, $\mathbb{Y}$ is not aposyndetic at any point of $\mathbb{K}$.

Therefore, the Theorem follows from Theorem 5.1.27.

Q.E.D.

5.3.24. Theorem. *Let $\sigma \times 1_{\mathcal{Q}}\colon \mathbb{H} \times \mathcal{Q} \twoheadrightarrow \mathbb{F} \times \mathcal{Q}$ be the universal covering map of $\mathbb{F} \times \mathcal{Q}$. Let X be a nonaposyndetic homogeneous continuum essentially embedded in $W \times \mathcal{Q}$. Let $\widetilde{X} = (\sigma \times 1_{\mathcal{Q}})^{-1}(X)$ and let $\mathbb{K}$ be a component of $\widetilde{X}$. Let $\mathbb{Y} = Cl_{(\mathbb{H} \times \mathcal{Q}) \cup \mathcal{S}^1}(\mathbb{K})$. If $\tilde{x} \in \mathbb{K}$, then $(\sigma \times 1_{\mathcal{Q}})|_{\mathcal{T}_{\mathbb{Y}}(\{\tilde{x}\})}$ is one–to–one.*

Proof. Let $\tilde{x}_1$ and $\tilde{x}_2$ be distinct points of $\mathcal{T}_{\mathbb{Y}}(\{\tilde{x}\})$ such that $(\sigma \times 1_{\mathcal{Q}})(\tilde{x}_1) = (\sigma \times 1_{\mathcal{Q}})(\tilde{x}_2)$. By Proposição 9 (p. 132) of [17], there exists a covering homeomorphism $\varphi\colon \mathbb{H} \times \mathcal{Q} \twoheadrightarrow \mathbb{H} \times \mathcal{Q}$ such that $\varphi(\tilde{x}_1) = \tilde{x}_2$. Since φ does not fix (setwise) any compact set (Theorem 5.3.6), it follows that $\varphi(\mathcal{T}_{\mathbb{Y}}(\{\tilde{x}\})) \cap (\mathbb{Y} \setminus \mathcal{T}_{\mathbb{Y}}(\{\tilde{x}\})) \neq \emptyset$ and $\mathcal{T}_{\mathbb{Y}}(\{\tilde{x}\}) \cap (\mathbb{Y} \setminus \varphi(\mathcal{T}_{\mathbb{Y}}(\{\tilde{x}\}))) \neq \emptyset$. Since $\varphi(\mathcal{T}_{\mathbb{Y}}(\{\tilde{x}\})) \cap \mathcal{T}_{\mathbb{Y}}(\{\tilde{x}\}) \neq \emptyset$, this contradicts the fact that the decomposition, $\mathcal{G}$, of $\mathbb{Y}$ is terminal (Theorem 5.3.23).

Q.E.D.

5.3.25. Theorem. Let $\sigma \times 1_{\mathcal{Q}} \colon \mathbb{H} \times \mathcal{Q} \twoheadrightarrow \mathbb{F} \times \mathcal{Q}$ be the universal covering map of $\mathbb{F} \times \mathcal{Q}$. Let X be a nonaposyndetic homogeneous continuum essentially embedded in $W \times \mathcal{Q}$. Let $\widetilde{X} = (\sigma \times 1_{\mathcal{Q}})^{-1}(X)$ and let $\mathbb{K}_1$ and $\mathbb{K}_2$ be components of $\widetilde{X}$. Let $\mathbb{Y}_j = Cl_{(\mathbb{H} \times \mathcal{Q}) \cup \mathcal{S}^1}(\mathbb{K}_j)$ $j \in \{1, 2\}$. If $\tilde{x}_1 \in \mathbb{K}_1$ and $\tilde{x}_2 \in \mathbb{K}_2$ such that $(\sigma \times 1_{\mathcal{Q}})(\tilde{x}_1) = (\sigma \times 1_{\mathcal{Q}})(\tilde{x}_2)$, then $\mathcal{T}_{\mathbb{Y}_1}(\{\tilde{x}_1\})$ is homeomorphic to $\mathcal{T}_{\mathbb{Y}_2}(\{\tilde{x}_2\})$, and $(\sigma \times 1_{\mathcal{Q}})(\mathcal{T}_{\mathbb{Y}_1}(\{\tilde{x}_1\})) = (\sigma \times 1_{\mathcal{Q}})(\mathcal{T}_{\mathbb{Y}_2}(\{\tilde{x}_2\}))$.

Proof. Let $\varphi \colon \mathbb{H} \times \mathcal{Q} \twoheadrightarrow \mathbb{H} \times \mathcal{Q}$ be a covering homeomorphism such that $\varphi(\tilde{x}_1) = \tilde{x}_2$ (Proposição 9 (p. 132) of [17]). By Lemma 5.1.1,

$$\varphi(\mathcal{T}_{\mathbb{Y}_1}(\{\tilde{x}_1\})) = \mathcal{T}_{\mathbb{Y}_2}(\{\varphi(\tilde{x}_1)\}) = \mathcal{T}_{\mathbb{Y}_2}(\{\tilde{x}_2\}).$$

Hence, $\mathcal{T}_{\mathbb{Y}_1}(\{\tilde{x}_1\})$ is homeomorphic to $\mathcal{T}_{\mathbb{Y}_2}(\{\tilde{x}_2\})$, and

$$(\sigma \times 1_{\mathcal{Q}})(\mathcal{T}_{\mathbb{Y}_1}(\{\tilde{x}_1\})) = (\sigma \times 1_{\mathcal{Q}})(\mathcal{T}_{\mathbb{Y}_2}(\{\tilde{x}_2\})).$$

$$\text{Q.E.D.}$$

5.3.26. Lemma. Let $\sigma \times 1_{\mathcal{Q}} \colon \mathbb{H} \times \mathcal{Q} \twoheadrightarrow \mathbb{F} \times \mathcal{Q}$ be the universal covering map of $\mathbb{F} \times \mathcal{Q}$. Let X be a nonaposyndetic homogeneous continuum essentially embedded in $W \times \mathcal{Q}$. Let $\widetilde{X} = (\sigma \times 1_{\mathcal{Q}})^{-1}(X)$. If $\widetilde{Z}$ is a subcontinuum of $\widetilde{X}$ such that $(\sigma \times 1_{\mathcal{Q}})|_{\widetilde{Z}}$ is one–to–one, then there exists an open subset $\widetilde{U}$ of $\mathbb{H} \times \mathcal{Q}$ such that $\widetilde{Z} \subset \widetilde{U}$ and $(\sigma \times 1_{\mathcal{Q}})|_{\widetilde{U}}$ is one–to–one.

Proof. Suppose the lemma is not true. Then for each $n \in \mathbb{N}$, there exist $\tilde{x}_n, \tilde{y}_n \in \mathcal{V}_{\frac{1}{n}}^{\tilde{d}}(\widetilde{Z})$ such that $\tilde{x}_n \neq \tilde{y}_n$ and $(\sigma \times 1_{\mathcal{Q}})(\tilde{x}_n) = (\sigma \times 1_{\mathcal{Q}})(\tilde{y}_n)$. Since $\mathbb{F} \times \mathcal{Q}$ is compact and $\sigma \times 1_{\mathcal{Q}}$ is a covering map, without loss of generality, we assume that the sequences $\{\tilde{x}_n\}_{n=1}^{\infty}$ and $\{\tilde{y}_n\}_{n=1}^{\infty}$ converge to $\tilde{x}_0$ and $\tilde{y}_0$, respectively. Note that $x_0, y_0 \in \widetilde{Z}$.

Since $\sigma \times 1_{\mathcal{Q}}$ is continuous, the sequences $\{(\sigma \times 1_{\mathcal{Q}})(\tilde{x}_n)\}_{n=1}^{\infty}$ and $\{(\sigma \times 1_{\mathcal{Q}})(\tilde{y}_n)\}_{n=1}^{\infty}$ converge to $(\sigma \times 1_{\mathcal{Q}})(\tilde{x}_0)$ and $(\sigma \times 1_{\mathcal{Q}})(\tilde{y}_0)$, respectively. Since for each $n \in \mathbb{N}$, $(\sigma \times 1_{\mathcal{Q}})(\tilde{x}_n) = (\sigma \times 1_{\mathcal{Q}})(\tilde{y}_n)$, we have that $(\sigma \times 1_{\mathcal{Q}})(\tilde{x}_0) = (\sigma \times 1_{\mathcal{Q}})(\tilde{y}_0)$. Hence, $\tilde{x}_0 = \tilde{y}_0$ ($(\sigma \times 1_{\mathcal{Q}})|_{\widetilde{Z}}$ is one–to–one).

Let V be an evenly covered open subset of $\mathbb{F} \times \mathcal{Q}$ such that $(\sigma \times 1_{\mathcal{Q}})(\tilde{x}_0) \in V$. Let $\widetilde{V}$ be the open subset of $(\sigma \times 1_{\mathcal{Q}})^{-1}(V)$ such that $\tilde{x}_0 \in \widetilde{V}$ and $(\sigma \times 1_Q)|_{\widetilde{V}} \colon \widetilde{V} \twoheadrightarrow V$ is a homeomorphism. Then $\widetilde{V}$ is an open subset of $\mathbb{H} \times Q$ containing $\tilde{x}_0$. Since $\{\tilde{x}_n\}_{n=1}^{\infty}$ and $\{\tilde{y}_n\}_{n=1}^{\infty}$ converge to $\tilde{x}_0$, there exists $N \in \mathbb{N}$ such that $\tilde{x}_n, \tilde{y}_n \in \widetilde{V}$ for every $n \geq N$. Hence, $(\sigma \times 1_Q)(\tilde{x}_n) \neq (\sigma \times 1_Q)(\tilde{y}_n)$ for each $n \geq N$ $((\sigma \times 1_Q)|_{\widetilde{V}} \colon \widetilde{V} \twoheadrightarrow V$ is a homeomorphism), a contradiction to the choices of $\tilde{x}_n$ and $\tilde{y}_n$.

Therefore, there exists an open subset $\widetilde{U}$ of $\mathbb{H} \times Q$ such that $\widetilde{Z} \subset \widetilde{U}$ and $(\sigma \times 1_Q)|_{\widetilde{U}}$ is one–to–one.

Q.E.D.

5.3.27. Theorem. *Let $\sigma \times 1_Q \colon \mathbb{H} \times Q \twoheadrightarrow \mathbb{F} \times Q$ be the universal covering map of $\mathbb{F} \times Q$. Let X be a nonaposyndetic homogeneous continuum essentially embedded in $W \times Q$. Let $\widetilde{X} = (\sigma \times 1_Q)^{-1}(X)$, and let $\mathbb{K}$ be a component of $\widetilde{X}$. Let $\mathbb{Y} = Cl_{(\mathbb{H} \times Q) \cup S^1}(\mathbb{K})$. If $\tilde{x} \in \mathbb{K}$, then $(\sigma \times 1_Q)(\mathcal{T}_{\mathbb{Y}}(\{\tilde{x}\}))$ is a maximal terminal proper, cell–like subcontinuum of X.*

Proof. Let $\tilde{x} \in \mathbb{K}$. Recall that $\mathcal{T}_{\mathbb{Y}}(\{\tilde{x}\})$ is cell–like ((3) of Theorem 5.3.23). It follows, from Theorem 5.3.24, that $(\sigma \times 1_Q)(\mathcal{T}_{\mathbb{Y}}(\{\tilde{x}\}))$ is cell–like. Hence, $(\sigma \times 1_Q)(\mathcal{T}_{\mathbb{Y}}(\{\tilde{x}\}))$ is a proper subcontinuum of X (there exists a map $f \colon X \twoheadrightarrow W$ that is not homotopic to a constant map).

Since $(\sigma \times 1_Q)|_{\mathcal{T}_{\mathbb{Y}}(\{\tilde{x}\})}$ is one–to–one (Theorem 5.3.24), there exists an open set $\widetilde{U}$ such that $\mathcal{T}_{\mathbb{Y}}(\{\tilde{x}\}) \subset \widetilde{U}$ and $(\sigma \times 1_Q)|_{\widetilde{U}}$ is one–to–one, by Lemma 5.3.26. Thus, $(\sigma \times 1_Q)|_{\widetilde{U}} \colon \widetilde{U} \twoheadrightarrow (\sigma \times 1_Q)(\widetilde{U})$ is a homeomorphism.

If M is a subcontinuum of X such that $M \cap (\sigma \times 1_Q)(\mathcal{T}_{\mathbb{Y}}(\{\tilde{x}\})) \neq \emptyset$ and $M \cap (X \setminus (\sigma \times 1_Q)(\mathcal{T}_{\mathbb{Y}}(\{\tilde{x}\}))) \neq \emptyset$, then we may assume, by taking a subcontinuum if necessary, that $M \subset (\sigma \times 1_Q)(\widetilde{U})$. Hence, M has a lift $\widetilde{M}$ such that $\widetilde{M} \cap \mathcal{T}_{\mathbb{Y}}(\{\tilde{x}\}) \neq \emptyset$ and $\widetilde{M} \cap (\mathbb{Y} \setminus \mathcal{T}_{\mathbb{Y}}(\{\tilde{x}\})) \neq \emptyset$. Since $\mathcal{T}_{\mathbb{Y}}(\{\tilde{x}\})$ is terminal, $\mathcal{T}_{\mathbb{Y}}(\{\tilde{x}\}) \subset \widetilde{M}$. Thus, $(\sigma \times 1_Q)(\mathcal{T}_{\mathbb{Y}}(\{\tilde{x}\})) \subset M$. Hence, $(\sigma \times 1_Q)(\mathcal{T}_{\mathbb{Y}}(\{\tilde{x}\}))$ is a terminal subcontinuum of X.

Now, we show that if T is a proper, terminal, cell–like subcontinuum of X such that $(\sigma \times 1_Q)(\mathcal{T}_{\mathbb{Y}}(\{\tilde{x}\})) \subset T$, then $T = (\sigma \times 1_Q)(\mathcal{T}_{\mathbb{Y}}(\{\tilde{x}\}))$. Since T is cell–like, the inclusion map $i \colon T \to \mathbb{F} \times Q$ lifts to a one–to–one map $\tilde{\imath} \colon T \to \mathbb{H} \times Q$ such that $\tilde{\imath}((\sigma \times 1_Q)(\tilde{x})) = \tilde{x}$. Hence, $\tilde{\imath}(T) \cap \mathcal{T}_{\mathbb{Y}}(\{\tilde{x}\}) \neq \emptyset$. Let $\widetilde{T} = \tilde{\imath}(T)$.

Since $\tilde{\imath}$ is one–to–one, $(\sigma \times 1_\mathcal{Q})|_{\tilde{T}}$ is one–to–one. Thus, there exists an open set $\tilde{V}$ such that $\tilde{T} \subset \tilde{V}$ and $(\sigma \times 1_\mathcal{Q})|_{\tilde{V}}$ is one–to–one, by Lemma 5.3.26. Since $\mathcal{T}_\mathbb{Y}(\{\tilde{x}\})$ is a maximal terminal subcontinuum of $\mathbb{Y}$, either $\tilde{T}$ is not terminal or $\tilde{T} = \mathcal{T}_\mathbb{Y}(\{\tilde{x}\})$. Let $\tilde{N}$ be a proper subcontinuum of $\mathbb{Y}$ such that $\tilde{N} \cap \tilde{T} \neq \emptyset$ and $\tilde{N} \cap \mathbb{Y} \setminus \tilde{T} \neq \emptyset$. Again, we assume that $\tilde{N} \subset \tilde{V}$. Hence, $(\sigma \times 1_\mathcal{Q})(\tilde{N}) \cap T \neq \emptyset$ and $(\sigma \times 1_\mathcal{Q})(\tilde{N}) \cap (X \setminus T) \neq \emptyset$. Since T is terminal, $T \subset (\sigma \times 1_\mathcal{Q})(\tilde{N})$. Thus, $\tilde{T} \subset \tilde{N}$, and so $\tilde{T}$ is terminal. Hence, $\tilde{T} = \mathcal{T}_\mathbb{Y}(\{\tilde{x}\})$, and $T = (\sigma \times 1_\mathcal{Q})(\mathcal{T}_\mathbb{Y}(\{\tilde{x}\}))$.

Q.E.D.

We are ready to state and prove Rogers's Terminal Decomposition Theorem.

5.3.28. Theorem. *Let $\sigma \times 1_\mathcal{Q} \colon \mathbb{H} \times \mathcal{Q} \twoheadrightarrow \mathbb{F} \times \mathcal{Q}$ be the universal covering map of $\mathbb{F} \times \mathcal{Q}$. Let X be a homogeneous continuum that admits a map into the figure eight W that is not homotopic to a constant map. Hence, we may consider X essentially embedded in $\mathbb{F} \times \mathcal{Q}$. Let $\tilde{X} = (\sigma \times 1_\mathcal{Q})^{-1}(X)$. If $\mathcal{G} = \{(\sigma \times 1_\mathcal{Q})(\mathcal{T}_\mathbb{Y}(\{\tilde{x}\})) \mid \tilde{x} \in \mathbb{K}$, where $\mathbb{K}$ is a component of $\tilde{X}$, and $\mathbb{Y} = Cl_{(\mathbb{H} \times \mathcal{Q}) \cup \mathcal{S}^1}(\mathbb{K})\}$, then $\mathcal{G}$ is a continuous decomposition such that the following hold:*

(1) $\mathcal{G}$ is a monotone and terminal decomposition of X.

(2) The elements of $\mathcal{G}$ are mutually homeomorphic, indecomposable, cell–like terminal, homogeneous continua.

(3) The quotient space, $X/\mathcal{G}$, is a homogeneous continuum.

(4) $X/\mathcal{G}$ does not contain any proper, nondegenerate terminal subcontinuum.

(5) If X is decomposable, then $X/\mathcal{G}$ is an aposyndetic continuum; in fact, the decomposition $\mathcal{G}$ is Jones's decomposition.

(6) If the elements of $\mathcal{G}$ are nondegenerate, then they have the same dimension as X and $X/\mathcal{G}$ is one–dimensional.

Proof. Note that, by Theorem 5.3.27, $\mathcal{G}$ is a collection of maximal proper terminal cell–like subcontinua of X.

We show that $\mathcal{G}$ is a decomposition of X. Let $\tilde{x}_1, \tilde{x}_2 \in \tilde{X}$ such that $(\sigma \times 1_\mathcal{Q})(\mathcal{T}_{\mathbb{Y}_1}(\{\tilde{x}_1\})) \cap (\sigma \times 1_\mathcal{Q})(\mathcal{T}_{\mathbb{Y}_2}(\{\tilde{x}_2\})) \neq \emptyset$ ($\mathbb{K}_j$ is the component of $\tilde{X}$ such that $\tilde{x}_j \in \mathbb{K}_j$ and $\mathbb{Y}_j = Cl_{(\mathbb{H} \times \mathcal{Q}) \cup \mathcal{S}^1}(\mathbb{K}_j)$,

$j \in \{1, 2\}$). Let $z \in (\sigma \times 1_{\mathcal{Q}})(\mathcal{T}_{\mathbb{Y}_1}(\{\tilde{x}_1\})) \cap (\sigma \times 1_{\mathcal{Q}})(\mathcal{T}_{\mathbb{Y}_2}(\{\tilde{x}_2\}))$. Then there exists $\tilde{z}_j \in \mathcal{T}_{\mathbb{Y}_j}(\{\tilde{x}_j\})$ such that $(\sigma \times 1_{\mathcal{Q}})(\tilde{z}_j) = z$, $j \in \{1, 2\}$. Hence, by Theorem 5.3.25, $\mathcal{T}_{\mathbb{Y}_1}(\{\tilde{x}_1\})$ is homeomorphic to $\mathcal{T}_{\mathbb{Y}_2}(\{\tilde{x}_2\})$ and $(\sigma \times 1_{\mathcal{Q}})(\mathcal{T}_{\mathbb{Y}_1}(\{\tilde{x}_1\})) = (\sigma \times 1_{\mathcal{Q}})(\mathcal{T}_{\mathbb{Y}_2}(\{\tilde{x}_2\}))$. Since $\sigma \times 1_{\mathcal{Q}}$ is a covering map,

$$X = \bigcup \{(\sigma \times 1_{\mathcal{Q}})(\mathcal{T}_{\mathbb{Y}}(\{\tilde{x}\})) \mid \tilde{x} \in \mathbb{K}, \text{ where } \mathbb{K} \text{ is a component of}$$

$$\widetilde{X}, \text{ and } \mathbb{Y} = Cl_{(\mathbb{H} \times \mathcal{Q}) \cup \mathcal{S}^1}(\mathbb{K})\}.$$

Therefore, $\mathcal{G}$ is a decomposition of X.

Now, we prove that the homeomorphism group of X, $\mathcal{H}(X)$, respects $\mathcal{G}$, and then, apply Theorem 5.1.4 to conclude that $\mathcal{G}$ is continuous.

Let $h \in \mathcal{H}(X)$, and let $\tilde{x}_1, \tilde{x}_2 \in \widetilde{X}$ such that

$$h((\sigma \times 1_{\mathcal{Q}})(\mathcal{T}_{\mathbb{Y}_1}(\{\tilde{x}_1\}))) \cap (\sigma \times 1_{\mathcal{Q}})(\mathcal{T}_{\mathbb{Y}_2}(\{\tilde{x}_2\})) \neq \emptyset.$$

Since $(\sigma \times 1_{\mathcal{Q}})(\mathcal{T}_{\mathbb{Y}_1}(\{\tilde{x}_1\}))$ is terminal, $h((\sigma \times 1_{\mathcal{Q}})(\mathcal{T}_{\mathbb{Y}_1}(\{\tilde{x}_1\})))$ is terminal (Corollary 5.1.12). Thus, since $(\sigma \times 1_{\mathcal{Q}})(\mathcal{T}_{\mathbb{Y}_2}(\{\tilde{x}_2\}))$ is a maximal terminal subcontinuum,

$$h((\sigma \times 1_{\mathcal{Q}})(\mathcal{T}_{\mathbb{Y}_1}(\{\tilde{x}_1\}))) \subset (\sigma \times 1_{\mathcal{Q}})(\mathcal{T}_{\mathbb{Y}_2}(\{\tilde{x}_2\})).$$

This implies that $(\sigma \times 1_{\mathcal{Q}})(\mathcal{T}_{\mathbb{Y}_1}(\{\tilde{x}_1\})) \subset h^{-1}((\sigma \times 1_{\mathcal{Q}})(\mathcal{T}_{\mathbb{Y}_2}(\{\tilde{x}_2\})))$. Hence, $(\sigma \times 1_{\mathcal{Q}})(\mathcal{T}_{\mathbb{Y}_1}(\{\tilde{x}_1\})) = h^{-1}((\sigma \times 1_{\mathcal{Q}})(\mathcal{T}_{\mathbb{Y}_2}(\{\tilde{x}_2\})))$. Consequently, $h((\sigma \times 1_{\mathcal{Q}})(\mathcal{T}_{\mathbb{Y}_1}(\{\tilde{x}_1\}))) = (\sigma \times 1_{\mathcal{Q}})(\mathcal{T}_{\mathbb{Y}_2}(\{\tilde{x}_2\}))$. Therefore, $\mathcal{H}(X)$ respects $\mathcal{G}$.

Since $\mathcal{H}(X)$ respects $\mathcal{G}$, by Theorem 5.1.4, the elements of $\mathcal{G}$ are mutually homeomorphic homogeneous continua, and the quotient space $X/\mathcal{G}$ is homogeneous. The fact that the elements of $\mathcal{G}$ have the same dimension as X follows from Theorem 8 of [32]. The proof of the indecomposability of the elements of $\mathcal{G}$ is similar to the one given in Theorem 5.1.19. The fact that $X/\mathcal{G}$ is one–dimensional may be found in Theorem 6 of [35].

Next, we show that $X/\mathcal{G}$ does not contain nondegenerate proper terminal subcontinua. To this end, let Γ be a nondegenerate terminal proper subcontinuum of $X/\mathcal{G}$. Let $q \colon X \twoheadrightarrow X/\mathcal{G}$ be the quotient map. Then q is a monotone map. Thus, $q^{-1}(\Gamma)$ is a subcontinuum of X (Lemma 2.1.12). Note that

$$q^{-1}(\Gamma) = \bigcup \{(\sigma \times 1_{\mathcal{Q}})(\mathcal{T}_{\mathbb{Y}}(\{\tilde{x}\})), \mid \tilde{x} \in \mathbb{K}, \text{ where } \mathbb{K} \text{ is a component}$$

of $\widetilde{X}$, $\mathbb{Y} = Cl_{(\mathbb{H} \times \mathcal{Q}) \cup \mathcal{S}^1}(\mathbb{K})$ and $q((\sigma \times 1_\mathcal{Q})(\mathcal{T}_\mathbb{Y}(\{\tilde{x}\}))) \in \Gamma\}$.

Since Γ is a nondegenerate proper subcontinuum of $X/\mathcal{G}$, $q^{-1}(\Gamma)$ is a proper subcontinuum of X and if $q((\sigma \times 1_\mathcal{Q})(\mathcal{T}_\mathbb{Y}(\{\tilde{x}\}))) \in \Gamma$, $(\sigma \times 1_\mathcal{Q})(\mathcal{T}_\mathbb{Y}(\{\tilde{x}\})) \subsetneq q^{-1}(\Gamma)$. Hence, $q^{-1}(\Gamma)$ is not a terminal subcontinuum of X (each element $(\sigma \times 1_\mathcal{Q})(\mathcal{T}_\mathbb{Y}(\{\tilde{x}\}))$ of $\mathcal{G}$ is a maximal terminal proper subcontinuum of X). Thus, there exists a subcontinuum Z of X such that $Z \cap q^{-1}(\Gamma) \neq \emptyset$, $Z \setminus q^{-1}(\Gamma) \neq \emptyset$ and $q^{-1}(\Gamma) \setminus Z \neq \emptyset$. Since $q(Z \cap q^{-1}(\Gamma)) = q(Z) \cap \Gamma$, $q(Z)$ is a subcontinuum of $X/\mathcal{G}$ such that $q(Z) \cap \Gamma \neq \emptyset$. Since Γ is a terminal subcontinuum of $X/\mathcal{G}$, either $q(Z) \subset \Gamma$ or $\Gamma \subset q(Z)$.

If $q(Z) \subset \Gamma$, then $q^{-1}(q(Z)) \subset q^{-1}(\Gamma)$. Hence, $Z \subset q^{-1}(\Gamma)$, a contradiction. Suppose, then, that $\Gamma \subset q(Z)$. This implies that $q^{-1}(\Gamma) \subset q^{-1}(q(Z))$. Thus, for each $x \in q^{-1}(\Gamma)$, there exists $z \in Z$ such that $q(z) = q(x)$. Hence, $q^{-1}(q(x)) \cap Z \neq \emptyset$. Since $q^{-1}(q(x))$ is a terminal subcontinuum of X, either $Z \subset q^{-1}(q(x))$ or $q^{-1}(q(x)) \subset Z$. If $Z \subset q^{-1}(q(x))$, then $Z \subset q^{-1}(\Gamma)$, a contradiction. Thus, $q^{-1}(q(x)) \subset Z$. Since x was an arbitrary point of $q^{-1}(\Gamma)$, $q^{-1}(\Gamma) \subset Z$, a contradiction. Therefore, $X/\mathcal{G}$ does not contain nondegenerate terminal proper subcontinua.

Finally, suppose X is decomposable. Let $\mathcal{G}' = \{\mathcal{T}_X(\{x\}) \mid x \in X\}$ be Jones's decomposition (Theorem 5.1.19). We show that $\mathcal{G} = \mathcal{G}'$. Thus, by Theorem 5.1.19, $X/\mathcal{G}$ is aposyndetic.

Let $(\sigma \times 1_\mathcal{Q})(\mathcal{T}_\mathbb{Y}(\{\tilde{x}\})) \in \mathcal{G}$. Let $x \in (\sigma \times 1_\mathcal{Q})(\mathcal{T}_\mathbb{Y}(\{\tilde{x}\}))$. Then $(\sigma \times 1_\mathcal{Q})(\mathcal{T}_\mathbb{Y}(\{\tilde{x}\})) \cap \mathcal{T}_X(\{x\}) \neq \emptyset$. Since both of these continua are maximal terminal subcontinua of X (Corollary 5.1.20), $(\sigma \times 1_\mathcal{Q})(\mathcal{T}_\mathbb{Y}(\{\tilde{x}\})) = \mathcal{T}_X(\{x\})$. Thus, $(\sigma \times 1_\mathcal{Q})(\mathcal{T}_\mathbb{Y}(\{\tilde{x}\})) \in \mathcal{G}'$. Hence, $\mathcal{G} \subset \mathcal{G}'$.

Let $\mathcal{T}_X(\{x\}) \in \mathcal{G}'$. Since $\sigma \times 1_\mathcal{Q}$ is a covering map, there exists $\tilde{x} \in \widetilde{X}$ such that $(\sigma \times 1_\mathcal{Q})(\tilde{x}) = x$. Let $\mathbb{K}$ be the component of $\widetilde{X}$ such that $\tilde{x} \in \mathbb{K}$. Then $(\sigma \times 1_\mathcal{Q})(\mathcal{T}_\mathbb{Y}(\{\tilde{x}\})) \cap \mathcal{T}_X(\{x\}) \neq \emptyset$. Since both of these continua are maximal terminal subcontinua of X, $(\sigma \times 1_\mathcal{Q})(\mathcal{T}_\mathbb{Y}(\{\tilde{x}\})) = \mathcal{T}_X(\{x\})$. Thus, $\mathcal{T}_X(\{x\}) \in \mathcal{G}$. Hence, $\mathcal{G}' \subset \mathcal{G}$.

Therefore, $\mathcal{G} = \mathcal{G}'$.

Q.E.D.

5.4 Case and Minc–Rogers Continua

Aposyndetic homogeneous continua of dimension greater than one are very easy to construct, since the product of homogeneous continua is homogeneous and aposyndetic (Corollary 3.3.9). It is difficult to find one–dimensional aposyndetic homogeneous continua. In fact, R. D. Anderson showed that the simple closed curve and the Menger universal curve are the only one–dimensional locally connected homogeneous continua (Theorem XIII of [5]). Hence, if the only one–dimensional aposyndetic homogeneous continua were locally connected, then Theorem 5.1.19 would imply that it would be enough to study one–dimensional indecomposable homogeneous continua.

We present a collection of one–dimensional aposyndetic homogeneous continua which are not locally connected. We use inverse limits to construct such continua. First, we present the construction of J. T. Rogers, Jr. [29] of a continuum originally constructed by J. H. Case [7]. Afterwards, we give a sketch of the generalization of Rogers's construction made by P. Minc and J. T. Rogers, Jr. [23].

The *Sierpiński universal plane curve*, $\mathbb{S}$, can be constructed by taking the unit square with boundary B, deleting the interior of the middle–ninth of that square (leaving the boundary of such a square, C), then deleting the interiors of the middle–ninths of each of the eight squares remaining. Let us continue in this manner step by step. The points which have not been removed constitute the required curve.

The next Theorem gives a characterization of $\mathbb{S}$; a proof of it may be found in Theorem 4 of [37]:

5.4.1. Theorem. *In order that a plane one–dimensional locally connected continuum Z be the Sierpiński universal plane curve $\mathbb{S}$ it is necessary and sufficient that for each open connected subset U of Z and each point $z \in U$, $U \setminus \{z\}$ is connected.*

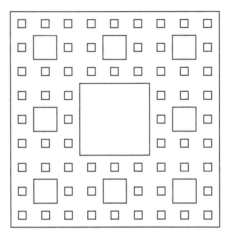

The *Menger universal curve*, $\mathbb{M}$, can be described as the set of all points of the unit cube, $[0,1]^3$, that project, in each of the x, y and z directions, onto Sierpiński universal plane curves on the faces of the cube.

5.4.2. Lemma. *Each point of the Menger universal curve has arbitrarily small neighborhoods with connected boundary.*

Proof. Note that the point $(1,0,0) \in \mathbb{M}$ has arbitrarily small neighborhoods homeomorphic to $\mathbb{M}$. Since $\mathbb{M}$ is homogeneous (Theorem III of [4]), this is true for all points of $\mathbb{M}$.

Q.E.D.

The next Theorem presents a characterization of $\mathbb{M}$; a proof of it may be found in Theorem XII of [5]:

5.4.3. Theorem. *In order that a one–dimensional locally connected continuum Z be the Menger universal curve $\mathbb{M}$ it is necessary and sufficient that for each open connected subset U of Z, U is not embeddable in the plane, and for each point $z \in U$, $U \setminus \{z\}$ is connected.*

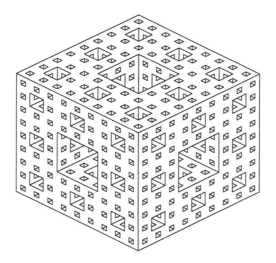

5.4.4. Definition. A covering map $\sigma\colon \widetilde{X} \twoheadrightarrow X$ is *regular* provided that for each point $x \in X$ and each pair of points $\tilde{x}_1, \tilde{x}_2 \in \sigma^{-1}(x)$, there exists a covering homeomorphism $\varphi\colon \widetilde{X} \twoheadrightarrow \widetilde{X}$ such that $\varphi(\tilde{x}_1) = \tilde{x}_2$.

5.4.5. Definition. A *solenoidal sequence* is an inverse sequence $\{X_n, f_n^{n+1}\}$ of continua such that each bonding map $f_n^m\colon X_m \twoheadrightarrow X_n$ is a regular covering map. The inverse limit of a solenoidal sequence is called a *solenoidal space*.

5.4.6. Theorem. *If the points $p = (p_n)_{n=1}^\infty$ and $q = (q_n)_{n=1}^\infty$ of the solenoidal space $X_\infty = \varprojlim\{X_n, f_n^{n+1}\}$ have the same first coordinate, then there exists a homeomorphism of X_∞ onto itself mapping p to q.*

Proof. For each $n \in \mathbb{N}$, we construct a homeomorphism $h_n\colon X_n \twoheadrightarrow X_n$ such that $h_n(p_n) = q_n$ and $f_n^{n+1} \circ h_{n+1} = h_n \circ f_n^{n+1}$. Then Theorems 2.1.46, 2.1.47 and 2.1.48 imply that $h = \varprojlim\{h_n\}$ is a homeomorphism of X_∞ onto itself such that $h(p) = q$.

Since $p_1 = q_1$, we take $h_1 = 1_{X_1}$. Note that $p_2, q_2 \in (f_1^2)^{-1}(p_1)$. Since f_1^2 is a regular covering map, there exists a covering homeomorphism $h_2\colon X_2 \twoheadrightarrow X_2$ such that $h_2(p_2) = q_2$ and $f_1^2 \circ h_2 = f_1^2 = h_1 \circ f_1^2$.

Note that $p_3, q_3 \in (f_1^3)^{-1}(p_1)$. Since f_1^3 is a regular covering map, there exists a covering homeomorphism $h_3 \colon X_3 \twoheadrightarrow X_3$ such that $h_3(p_3) = q_3$ and $f_1^3 \circ h_3 = f_1^3 = h_1 \circ f_1^3$. Observe that both $h_2 \circ f_2^3$ and $f_2^3 \circ h_3$ are liftings of f_1^3 such that $h_2 \circ f_2^3(p_3) = h_2(p_2) = q_2$ and $f_2^3 \circ h_3(p_3) = f_2^3(q_3) = q_2$. Hence, $h_2 \circ f_2^3 = f_2^3 \circ h_3$, by Proposition 5.2.5.

Suppose we have defined a covering homeomorphism $h_k \colon X_k \twoheadrightarrow X_k$ for $f_1^k \colon X_k \twoheadrightarrow X_1$ such that $h_k(p_k) = q_k$, $f_1^k \circ h_k = f_1^k$ and $h_{k-1} \circ f_{k-1}^k = f_{k-1}^k \circ h_k$ for each $k \in \{1, \dots, n-1\}$.

Since $p_n, q_n \in (f_1^n)^{-1}(p_1)$ and f_1^n is a regular covering map, there exists a covering homeomorphism $h_n \colon X_n \twoheadrightarrow X_n$ such that $h_n(p_n) = q_n$ and $f_1^n \circ h_n = f_1^n$. Again, both $h_{n-1} \circ f_{n-1}^n$ and $f_{n-1}^n \circ h_n$ are liftings of f_1^n such that $h_{n-1} \circ f_{n-1}^n(p_n) = h_{n-1}(p_{n-1}) = q_{n-1}$ and $f_{n-1}^n \circ h_n(p_n) = f_{n-1}^n(q_n) = q_{n-1}$. Hence, $h_{n-1} \circ f_{n-1}^n = f_{n-1}^n \circ h_n$ (Proposition 5.2.5).

$$\textbf{Q.E.D.}$$

5.4.7. Definition. Let V be an open subset of a metric space Y. Call Y *stably homogeneous over* V if for each pair of points $p, q \in V$, there exists a homeomorphism $h \colon Y \twoheadrightarrow Y$ such that $h(p) = q$ and $h|_{(Y \setminus V)} = 1_{(Y \setminus V)}$.

5.4.8. Definition. A metric space Y is *strongly locally homogeneous* if for each point $y \in Y$ and each open subset U of Y such that $y \in U$, there exists an open subset V of Y such that $y \in V \subset U$ and such that Y is stably homogeneous over V.

5.4.9. Definition. Let $\{X_n, f_n^{n+1}\}$ be an inverse sequence of continua. If $X_\infty = \varprojlim\{X_n, f_n^{n+1}\}$, we say that X_∞ is *locally a product of an open subset of X_1 and the Cantor set*, $\mathcal{C}$, provided that for each point $x \in X_\infty$, there exists an open subset U_1 of X_1 such that $x \in f_1^{-1}(U_1)$ and $f_1^{-1}(U_1)$ is homeomorphic to $U_1 \times \mathcal{C}$.

5.4.10. Theorem. *Let $\{X_n, f_n^{n+1}\}$ be a solenoidal sequence such that X_1 is strongly locally homogeneous. If $X_\infty = \varprojlim\{X_n, f_n^{n+1}\}$ and X_∞ is locally a product of an open subset of X_1 and the Cantor set, C, then X_∞ is homogeneous.*

Proof. Let $p = (p_n)_{n=1}^\infty \in X_\infty$. Let $\mathbb{H}$ be the set of points $q_1 \in X_1$ such that there exists a homeomorphism of X_∞ onto itself such that takes p to a point whose first coordinate is q_1. We show that $\mathbb{H}$ is both open and closed.

Let $q_1 \in \mathbb{H}$. Then there exists a homeomorphism $g \colon X_\infty \twoheadrightarrow X_\infty$ such that $f_1(g(p)) = q_1$ (recall that $f_1 \colon X_\infty \to X_1$ is the projection map).

Since X_∞ is locally a product of an open subset of X_1 and the Cantor set, there exists an open set U_1 of X_1 containing q_1 such that $f_1^{-1}(U_1)$ is homeomorphic to $U_1 \times C$.

Since X_1 is strongly locally homogeneous, there exists an open set V of X_1 such that $q_1 \in V \subset U_1$ and X_1 is stably homogeneous over V. If $r_1 \in V$, then there exists a homeomorphism $h_1 \colon X_1 \twoheadrightarrow X_1$ such that $h_1(q_1) = r_1$ and $h_1|_{(X_1 \setminus V)} = 1_{(X_1 \setminus V)}$.

Define the map $h \colon X_\infty \twoheadrightarrow X_\infty$ as follows:

$$h|_{(X_\infty \setminus f_1^{-1}(V))} = 1_{(X_\infty \setminus f_1^{-1}(V))}$$

and

$$h|_{f_1^{-1}(V)} = h_1 \times 1_C.$$

Then h is a homeomorphism, and $f_1(h(g(p))) = r_1$. Hence, $r_1 \in \mathbb{H}$. Thus, $\mathbb{H}$ is open.

A similar argument shows that $\mathbb{H}$ is closed. Hence, $\mathbb{H} = X_1$. Then, by Theorem 5.4.6, X_∞ is homogeneous.

Q.E.D.

5.4.11. Lemma. *View the Sierpiński universal plane curve $\mathbb{S}$ as a subset of an annulus $\mathbb{D}$ with boundary $B \cup C$. If $g \colon \mathbb{D} \twoheadrightarrow \mathbb{D}$ is a double–covering map, i.e., g is a covering map such that its fibres have exactly two points, such that $g(B) = B$ and $g(C) = C$, then $g^{-1}(\mathbb{S})$ is homeomorphic to $\mathbb{S}$.*

Proof. First, we show that $g^{-1}(\mathbb{S})$ is connected. Let $p_2 \in g^{-1}(\mathbb{S})$, and let $p_1 \in \mathbb{S}$ such that $g(p_2) = p_1$. Let A be an arc in $\mathbb{S}$ from p_1

to a point of B. The arc A lifts to an arc $\widetilde{A}$ in $g^{-1}(\mathbb{S})$ from p_2 to a point of B, by a proof similar to the one given in Theorem 1.3.31. Hence, $g^{-1}(\mathbb{S})$ is arcwise connected.

Since g is a covering map, g preserves local properties. Hence, by Theorem 5.4.1, $g^{-1}(\mathbb{S})$ is homeomorphic to $\mathbb{S}$.

$$\textbf{Q.E.D.}$$

5.4.12. Lemma. *View the Menger universal curve $\mathbb{M}$ as a subset of $\mathbb{D} \times [0,1]$, where $\mathbb{D}$ is an annulus with boundary $B \cup C$. If $f \colon \mathbb{D} \times [0,1] \twoheadrightarrow \mathbb{D} \times [0,1]$ is given by $f = g \times 1_{[0,1]}$, where $g \colon \mathbb{D} \twoheadrightarrow \mathbb{D}$ is a double–covering map such that $g(B) = B$ and $g(C) = C$, then $f^{-1}(\mathbb{M})$ is homeomorphic to $\mathbb{M}$.*

Proof. A proof similar to that of Lemma 5.4.11 shows that $f^{-1}(\mathbb{M})$ is arcwise connected.

Since f is a covering map, f preserves local properties. Hence, by Theorem 5.4.3, $f^{-1}(\mathbb{M})$ is homeomorphic to $\mathbb{M}$.

$$\textbf{Q.E.D.}$$

The following Theorem is a consequence of Theorem 5.6 of [22]:

5.4.13. Theorem. *If $\{X_n, f_n^{n+1}\}$ is a solenoidal sequence of manifolds, then $X_\infty = \varprojlim \{X_n, f_n^{n+1}\}$ is locally a product of an open subset of X_1 and the Cantor set $\mathcal{C}$.*

5.4.14. Definition. A continuum X is said to be *colocally connected at $x \in X$* provided that for each open subset U of X such that $x \in U$, there exists an open subset V of X such that $x \in V \subset U$ and $X \setminus V$ is connected The continuum X is *colocally connected* if it is colocally connected at each of its points.

5.4.15. Remark. Note that each colocally connected continuum is aposyndetic.

The continuum, $\mathbb{C}$, constructed in the following Theorem is known as the *Case continuum*.

5.4.16. Theorem. *Let* $f \colon \mathbb{D} \times [0,1] \twoheadrightarrow \mathbb{D} \times [0,1]$ *be the covering map of Lemma 5.4.12. If* $X_1 = \mathbb{M}$, $X_2 = f^{-1}(\mathbb{M}), \ldots, X_n = (f^n)^{-1}(\mathbb{M})$, *and if for each* $n \in \mathbb{N}$, $f_n^{n+1} = f|_{X_{n+1}}$, *then* $\{X_n, f_n^{n+1}\}$ *is a solenoidal sequence of Menger curves and* $\mathbb{C} = \varprojlim\{X_n, f_n^{n+1}\}$ *is an aposyndetic, homogeneous, one–dimensional solenoidal space that is not locally connected.*

Proof. Since the solenoidal sequence $\{X_n, f_n^{n+1}\}$ is obtained as the restriction of a solenoidal sequence of manifolds, $\mathbb{C}$ is locally homeomorphic to the product of an open subset of $X_1 = \mathbb{M}$ and the Cantor set (Theorem 5.4.13). Hence, $\mathbb{C}$ is not locally connected. Since $\mathbb{M}$ is one–dimensional, $\mathbb{C}$ is one–dimensional (15.6 of [25]).

By Theorem XVI of [5], $\mathbb{M}$ is strongly locally homogeneous. Hence, $\mathbb{C}$ is homogeneous, by Theorem 5.4.10.

To see $\mathbb{C}$ is aposyndetic, we show that $\mathbb{C}$ is colocally connected.

Since $\mathbb{C}$ is locally homeomorphic to the product of an open subset of $X_1 = \mathbb{M}$ and the Cantor set, by Proposition 2.1.9 and Lemma 5.4.2, each point of $\mathbb{C}$ has arbitrarily small neighborhoods of the form $U = f_n^{-1}(U_n)$, where U_n is an open subset of X_n with connected boundary B_n, and $Cl_{\mathbb{C}}(U)$ is homeomorphic to $Cl_{X_n}(U_n) \times \mathcal{C}$. Let $U = f_n^{-1}(U_n)$ be one of such neighborhoods, and suppose $\mathbb{C} \setminus U$ is not connected. Then there exist two disjoint closed subsets C and D of $\mathbb{C}$ such that $\mathbb{C} \setminus U = C \cup D$. Let B_n be the boundary of U_n. Note that no component of $f_n^{-1}(B_n)$ intersects both C and D. Define C' to be the union of all the components K of U such that $Cl_{\mathbb{C}}(K) \cap C \neq \emptyset$. Define D' in a similar way. Note that $C \cup C'$ and $D \cup D'$ are disjoint closed subsets of $\mathbb{C}$ whose union is $\mathbb{C}$. Since $\mathbb{C}$ is connected, either $C \cup C'$ or $D \cup D'$ is empty. Hence, $\mathbb{C} \setminus U$ is connected.

Q.E.D.

To finish this section, we sketch the construction of the Minc–Rogers continua.

Fix $n \in \mathbb{N}$ such that $n \geq 3$. Let $\mathbb{T}^n = \prod_{k=1}^{n} \mathcal{S}^1$ be the n–

dimensional torus. Let $\mathbb{Z}_\ell = \prod_{k=1}^{\ell-1}\{e\} \times \mathcal{S}^1 \times \prod_{k=\ell+1}^{n}\{e\}$, for $\ell \in \{1,\dots,n\}$, where $e = (1,0)$. One can assume that the Menger universal curve $\mathbb{M}$ is embedded in $\mathbb{T}^n$ in such a way that $\mathbb{Z}_\ell \subset \mathbb{M}$ for every $\ell \in \{1,\dots,n\}$.

For each $\boldsymbol{\alpha} \in \mathbb{Z}^n$, $\boldsymbol{\alpha} = (\alpha_1,\dots,\alpha_n)$, define $f_{\boldsymbol{\alpha}} \colon \mathbb{T}^n \to \mathbb{T}^n$ by

$$f_{\boldsymbol{\alpha}}((z_1,\dots,z_n)) = (z_1^{\alpha_1},\dots,z_n^{\alpha_n}).$$

The proof of the next Lemma is similar to the proof given for Lemma 5.4.12.

5.4.17. Lemma. *With the above notation, $f_{\boldsymbol{\alpha}}^{-1}(\mathbb{M})$ is homeomorphic to $\mathbb{M}$ containing $\mathbb{Z}_\ell$ for every $\ell \in \{1,\dots,n\}$.*

The proof of the following Theorem is similar to the one given in Theorem 5.4.16. The spaces obtained from Theorem 5.4.18 are known as *Minc–Rogers continua*.

5.4.18. Theorem. *With the above notation. Let $\Lambda = \{\boldsymbol{\alpha}^k\}_{k=1}^{\infty}$ be a sequence of elements of $\mathbb{Z}^n$. Assume that $\boldsymbol{\alpha}^k = (\alpha_1^k,\dots,\alpha_n^k)$. Let $\mathbb{M}_1^\Lambda = \mathbb{M}$, and $\mathbb{M}_k^\Lambda = f_{\boldsymbol{\alpha}^k}^{-1}(\mathbb{M}_{k-1}^\Lambda)$. If $\mathbb{M}^\Lambda = \varprojlim\{\mathbb{M}_k^\Lambda, f_{\boldsymbol{\alpha}^k}|_{\mathbb{M}_{k+1}^\Lambda}\}$, then $\mathbb{M}^\Lambda$ is a one–dimensional colocally connected homogeneous continuum.*

5.5 Covering Spaces of Some Homogeneous Continua

We study covering spaces of certain homogeneous continua.

Let us note first that, by Corollary 1.3.30, we have that $\mathbb{R}$ is a covering space of $\mathcal{S}^1$ since the exponential map is a covering map. It is easy to see that, for a given $n \in \mathbb{N}$, the map $p\colon \mathcal{S}^1 \twoheadrightarrow \mathcal{S}^1$ given by $p(z) = z^n$, where $\mathcal{S}^1$ is considered as a subset of the set of complex numbers $\mathbb{C}$, is a covering map. Hence, $\mathcal{S}^1$ is a covering space of itself in a nontrivial way.

Let X be a continuum for which we can define a map $f_{\mathbb{H}}\colon X \to W$, where W is the figure eight $C_1 \cup C_2$ that is not homotopic to a constant map. Let $g\colon X \to \mathcal{Q}$ be an embedding of X into the Hilbert cube (Theorem 1.1.16). Then the map

$$(f_{\mathbb{H}}, g)\colon X \to \mathcal{W} \times \mathcal{Q}$$

given by

$$(f_{\mathbb{H}}, g)(x) = (f_{\mathbb{H}}(x), g(x))$$

is an embedding that is not homotopic to a constant map. Let $\sigma \times 1_{\mathcal{Q}}\colon \widetilde{W} \times \mathcal{Q} \twoheadrightarrow \mathcal{W} \times \mathcal{Q}$ be the universal covering map of $W \times \mathcal{Q}$, and define $\widetilde{X}_{\mathbb{H}} = (\sigma \times 1_{\mathcal{Q}})^{-1}(X)$. We are going to study $\widetilde{X}_{\mathbb{H}}$ for several continua.

5.5.1. Definition. *A simple triod is a continuum which is the union of three arcs having only one end point in common.*

5.5.2. Theorem. *Let* $\mathbf{n} = \{n_k\}_{k=1}^{\infty}$ *be a sequence of positive integers. Let*

$$f_{\mathbb{H}}\colon \Sigma \to W$$

be a map that is not homotopic to a constant map, where Σ *is the* $\mathbf{n}$*–solenoid. Then* $\widetilde{\Sigma}_{\mathbb{H}}$ *is homeomorphic to* $\mathbb{R} \times \mathcal{C}^*$, *where* $\mathcal{C}^*$ *is the Cantor set minus one point.*

Proof. Let $f_{\mathbb{H}}\colon \Sigma \to W$ be an essential map. By Theorem of [2], we know that $\widetilde{\Sigma}_{\mathbb{H}}$ is homeomorphic to $\mathbb{K}_{\mathbb{H}} \times \widetilde{\mathbb{B}}_{\mathbb{H}}$, where $\mathbb{K}_{\mathbb{H}}$ is a component of $\widetilde{\Sigma}_{\mathbb{H}}$ and $\widetilde{\mathbb{B}}_{\mathbb{H}}$ is a zero–dimensional locally compact homogeneous space. Hence, $\mathbb{B}_{\mathbb{H}}$ is homeomorphic to $\mathcal{C}^*$.

We divide the proof in eight steps.

Step 1. $\widetilde{\Sigma}_{\mathbb{H}}$ *does not contain simple triods.*

It is known that each proper subcontinuum of Σ is an arc (Theorem 2 of [11]). Hence, Σ does not contain simple triods. Since $\sigma \times 1_{\mathcal{Q}}$ is a local homeomorphism, $\widetilde{\Sigma}_{\mathbb{H}}$ does not contain simple triods.

Step 2. $\widetilde{\Sigma}_{\mathbb{H}}$ *does not contain simple closed curves.*

Suppose $\widetilde{\Sigma}_{\mathbb{H}}$ contains a simple closed curve $\widetilde{C}$. Then $(\sigma \times 1_{\mathcal{Q}})(\widetilde{C})$ is a subcontinuum of Σ. Hence, $(\sigma \times 1_{\mathcal{Q}})(\widetilde{C})$ is an arc A (Theorem 2 of [11]). Let a be one of the end points of A. Let U be an evenly covered open set of Σ containing a. Let $x \in U \setminus A$ be in the same arc component of U containing a; so there is an arc xa from x to a contained in U, such that $xa \cap A = \{a\}$. Let $\tilde{a} \in \widetilde{C}$ be such that $(\sigma \times 1_{\mathcal{Q}})(\tilde{a}) = a$, and let $\widetilde{U}$ be an open set of $\widetilde{\Sigma}_{\mathbb{H}}$ containing $\tilde{a}$ so that $(\sigma \times 1_{\mathcal{Q}})|_{\widetilde{U}} \colon \widetilde{U} \twoheadrightarrow (\sigma \times 1_{\mathcal{Q}})(\widetilde{U}) = U$ is a homeomorphism. Let $\tilde{x} \in \widetilde{U} \cap (\sigma \times 1_{\mathcal{Q}})^{-1}(x)$. Then the arc xa can be lifted to an arc $\tilde{x}\tilde{a}$, and $\tilde{x}\tilde{a} \cap \widetilde{C} = \{\tilde{a}\}$. This implies that $\widetilde{\Sigma}_{\mathbb{H}}$ contains a simple triod, a contradiction to Step 1. Therefore, $\widetilde{\Sigma}_{\mathbb{H}}$ does not contain a simple closed curve.

Given two points $\tilde{p}$ and $\tilde{q}$ in the same arc component of $\widetilde{\Sigma}_{\mathbb{H}}$, the union of all the arcs in $\widetilde{\Sigma}_{\mathbb{H}}$ that have $\tilde{p}$ as an end point and contain $\tilde{q}$ is called a *ray* starting at $\tilde{p}$.

Step 3. *A ray $\widetilde{R}$ in $\widetilde{\Sigma}_{\mathbb{H}}$ is the union of a countable number of arcs.*

Let $\tilde{p}$ be the starting point of $\widetilde{R}$, and let $\{\tilde{p}_n\}_{n=1}^{\infty}$ be a countable dense subset of $\widetilde{R}$. We assert that $\widetilde{R}$ is the union of the arcs $\{\tilde{p}\tilde{p}_n\}_{n=1}^{\infty}$. Suppose there exists a point $\tilde{r}$ in $\widetilde{R} \setminus \bigcup_{n=1}^{\infty} \tilde{p}\tilde{p}_n$. Consider the arc $\tilde{p}\tilde{r}$. Now, $\tilde{r}$ must be in the interior of an arc. Thus, $\tilde{p}\tilde{r}$ may be extended to an arc $\tilde{p}\tilde{s}$ so that $\tilde{r}$ is contained in the relative interior of $\tilde{p}\tilde{s}$. Since $\widetilde{\Sigma}_{\mathbb{H}}$ contains no simple triods, each $\tilde{p}_n$ belongs to the arc $\tilde{p}\tilde{r}$. This implies that $\{\tilde{p}_n\}_{n=1}^{\infty}$ is not dense in $\widetilde{R}$, since no $\tilde{p}_n$ is near $\tilde{s}$, a contradiction. Therefore, $\widetilde{R} = \bigcup_{n=1}^{\infty} \tilde{p}\tilde{p}_n$.

Step 4. *For each point $\tilde{p}$ of an arc component $\widetilde{A}$ of $\widetilde{\Sigma}_{\mathbb{H}}$, $\widetilde{A}$ is the union of two rays $\widetilde{R}_1$ and $\widetilde{R}_2$ starting at $\tilde{p}$ such that $\widetilde{R}_1 \cap \widetilde{R}_2 = \{\tilde{p}\}$.*

We know that $\tilde{p}$ is in the interior of an arc $\tilde{a}\tilde{b}$. Since $\widetilde{\Sigma}_{\mathbb{H}}$ does not contain simple triods, we have that $\widetilde{\mathcal{A}}$ is the union of two rays starting at $\tilde{p}$ going through $\tilde{a}$ and $\tilde{b}$, respectively. Since $\widetilde{\Sigma}_{\mathbb{H}}$ does not contain simple closed curves, these rays intersect only at $\tilde{p}$.

Step 5. $\widetilde{\Sigma}_{\mathbb{H}}$ *has uncountably many arc components.*

Suppose $\widetilde{\Sigma}_{\mathbb{H}}$ has only countably many arc components. Then $\widetilde{\Sigma}_{\mathbb{H}}$ is the union of countably many arcs. By Theorem 1.5.12, one of these arcs contains an open set of $\widetilde{\Sigma}_{\mathbb{H}}$, a contradiction to the fact that $\widetilde{\Sigma}_{\mathbb{H}}$ is not locally connected.

Step 6. *The arc components of* $\widetilde{\Sigma}_{\mathbb{H}}$ *are unbounded.*

Let $\widetilde{\mathcal{A}}$ be an arc component of $\widetilde{\Sigma}_{\mathbb{H}}$, and let $\tilde{p} \in Cl_{\widetilde{\Sigma}_{\mathbb{H}}}(\widetilde{\mathcal{A}})$. We show that $Cl_{\widetilde{\Sigma}_{\mathbb{H}}}(\widetilde{\mathcal{A}}) = Cl_{\widetilde{\Sigma}_{\mathbb{H}}}(\widetilde{\mathcal{A}}_{\tilde{p}})$, where $\widetilde{\mathcal{A}}_{\tilde{p}}$ is the arc component of $\widetilde{\Sigma}_{\mathbb{H}}$ containing $\tilde{p}$. Suppose there exists a point $\tilde{q} \in Cl_{\widetilde{\Sigma}_{\mathbb{H}}}(\widetilde{\mathcal{A}}_{\tilde{p}}) \setminus Cl_{\widetilde{\Sigma}_{\mathbb{H}}}(\widetilde{\mathcal{A}})$. Let $\varepsilon \leq \tilde{d}(\tilde{q}, Cl_{\widetilde{\Sigma}_{\mathbb{H}}}(\widetilde{\mathcal{A}}))$ such that each ε–homeomorphism may be lifted to $\widetilde{\Sigma}_{\mathbb{H}}$ (Theorem 5.2.13). Hence, by Theorem 4.2.31, there exists an Effros number $\delta > 0$ for this ε. Let $\tilde{a} \in \widetilde{\mathcal{A}}$ such that $\tilde{d}(\tilde{a}, \tilde{p}) < \delta$. Hence, there exists an ε–homeomorphism $\tilde{h} \colon \widetilde{\Sigma}_{\mathbb{H}} \twoheadrightarrow \widetilde{\Sigma}_{\mathbb{H}}$ such that $\tilde{h}(\tilde{p}) = \tilde{a}$. Note that $\tilde{h}(\tilde{q}) \in Cl_{\widetilde{\Sigma}_{\mathbb{H}}}(\widetilde{\mathcal{A}})$; this is a contradiction. Thus, $Cl_{\widetilde{\Sigma}_{\mathbb{H}}}(\widetilde{\mathcal{A}}_{\tilde{p}}) \subset Cl_{\widetilde{\Sigma}_{\mathbb{H}}}(\widetilde{\mathcal{A}})$.

A symmetric argument shows that $Cl_{\widetilde{\Sigma}_{\mathbb{H}}}(\widetilde{\mathcal{A}}) \subset Cl_{\widetilde{\Sigma}_{\mathbb{H}}}(\widetilde{\mathcal{A}}_{\tilde{p}})$.

Therefore, $Cl_{\widetilde{\Sigma}_{\mathbb{H}}}(\widetilde{\mathcal{A}}) = Cl_{\widetilde{\Sigma}_{\mathbb{H}}}(\widetilde{\mathcal{A}}_{\tilde{p}})$.

Let

$$\mathcal{G} = \{Cl_{\widetilde{\Sigma}_{\mathbb{H}}}(\widetilde{\mathcal{A}}) \mid \widetilde{\mathcal{A}} \text{ is an arc component of } \widetilde{\Sigma}_{\mathbb{H}}\}.$$

The above argument shows that the elements of $\mathcal{G}$ are pairwise disjoint.

Suppose that one arc component of $\widetilde{\Sigma}_{\mathbb{H}}$ is bounded. Then all of them are. Hence, $\mathcal{G}$ is an uncountable collection of continua (Step 5). Since each arc component of Σ is dense, $\sigma \times 1_{\mathcal{Q}}$ sends each element of $\mathcal{G}$ onto Σ, and this contradicts the fact that the fibres of $\sigma \times 1_{\mathcal{Q}}$ are countable.

Step 7. *Each arc component of* $\widetilde{\Sigma}_{\mathbb{H}}$ *is a closed subset of* $\widetilde{\Sigma}_{\mathbb{H}}$ *which is homeomorphic to* $\mathbb{R}$.

Since Σ does not contain simple closed curves, by Corollary 8 of [31], there is a one–to–one map $j\colon \mathbb{R} \to \Sigma$ such that $j(\mathbb{R})$ is a dense arc component of Σ. If $\mathbb{K}_{\mathbb{H}}$ is a component of $\widetilde{\Sigma}_{\mathbb{H}}$, then let $\tilde{j}\colon \mathbb{R} \to \mathbb{K}_{\mathbb{H}}$ be a lift of j so that $\tilde{j}(\mathbb{R}) \subset \mathbb{K}_{\mathbb{H}}$ (Lemma 79.1 of [24]).

Let $\varepsilon > 0$ be a positive number such that each ε–homeomorphism of Σ lifts to an ε–homeomorphism of $\widetilde{\Sigma}_{\mathbb{H}}$ (Theorem 5.2.13), and let $2\delta > 0$ be an Effros number for this ε (Theorem 4.2.31).

Let $v \in W$ be the common point of C_1 and C_2. Cover $\{v\} \times \mathcal{Q}$ with a finite collection $\mathcal{B}$ of δ–balls. Since the arc components of $\widetilde{\Sigma}_{\mathbb{H}}$ are unbounded (Step 6) and $\mathcal{B}$ is finite, we can find an arc $\widetilde{A}_0$, contained in $\tilde{j}(\mathbb{R})$, whose end points are in two different liftings, $\widetilde{B}_0$ and $\widetilde{B}_1$, of the same element B of $\mathcal{B}$.

Let $\widetilde{A}_0 = \tilde{a}_0\tilde{a}_1$, where $\tilde{a}_k \in \widetilde{B}_k$, $k \in \{0,1\}$. Let $(\varphi \times 1_{\mathcal{Q}})\colon \mathbb{H} \times \mathcal{Q} \to \mathbb{H} \times \mathcal{Q}$ be the covering homeomorphism that maps $\widetilde{B}_0$ onto $\widetilde{B}_1$. Since $(\varphi \times 1_{\mathcal{Q}})(\tilde{a}_0) \in \widetilde{B}_1$, there is an ε–homeomorphism $\tilde{h}\colon \widetilde{\Sigma}_{\mathbb{H}} \to \widetilde{\Sigma}_{\mathbb{H}}$ such that $\tilde{h} \circ (\varphi \times 1_{\mathcal{Q}})(\tilde{a}_0) = \tilde{a}_1$. Then $\widetilde{A}_1 = \widetilde{A}_0 \cup \tilde{h} \circ (\varphi \times 1)(\widetilde{A}_0)$ is an arc, because $\tilde{a}_1 \in \widetilde{A}_0 \cap \tilde{h} \circ (\varphi \times 1)(\widetilde{A}_0)$ and $\widetilde{\Sigma}_{\mathbb{H}}$ does not contain either simple triods or simple closed curves. If we call $\tilde{a}_2 = \tilde{h} \circ (\varphi \times 1_{\mathcal{Q}})(\tilde{a}_1)$, we may write $\widetilde{A}_1 = \tilde{a}_0\tilde{a}_1\tilde{a}_2$.

Now observe that $\tilde{h} \circ (\varphi \times 1_{\mathcal{Q}})(\widetilde{A}_1)$ is an arc, and $\widetilde{A}_1 \cap \tilde{h} \circ (\varphi \times 1_{\mathcal{Q}})(\widetilde{A}_1) = \tilde{a}_1\tilde{a}_2$. Thus, we have that $\widetilde{A}_2 = \widetilde{A}_1 \cup \tilde{h} \circ (\varphi \times 1_{\mathcal{Q}})(\widetilde{A}_1)$ is an arc. If $\tilde{a}_3 = \tilde{h} \circ (\varphi \times 1_{\mathcal{Q}})(\tilde{a}_2)$ we may write $\widetilde{A}_2 = \tilde{a}_0\tilde{a}_1\tilde{a}_2\tilde{a}_3$.

If we continue in this way, we obtain a sequence $\{\widetilde{A}_n\}_{n=1}^{\infty}$ of arcs so that the "right" end points of the arcs tend to the attracting point z^+ of $\varphi \times 1_{\mathcal{Q}}$.

Similarly, using $\left(\tilde{h} \circ (\varphi \times 1_{\mathcal{Q}})\right)^{-1}$, we can construct a sequence $\{\widetilde{A}_{-n}\}_{n=1}^{\infty}$ of arcs having their "left" end points tending to the repelling point z^- of $\varphi \times 1_{\mathcal{Q}}$.

Thus, $\tilde{j}(\mathbb{R})$ is closed, and each point of $\tilde{j}(\mathbb{R})$ has a neighborhood homeomorphic to an open interval. Therefore $\tilde{j}(\mathbb{R})$ is homeomorphic to $\mathbb{R}$.

Step 8. *The components of $\widetilde{\Sigma}_{\mathbb{H}}$ are homeomorphic to $\mathbb{R}$.*

We prove that each arc component of $\widetilde{\Sigma}_{\mathbb{H}}$ is a component of $\widetilde{\Sigma}_{\mathbb{H}}$. By Corollary of [2], we only need to show that each arc component is a quasicomponent, and for this, it is enough to prove that if $\widetilde{\mathcal{A}}_1$

and $\widetilde{\mathcal{A}}_2$ are two distinct arc components of $\widetilde{\Sigma}_{\mathbb{H}}$, then there is a closed and open subset of $\widetilde{\Sigma}_{\mathbb{H}}$ containing $\widetilde{\mathcal{A}}_1$ which is disjoint from $\widetilde{\mathcal{A}}_2$.

Let $\widetilde{\mathcal{A}}_1$ and $\widetilde{\mathcal{A}}_2$ be two different arc components of $\widetilde{\Sigma}_{\mathbb{H}}$. Let $\tilde{a}_1 \in \widetilde{\mathcal{A}}_1$. Since $\widetilde{\Sigma}_{\mathbb{H}}$ is locally homeomorphic to Σ, we can find a compact set of $\widetilde{\Sigma}_{\mathbb{H}}$ of the form $[r, s] \times \mathcal{C}$, such that $(r, s) \times \mathcal{C}$ is an open subset of $\widetilde{\Sigma}_{\mathbb{H}}$ containing $\tilde{a}_1$. Let $\widetilde{U}$ be a closed and open subset of $(r, s) \times \mathcal{C}$ containing $\tilde{a}_1$ such that $\widetilde{U} \cap \widetilde{\mathcal{A}}_2 = \emptyset$.

Let $\widetilde{V}$ be the union of all the arc components $\widetilde{\mathcal{A}}$ of $\widetilde{\Sigma}_{\mathbb{H}}$ such that $\widetilde{\mathcal{A}} \cap \widetilde{U} \neq \emptyset$. Clearly $\widetilde{\mathcal{A}}_1 \subset \widetilde{V}$ and $\widetilde{\mathcal{A}}_2 \cap \widetilde{V} = \emptyset$. We show that $\widetilde{V}$ is a closed and open subset of $\widetilde{\Sigma}_{\mathbb{H}}$.

First, we show that $\widetilde{V}$ is open. Let $\tilde{x} \in \widetilde{V}$ and let $\widetilde{\mathcal{A}}_{\tilde{x}}$ be the arc component of $\widetilde{\Sigma}_{\mathbb{H}}$ containing $\tilde{x}$. Let $\tilde{x}_1 \in \widetilde{\mathcal{A}}_{\tilde{x}} \cap \widetilde{U}$. Take $\varepsilon > 0$ such that each ε–homeomorphism of Σ may be lifted to $\widetilde{\Sigma}_{\mathbb{H}}$ (Theorem 5.2.13), and $\mathcal{V}_{\varepsilon}^{\tilde{d}}(\tilde{x}_1) \subset \widetilde{U}$. Let $\delta > 0$ be an Effros number for this ε (Theorem 4.2.31). Let $\tilde{x}_2 \in \mathcal{V}_{\delta}^{\tilde{d}}(\tilde{x})$. Then there is an ε–homeomorphism $\tilde{h} \colon \widetilde{\Sigma}_{\mathbb{H}} \twoheadrightarrow \widetilde{\Sigma}_{\mathbb{H}}$ such that $\tilde{h}(\tilde{x}) = \tilde{x}_2$. Hence, $\tilde{h}(\tilde{x}) \in \widetilde{\mathcal{A}}_{\tilde{x}_2}$, $\widetilde{\mathcal{A}}_{\tilde{x}_2}$ being the arc component of $\widetilde{\Sigma}_{\mathbb{H}}$ containing $\tilde{x}_2$. Therefore, $\tilde{d}\left(\tilde{h}(\tilde{x}_1), \tilde{x}_1\right) < \varepsilon$. Thus, $\widetilde{\mathcal{A}}_{\tilde{x}_2} \cap \widetilde{U} \neq \emptyset$. Consequently, $\tilde{x}_2 \in \widetilde{V}$. Therefore, $\widetilde{V}$ is open.

Now, we prove that $\widetilde{V}$ is closed. Let $\tilde{x} \in Cl_{\widetilde{\Sigma}_{\mathbb{H}}}(\widetilde{V})$, and let $\widetilde{\mathcal{A}}_{\tilde{x}}$ be the arc component of $\widetilde{\Sigma}_{\mathbb{H}}$ containing $\tilde{x}$. For every $n \in \mathbb{N}$, let $\delta_n > 0$ be an Effros number for $\dfrac{1}{n}$. Let $\tilde{x}_n \in \widetilde{V} \cap \mathcal{V}_{\delta_n}^{\tilde{d}}(\tilde{x})$. For each $n \in \mathbb{N}$, there is a $\dfrac{1}{n}$–homeomorphism $\tilde{h}_n \colon \widetilde{\Sigma}_{\mathbb{H}} \twoheadrightarrow \widetilde{\Sigma}_{\mathbb{H}}$ so that $\tilde{h}_n(\tilde{x}_n) = \tilde{x}$.

Let $\widetilde{\mathcal{A}}_{\tilde{x}_n}$ be the arc component of $\widetilde{\Sigma}_{\mathbb{H}}$ containing $\tilde{x}_n$. Let $\tilde{x}_n' \in \widetilde{\mathcal{A}}_{\tilde{x}_n} \cap \widetilde{U}$. Then for each $n \in \mathbb{N}$, $\tilde{h}_n(\tilde{x}_n') \in \widetilde{\mathcal{A}}_{\tilde{x}}$, and $\tilde{d}\left(\tilde{x}_n', \tilde{h}_n(\tilde{x}_n')\right) < \dfrac{1}{n}$.

Since $\widetilde{U}$ is compact, without loss of generality, we assume that the sequence, $\{\tilde{x}_n'\}_{n=1}^{\infty}$, converges to $\tilde{x}'$, where $\tilde{x}' \in \widetilde{U}$. Let $\varepsilon > 0$ be given, and let $n \in \mathbb{N}$ be such that $\dfrac{1}{n} < \dfrac{\varepsilon}{2}$ and $d(\tilde{x}_n', \tilde{x}') < \dfrac{\varepsilon}{2}$. Then

$$\tilde{d}\left(\tilde{x}', \tilde{h}_n(\tilde{x}_n')\right) \leq d(\tilde{x}', \tilde{x}_n') + \tilde{d}\left(\tilde{x}_n', \tilde{h}_n(\tilde{x}_n')\right) < \frac{\varepsilon}{2} + \frac{1}{n} < \varepsilon.$$

Hence, for every $\varepsilon > 0$, $\mathcal{V}_{\varepsilon}^{\tilde{d}}(\tilde{x}') \cap \widetilde{\mathcal{A}}_{\tilde{x}} \neq \emptyset$. Therefore, $\tilde{x}' \in Cl_{\widetilde{\Sigma}_{\mathbb{H}}}(\widetilde{\mathcal{A}}_{\tilde{x}}) =$

$\widetilde{\mathcal{A}}_{\tilde{x}}$ (Step 7), and $\widetilde{\mathcal{A}}_{\tilde{x}} \cap \widetilde{U} \neq \emptyset$. Thus $\tilde{x} \in \widetilde{V}$, and $\widetilde{V}$ is closed.

Therefore, $\widetilde{\Sigma}_{\mathbb{H}}$ is homeomorphic to $\mathbb{R} \times \mathcal{C}^*$.

Q.E.D.

5.5.3. Theorem. *Let $\mathbb{M}$ be the Menger universal curve. If $f_{\mathbb{H}} \colon \mathbb{M} \to W$ is a map that is not homotopic to a constant map, then $\widetilde{\mathbb{M}}_{\mathbb{H}}$ has countably many components, and for each component $\mathbb{K}_{\mathbb{H}}$ of $\widetilde{\mathbb{M}}_{\mathbb{H}}$, $\mathbb{Y}_{\mathbb{K}_{\mathbb{H}}} = Cl_{(\mathbb{H} \times \mathcal{Q}) \cup \mathcal{S}^1}(\mathbb{K}_{\mathbb{H}})$ is homeomorphic to $\mathbb{M}$.*

Proof. By Theorem of [2], $\widetilde{\mathbb{M}}_{\mathbb{H}}$ is homeomorphic to $\mathbb{K}_{\mathbb{H}} \times \widetilde{\mathbb{B}}_{\mathbb{H}}$. Since $\widetilde{\mathbb{M}}_{\mathbb{H}}$ is locally homeomorphic to $\mathbb{M}$, and $\mathbb{M}$ is locally connected, $\widetilde{\mathbb{B}}_{\mathbb{H}}$ is an at most countable and discrete space. Therefore, $\mathbb{K}_{\mathbb{H}}$ is locally homeomorphic to $\mathbb{M}$. Thus, $\mathbb{K}_{\mathbb{H}}$ is connected, locally arcwise connected (hence arcwise connected), and locally compact. Note that for each connected subset $\widetilde{U}$ of $\mathbb{K}_{\mathbb{H}}$ and each point $\tilde{z} \in \widetilde{U}$, $\widetilde{U} \setminus \{\tilde{z}\}$ is connected, and no open set of $\mathbb{K}_{\mathbb{H}}$ can be embedded in the plane.

By Theorem 5.3.21, $\mathbb{Y}_{\mathbb{K}_{\mathbb{H}}}$ is connected im kleinen at each point of $\mathbb{E}(\mathbb{K}_{\mathbb{H}}) = \mathbb{Y}_{\mathbb{K}_{\mathbb{H}}} \setminus \mathbb{K}_{\mathbb{H}}$. Hence $\mathbb{Y}_{\mathbb{K}_{\mathbb{H}}}$ is locally connected.

Now, we show that for each point $\tilde{z}$ of point of $\mathbb{E}(\mathbb{K}_{\mathbb{H}})$ and each connected subset $\widetilde{U}$ of $\mathbb{Y}_{\mathbb{K}_{\mathbb{H}}}$, such that $\tilde{z} \in \widetilde{U}$, $\widetilde{U} \setminus \{\tilde{z}\}$ is connected.

Let $\tilde{z} \in \mathbb{E}(\mathbb{K}_{\mathbb{H}})$ and let $\widetilde{U}$ be a connected open set of $\mathbb{Y}_{\mathbb{K}_{\mathbb{H}}}$ containing $\tilde{z}$. Observe that $\widetilde{U}$ is arcwise connected. Let $\widetilde{U}'$ be an open connected subset of $\mathbb{Y}_{\mathbb{K}_{\mathbb{H}}}$ containing $\tilde{z}$ such that $Cl_{\mathbb{Y}_{\mathbb{K}_{\mathbb{H}}}}(\widetilde{U}') \subset \widetilde{U}$. The projection of $\widetilde{U}' \setminus \mathbb{E}(\mathbb{K}_{\mathbb{H}})$ into $\widetilde{W}$ contains an infinite number of vertices of $\widetilde{W}$. Let $\widetilde{V}'$ be the set of all vertices $\tilde{v}$ of $\widetilde{W}$ contained in the projection of $\widetilde{U}' \setminus \mathbb{E}(\mathbb{K}_{\mathbb{H}})$ into $\widetilde{W}$ for which $\mathcal{Q}_{\tilde{v}} \cap \mathbb{K}_{\mathbb{H}} \subset \widetilde{U}'$, where $\mathcal{Q}_{\tilde{v}} = \{\tilde{v}\} \times \mathcal{Q}$.

Let $v \in W$ be the common point of C_1 and C_2. Cover $(\{v\} \times \mathcal{Q}) \cap \mathbb{M}$ with a finite open cover $\mathcal{B}$, where each element of $\mathcal{B}$ is evenly covered and arcwise connected. We take the elements of $\mathcal{B}$ small enough such that for each $\tilde{v} \in \widetilde{V}'$, the lifting $\mathcal{B}_{\tilde{v}}$ of $\mathcal{B}$, covering $\mathcal{Q}_{\tilde{v}} \cap \mathbb{K}_{\mathbb{H}}$, is such that $\bigcup \mathcal{B}_{\tilde{v}} \subset \widetilde{U}$.

Suppose that $\widetilde{U} \setminus \{\tilde{z}\}$ is not connected. Then $\widetilde{U} \setminus \{\tilde{z}\}$ can be written as $\widetilde{U}_1 \cup \widetilde{U}_2$, where $\widetilde{U}_1$ and $\widetilde{U}_2$ are disjoint open sets of $\widetilde{U}$.

We assert that there is an infinite subset $\widetilde{V}$ of $\widetilde{V}'$ with the property that if $\tilde{v} \in \widetilde{V}$, then $\bigcup \mathcal{B}_{\tilde{v}}$ is totally contained in $\widetilde{U}_1$ or in $\widetilde{U}_2$. To show this, let $\tilde{v}' \in \widetilde{V}'$, and suppose that $\mathcal{B}_{\tilde{v}'} = \{B_{\tilde{v}',1}, \ldots, B_{\tilde{v}',\ell}\}$. Let $\tilde{x}_{\tilde{v}',j} \in B_{\tilde{v}',j}$. Since $\mathbb{K}_\mathbb{H}$ is arcwise connected, for each $j \in \{1, \ldots, \ell - 1\}$, there is an arc $\alpha_j \colon [0,1] \to \mathbb{K}_\mathbb{H}$ such that $\alpha_j(0) = \tilde{x}_{\tilde{v}',1}$ and $\alpha_j(1) = \tilde{x}_{\tilde{v}',j+1}$. Since $\bigcup_{j=1}^{\ell-1} \alpha_j([0,1])$ is compact, there is a covering homeomorphism $(\varphi \times 1_\mathcal{Q}) \colon \mathbb{K}_\mathbb{H} \to \mathbb{K}_\mathbb{H}$ such that

$$(\varphi \times 1_\mathcal{Q}) \left(\bigcup_{j=1}^{\ell-1} \alpha_j([0,1]) \right) \subset \widetilde{U}.$$

Let $(\varphi \times 1_\mathcal{Q})(\tilde{v}') = \tilde{v}$. Since $\bigcup \mathcal{B}_{\tilde{v}} \cup (\varphi \times 1_\mathcal{Q}) \left(\bigcup_{j=1}^{\ell-1} \alpha_j([0,1]) \right)$ is connected, it is contained in $\widetilde{U}_1$ or in $\widetilde{U}_2$. Hence $\bigcup \mathcal{B}_{\tilde{v}}$ is totally contained in $\widetilde{U}_1$ or in $\widetilde{U}_2$. Since this was true for every $\tilde{v}' \in \widetilde{V}'$, such a subset $\widetilde{V}$ of $\widetilde{V}'$ exists.

Now, we assert that $\bigcup_{\tilde{v} \in \widetilde{V}} \left(\bigcup \mathcal{B}_{\tilde{v}} \right)$ is contained in $\widetilde{U}_1$ or in $\widetilde{U}_2$. To see this, let $\tilde{v}_1$ and $\tilde{v}_2$ in $\widetilde{V}$ and suppose that $\bigcup \mathcal{B}_{\tilde{v}_1} \subset \widetilde{U}_1$ and $\bigcup \mathcal{B}_{\tilde{v}_2} \subset \widetilde{U}_2$. Let $\tilde{x}_k \in \bigcup \mathcal{B}_{\tilde{v}_k}$, $k \in \{1,2\}$, then $\tilde{x}_k = (\tilde{w}_k, q_k)$, where $\tilde{w}_k \in \widetilde{W}$ and $q_k \in \mathcal{Q}$. Now, $\tilde{w}_k$ and $\tilde{z}$ determine a unique arc $[\tilde{w}_k, \tilde{z}]$ in $Cl_\mathbb{H}(\widetilde{W})$, from $\tilde{w}_k$ to $\tilde{z}$ (p. 284 of [34]). Now $[\tilde{w}_1, \tilde{z}] \cap [\tilde{w}_2, \tilde{z}]$ contains an element $\tilde{v}$ of $\widetilde{V}$. Without loss of generality, we assume that $\bigcup \mathcal{B}_{\tilde{v}}$ is contained in $\widetilde{U}_2$. Since $\widetilde{U}$ is arcwise connected, there exists an $\alpha \colon [0,1] \to \widetilde{U}$ such that $\alpha(0) = \tilde{x}_1$ and $\alpha(1) = \tilde{x}_2$. Since $\tilde{x}_k \in \widetilde{U}_k$, there is a $t \in [0,1]$ such that $\alpha(t) = \tilde{z}$. On the other hand, any arc from $\tilde{x}_1$ to $\tilde{z}$ must intersect $\mathcal{Q}_{\tilde{v}} \cap \mathbb{K}_\mathbb{H}$, which is a contradiction. Therefore, $\bigcup_{\tilde{v} \in \widetilde{V}} \left(\bigcup \mathcal{B}_{\tilde{v}} \right)$ is contained in $\widetilde{U}_1$ or in $\widetilde{U}_2$, say $\widetilde{U}_1$.

A similar argument shows that $\widetilde{U}_2$ is empty. Hence, $\widetilde{U} \setminus \{\tilde{z}\}$ is connected. Therefore, by Theorem 5.4.3, $\mathbb{Y}_{\mathbb{K}_\mathbb{H}}$ is homeomorphic to $\mathbb{M}$.

Q.E.D.

In the following Theorem we construct a covering space of the Case continuum whose components are locally connected.

5.5.4. Theorem. *The Case continuum $\mathbb{C}$ has a covering space with locally connected components.*

Proof. Note that, by Theorem 5.4.16, $\mathbb{C} \subset \Sigma \times [0,1]^2$, where Σ is the dyadic solenoid.

First, we use inverse limits to construct a covering space of $\Sigma \times [0,1]^2$.

Let $Y_n = \mathcal{S}^1 \times [0,1]^2$, and let $\widetilde{Y}_n = \mathbb{R} \times [0,1]^2 \times \{2^{n-1}\text{th roots of unity}\}$. Define the bonding maps $f_n^{n+1} \colon Y_{n+1} \twoheadrightarrow Y_n$ and $\tilde{f}_n^{n+1} \colon \widetilde{Y}_{n+1} \twoheadrightarrow \widetilde{Y}_n$ by $f_n^{n+1}((z,x)) = (z^2, x)$, and $\tilde{f}_n^{n+1}((r,x,t)) = (2r, x, t^2)$, respectively. Let $q_n \colon \widetilde{Y}_n \to Y_n$ be the covering map given by $q_n((r,x,t)) = (t\exp(2\pi r), x)$. Thus we can consider the following infinite ladder:

$$
\begin{array}{ccccccccc}
\cdots & \longleftarrow & \widetilde{Y}_{n-1} & \overset{f_{n-1}^n}{\longleftarrow} & \widetilde{Y}_n & \overset{f_n^{n+1}}{\longleftarrow} & \widetilde{Y}_{n+1} & \longleftarrow & \cdots \quad : \widetilde{Y}_\infty \\
& & \downarrow q_{n-1} & & \downarrow q_n & & \downarrow q_{n+1} & & \downarrow q_\infty \\
\cdots & \longleftarrow & Y_{n-1} & \underset{\tilde{f}_{n-1}^n}{\longleftarrow} & Y_n & \underset{\tilde{f}_n^{n+1}}{\longleftarrow} & Y_{n+1} & \longleftarrow & \cdots \quad : Y_\infty
\end{array}
$$

where $\widetilde{Y}_\infty = \varprojlim\{\widetilde{Y}_n, \tilde{f}_n^{n+1}\}$, $Y_\infty = \varprojlim\{Y_n, f_n^{n+1}\}$ and $q_\infty = \varprojlim\{q_n\}$. Note that $\widetilde{Y}_\infty$ is homeomorphic to $\mathbb{R} \times [0,1]^2 \times \mathcal{C}$ and Y_∞ is homeomorphic to $\Sigma \times [0,1]^2$.

Let $(\mathbb{R} \times [0,1]^2 \times \{t\})_{n+1}$ be a component of $\widetilde{Y}_{n+1}$. Note that the restriction of the map $\tilde{f}_n^{n+1}$ to this component, $\tilde{f}_n^{n+1}|_{(\mathbb{R}\times[0,1]^2\times\{t\})_{n+1}} \colon (\mathbb{R} \times [0,1]^2 \times \{t\})_{n+1} \subset \widetilde{Y}_{n+1} \twoheadrightarrow (\mathbb{R} \times [0,1]^2 \times \{t^2\})_n \subset \widetilde{Y}_n$, is a homeomorphism, since the map $g_{n+1}^n \colon (\mathbb{R} \times [0,1]^2 \times \{t^2\})_n \subset \widetilde{Y}_n \twoheadrightarrow (\mathbb{R} \times [0,1]^2 \times \{t\})_{n+1} \subset \widetilde{Y}_{n+1}$ given by $g_{n+1}^n((r,x,t^2)) = (\frac{1}{2}r, x, t)$ is its inverse.

Let $\tilde{t} = (t_n)_{n=1}^\infty \in \mathcal{C}$, and let $(\mathbb{R} \times [0,1]^2 \times \{\tilde{t}\})_\infty$ be a component of $\widetilde{Y}_\infty$. Consider the map $\tilde{f}_n|_{(\mathbb{R}\times[0,1]^2\times\{\tilde{t}\})_\infty} \colon (\mathbb{R} \times [0,1]^2 \times \{\tilde{t}\})_\infty \subset \widetilde{Y}_\infty \twoheadrightarrow (\mathbb{R} \times [0,1]^2 \times \{t_n\})_n \subset \widetilde{Y}_n$. We claim that this map is a homeomorphism. To see this, let us define the map $g^n \colon (\mathbb{R} \times [0,1]^2 \times \{t_n\})_n \subset \widetilde{Y}_n \twoheadrightarrow (\mathbb{R} \times [0,1]^2 \times \{\tilde{t}\})_\infty \subset \widetilde{Y}_\infty$ by

$$g^n((r, x, t_n)) = (f_1^n((r, x, t_n)), \dots, f_{n-1}^n((r, x, t_n)), (r, x, t_n),$$

$$g_{n+1}^n((r, x, t_n)), \dots, g_{n+m}^n((r, x, t_n)), \dots)$$

where $g_{n+m}^n = g_{n+m}^{n+m-1} \circ \cdots \circ g_{n+1}^n$. Since $g^n \circ \tilde{f}_n|_{(\mathbb{R} \times [0,1]^2 \times \{\tilde{t}\})_\infty} = 1_{(\mathbb{R} \times [0,1]^2 \times \{\tilde{t}\})_\infty}$, and $\tilde{f}_n|_{(\mathbb{R} \times [0,1]^2 \times \{\tilde{t}\})_\infty} \circ g^n = 1_{(\mathbb{R} \times [0,1]^2 \times \{t_n\})_n}$, we have that $\tilde{f}_n|_{(\mathbb{R} \times [0,1]^2 \times \{\tilde{t}\})_\infty}$ is a homeomorphism.

Now, assume that the Menger curve $\mathbb{M}$ is embedded in Y_1, as described in Lemma 5.4.12. Since $(f_1^2)^{-1}(\mathbb{M})$ is homeomorphic to $\mathbb{M}$, setting $X_1 = \mathbb{M}$, $X_2 = (f_1^2)^{-1}(\mathbb{M}), \dots, X_n = (f_1^n)^{-1}(\mathbb{M})$, we obtain that $\mathbb{C} = \varprojlim\{X_n, f_n^{n+1}\}$.

Let $\tilde{X}_n = q_n^{-1}(X_n)$. Note that $\tilde{X}_n$ is a locally compact, metric space, with 2^{n-1} components, and each component is locally homeomorphic to $\mathbb{M}$.

Let $\tilde{\mathbb{C}} = \varprojlim\{\tilde{X}_n, \tilde{f}_n^{n+1}\}$. Then $\tilde{\mathbb{C}}$ is a locally compact, metric space. Let $\mathbb{K}$ be a component of $\tilde{\mathbb{C}}$. To see that $\mathbb{K}$ is locally connected, note that $\mathbb{K}$ is contained in one of the components of $\tilde{Y}_\infty$. Let $(\mathbb{R} \times [0,1]^2 \times \{\tilde{t}\})_\infty$ be the component of $\tilde{Y}_\infty$ containing $\mathbb{K}$. Since $\tilde{f}_1|_{(\mathbb{R} \times [0,1]^2 \times \{\tilde{t}\})_\infty}$ is a homeomorphism, we have that $\tilde{f}_1|_{\mathbb{K}}$ is also a homeomorphism onto $\tilde{f}_1(\mathbb{K})$, but $\tilde{f}_1(\mathbb{K}) = \tilde{X}_1$. Since $\tilde{X}_1$ is locally homeomorphic to $\mathbb{M}$, $\mathbb{K}$ is locally homeomorphic to $\mathbb{M}$. Thus $\mathbb{K}$ is locally connected, because $\mathbb{M}$ is locally connected.

Q.E.D.

To finish this chapter, we show two different covering spaces of one of the Minc–Rogers continua (Theorem 5.4.18).

5.5.5. Theorem. *Let Λ be the constant sequence $\{(2, 2, 1)\}$. If $\mathbb{MR} = \mathbb{M}^\Lambda$, then $\mathbb{MR}$ has a covering space whose components are not locally connected.*

Proof. Note that, by Theorem 5.4.18, $\mathbb{MR} \subset \Sigma \times \Sigma \times \mathcal{S}^1$, where Σ is the dyadic solenoid.

First, we construct a covering space for $\Sigma \times \Sigma \times \mathcal{S}^1$ using inverse limits.

Let $Y_n = \mathcal{S}^1 \times \mathcal{S}^1 \times \mathcal{S}^1$, and $\widetilde{Y}_n = \mathbb{R} \times \mathcal{S}^1 \times \mathcal{S}^1 \times \{2^{n-1}\text{th roots of unity}\}$. Define the bonding maps

$$f_n^{n+1} \colon Y_{n+1} \twoheadrightarrow Y_n \text{ and } \tilde{f}_n^{n+1} \colon \widetilde{Y}_{n+1} \twoheadrightarrow \widetilde{Y}_n$$

by

$$f_n^{n+1}((z_1, z_2, z_3)) = (z_1^2, z_2^2, z_3) \text{ and } \tilde{f}_n^{n+1}((r, z_2, z_3, t)) = (2r, z_2^2, z_3, t^2),$$

respectively. Let $q_n \colon \widetilde{Y}_n \twoheadrightarrow Y_n$ be the map given by $q_n((r, z_2, z_3, t)) = (t \exp(2\pi r), z_2, z_3)$. Note that q_n is a covering map. Hence, we have the following infinite ladder:

$$
\begin{array}{ccccccccc}
\cdots & \longleftarrow & \widetilde{Y}_{n-1} & \xleftarrow{f_{n-1}^n} & \widetilde{Y}_n & \xleftarrow{f_n^{n+1}} & \widetilde{Y}_{n+1} & \longleftarrow \cdots & : \widetilde{Y}_\infty \\
 & & \Big\downarrow{\scriptstyle q_{n-1}} & & \Big\downarrow{\scriptstyle q_n} & & \Big\downarrow{\scriptstyle q_{n+1}} & & \Big\downarrow{\scriptstyle q_\infty} \\
\cdots & \longleftarrow & Y_{n-1} & \xleftarrow[\tilde{f}_{n-1}^n]{} & Y_n & \xleftarrow[\tilde{f}_n^{n+1}]{} & Y_{n+1} & \longleftarrow \cdots & : Y_\infty
\end{array}
$$

where $\widetilde{Y}_\infty = \varprojlim\{\widetilde{Y}_n, \tilde{f}_n^{n+1}\}$, $Y_\infty = \varprojlim\{Y_n, f_n^{n+1}\}$ and $q_\infty = \varprojlim\{q_n\}$. Note that $\widetilde{Y}_\infty$ is homeomorphic to $\mathbb{R} \times \Sigma \times \mathcal{S}^1 \times \mathcal{C}$ and Y_∞ is homeomorphic to $\Sigma \times \Sigma \times \mathcal{S}^1$.

Let $\tilde{t} = (t_n)_{n=1}^\infty \in \mathcal{C}$, and let $(\mathbb{R} \times \Sigma \times \mathcal{S}^1 \times \{\tilde{t}\})_\infty$ be a component of $\widetilde{Y}_\infty$. We assert that $q_\infty|_{(\mathbb{R} \times \Sigma \times \mathcal{S}^1 \times \{\tilde{t}\})_\infty}$ is one–to–one. To see this, let $(\tilde{r}, \tilde{s}, \tilde{x}, \tilde{t}) = ((r_n, s_n, x_n, t_n))_{n=1}^\infty$ and $(\tilde{r}', \tilde{s}', \tilde{x}', \tilde{t}) = ((r_n', s_n', x_n', t_n))_{n=1}^\infty$ be two points of $(\mathbb{R} \times \Sigma \times \mathcal{S}^1 \times \{\tilde{t}\})_\infty$ such that $q_\infty((\tilde{r}, \tilde{s}, \tilde{x}, \tilde{t})) = q_\infty((\tilde{r}', \tilde{s}', \tilde{x}', \tilde{t}))$. This equality implies that, for every $n \in \mathbb{N}$, $q_n((r_n, s_n, x_n, t_n)) = q_n((r_n', s_n', x_n', t_n))$. Then $(t_n \exp(2\pi r_n), s_n, x_n) = (t_n \exp(2\pi r_n'), s_n', x_n')$. Thus, for each $n \in \mathbb{N}$, $s_n = s_n'$, $x_n = x_n'$ and $t_n \exp(2\pi r_n) = t_n \exp(2\pi r_n')$. From this last equality we obtain that $\exp(2\pi r_n) = \exp(2\pi r_n')$. Hence, for every $n \in \mathbb{N}$, there is an $m_n \in \mathbb{Z}$ such that $2\pi r_n + 2\pi m_n = 2\pi r_n'$, from here, $r_n + m_n = r_n'$. For $n = 1$, we have that $r_1' = r_1 + m_1$. Then $r_2' = \frac{1}{2}r_1' = \frac{1}{2}r_1 + \frac{1}{2}m_1$, $r_3' = \frac{1}{2}r_2' = \frac{1}{4}r_1 + \frac{1}{4}m_1$, etc. Now, if $m_1 \neq 0$, there is a $k \in \mathbb{Z}$ such that $\frac{1}{2^k}m_1 \notin \mathbb{Z}$, and, on the other hand, $r_{k+1}' = \frac{1}{2^k}r_1 + \frac{1}{2^k}m_1$, a contradiction. Hence, $m_1 = 0$. Similarly, $m_n = 0$ for all $n \in \mathbb{N}$. Thus, $r_n = r_n'$, and $(\tilde{r}, \tilde{s}, \tilde{x}, \tilde{t}) = (\tilde{r}', \tilde{s}', \tilde{x}', \tilde{t})$.

Note that if $(\tilde{r}, \tilde{s}, \tilde{x}, \tilde{t}) \in (\mathbb{R} \times \Sigma \times \mathcal{S}^1 \times \mathcal{C})_\infty$, then $q_\infty((\tilde{r}, \tilde{s}, \tilde{x}, \tilde{t})) = (s', \tilde{s}, \tilde{x})$, i.e., q_∞ sends the second and third coordinates of a point in $\mathbb{R} \times \Sigma \times \mathcal{S}^1 \times \mathcal{C}$ identically to the second and third coordinates of its image in $\Sigma \times \Sigma \times \mathcal{S}^1$.

Let $\widetilde{U} = (\tilde{a}, \tilde{b}) \times ((\tilde{c}, \tilde{d}) \times \mathcal{C}') \times (\tilde{e}, \tilde{f}) \times \mathcal{C}''$ be a basic open set of $\widetilde{Y}_\infty$. Then $q_\infty(\widetilde{U}) = ((a, b) \times \mathcal{C}''') \times ((\tilde{c}, \tilde{d}) \times \mathcal{C}') \times (\tilde{e}, \tilde{f})$. Hence, if $\widetilde{U}_c = (\tilde{a}, \tilde{b}) \times ((\tilde{c}, \tilde{d}) \times \mathcal{C}') \times (\tilde{e}, \tilde{f}) \times \{\tilde{t}\}$ is an open set of a component of $\widetilde{Y}_\infty$, then $q_\infty(\widetilde{U}_c) = ((a, b) \times \{t\}) \times ((\tilde{c}, \tilde{d}) \times \mathcal{C}') \times (\tilde{e}, \tilde{f})$.

Assume that the Menger curve $\mathbb{M}$ is embedded in Y_1 in such a way that $\{e\} \times \mathcal{S}^1 \times \{e\} \subset \mathcal{M}$ and $\mathcal{S}^1 \times \{e\} \times \{e\} \subset \mathcal{M}$ (Lemma 5.4.17). Since $(f_1^2)^{-1}(\mathbb{M})$ is homeomorphic to $\mathbb{M}$, taking $X_1 = \mathbb{M}$, $X_2 = (f_1^2)^{-1}(\mathbb{M}), \ldots, X_n = (f_1^n)^{-1}(\mathbb{M})$, we obtain that $\mathbb{MR} = \varprojlim\{X_n, f_n^{n+1}\}$.

Let $\widetilde{X}_n = q_n^{-1}(X_n)$. Observe that $\widetilde{X}_n$ is a locally compact, metric space, with 2^{n-1} components, each of which is locally homeomorphic to $\mathbb{M}$. Let $\widetilde{\mathbb{MR}} = \varprojlim\{\widetilde{X}_n, \tilde{f}_n^{n+1}\}$. Let $\widetilde{U} = (\tilde{a}, \tilde{b}) \times ((\tilde{c}, \tilde{d}) \times \mathcal{C}') \times (\tilde{e}, \tilde{f}) \times \mathcal{C}''$ be a basic open set of $\widetilde{Y}_\infty$, then $\widetilde{U} \cap \widetilde{\mathbb{MR}}$ is a basic open set of $\widetilde{\mathbb{MR}}$. We can take $\widetilde{U}$ such that $q_\infty|_{\widetilde{U}}$ is a homeomorphism. Since $\widetilde{\mathbb{MR}} = q_\infty^{-1}(\mathbb{MR})$, we have that $q_\infty(\widetilde{U} \cap \widetilde{\mathbb{MR}}) = q_\infty(\widetilde{U}) \cap \mathbb{MR} = \left[((a, b) \times \mathcal{C}''') \times ((\tilde{c}, \tilde{d}) \times \mathcal{C}') \times (\tilde{e}, \tilde{f})\right] \cap \mathbb{MR}$. But this is homeomorphic to $\left[(a, b) \times (\tilde{c}, \tilde{d}) \times (\tilde{e}, \tilde{f}) \times \mathcal{C}' \times \mathcal{C}'''\right] \cap \mathbb{MR}$. Since $\mathbb{MR}$ is locally homeomorphic to the product of an open set of $\mathbb{M}$ and a Cantor set, there is an open set V of $\mathbb{MR}$ so that $\left[(a, b) \times (\tilde{c}, \tilde{d}) \times (\tilde{e}, \tilde{f}) \times \mathcal{C}' \times \mathcal{C}'''\right] \cap \mathbb{MR}$ is homeomorphic to $V \times \mathcal{C}' \times \mathcal{C}'''$. Hence $q_\infty([(\tilde{a}, \tilde{b}) \times ((\tilde{c}, \tilde{d}) \times \mathcal{C}') \times (\tilde{e}, \tilde{f}) \times \{\tilde{t}\}] \cap \widetilde{\mathbb{MR}})$ is homeomorphic to $V \times \mathcal{C}' \times \{t\}$. Then the components of $\widetilde{\mathbb{MR}}$ are locally homeomorphic to $V \times \mathcal{C}'$. In particular they are not locally connected.

Q.E.D.

5.5.6. Theorem. *Let Λ be the constant sequence $\{(2, 2, 1)\}$. If $\mathbb{MR} = \mathbb{M}^\Lambda$, then $\mathbb{MR}$ has a covering space whose components are locally connected.*

Proof. Let the inverse sequences $\{Y_n, f_n^{n+1}\}$ and $\{X_n, f_n^{n+1}\}$ be as in the proof of Theorem 5.5.5. Hence, $\mathbb{MR} = \varprojlim\{X_n, f_n^{n+1}\}$.

Let $\widetilde{Y}_n = \mathbb{R} \times \mathbb{R} \times \mathcal{S}^1 \times \{2^{n-1}\text{th roots of unity}\}$, and let $\tilde{f}_n^{n+1} \colon \widetilde{Y}_{n+1} \to \widetilde{Y}_n$ be given by $\tilde{f}_n^{n+1}((r_1, r_2, z_3, t)) = (2r_1, 2r_2, z_3, t^2)$. Define the covering map $q_n \colon \widetilde{Y}_n \to Y_n$ by

$$q_n((r_1, r_2, z_3, t)) = (t \exp(2\pi r_1), t \exp(2\pi r_2), z_3, t).$$

Thus, we may construct the following infinite and commutative ladder:

$$
\begin{array}{ccccccccc}
\cdots & \longleftarrow & \widetilde{Y}_{n-1} & \overset{f^n_{n-1}}{\longleftarrow} & \widetilde{Y}_n & \overset{f^{n+1}_n}{\longleftarrow} & \widetilde{Y}_{n+1} & \longleftarrow & \cdots \quad : \widetilde{Y}_\infty \\
 & & \downarrow {\scriptstyle q_{n-1}} & & \downarrow {\scriptstyle q_n} & & \downarrow {\scriptstyle q_{n+1}} & & \downarrow {\scriptstyle q_\infty} \\
\cdots & \longleftarrow & Y_{n-1} & \underset{\tilde{f}^n_{n-1}}{\longleftarrow} & Y_n & \underset{\tilde{f}^{n+1}_n}{\longleftarrow} & Y_{n+1} & \longleftarrow & \cdots \quad : Y_\infty
\end{array}
$$

where $\widetilde{Y}_\infty = \varprojlim\{\widetilde{Y}_n, \tilde{f}^{n+1}_n\}$ and $q_\infty = \varprojlim\{q_n\}$. Note that $\widetilde{Y}_\infty$ is homeomorphic to $\mathbb{R} \times \mathbb{R} \times \mathcal{S}^1 \times \mathcal{C}$.

Let $(\mathbb{R} \times \mathbb{R} \times \mathcal{S}^1 \times \{t\})_{n+1}$ be a component of $\widetilde{Y}_{n+1}$. Note that the restriction of bonding maps $\tilde{f}^{n+1}_n|_{(\mathbb{R}\times\mathbb{R}\times\mathcal{S}^1\times\{t\})_{n+1}} \colon (\mathbb{R} \times \mathbb{R} \times \mathcal{S}^1 \times \{t\})_{n+1} \subset \widetilde{Y}_{n+1} \to (\mathbb{R} \times \mathbb{R} \times \mathcal{S}^1 \times \{t^2\})_n \subset \widetilde{Y}_n$ is a homeomorphism, since the following map $g^n_{n+1} \colon (\mathbb{R}\times\mathbb{R}\times\mathcal{S}^1\times\{t^2\})_n \subset \widetilde{Y}_n \to (\mathbb{R}\times\mathbb{R}\times \mathcal{S}^1 \times \{t\})_{n+1} \subset \widetilde{Y}_{n+1}$ given by $g^n_{n+1}((r_1, r_2, z_3, t^2)) = (\frac{1}{2}r_1, \frac{1}{2}r_2, z_3, t)$ is its inverse.

Define $g^n_{n+m} = g^{n+m-1}_{n+m} \circ \cdots \circ g^n_{n+1}$. Let $\tilde{t} = (t_n)^\infty_{n=1} \in \mathcal{C}$, and let $(\mathbb{R} \times \mathbb{R} \times \mathcal{S}^1 \times \{\tilde{t}\})_\infty$ be a component of $\widetilde{Y}_\infty$. Consider the map $\tilde{f}_n|_{(\mathbb{R}\times\mathbb{R}\times\mathcal{S}^1\times\{\tilde{t}\})_\infty} \colon (\mathbb{R} \times \mathbb{R} \times \mathcal{S}^1 \times \{\tilde{t}\})_\infty \subset \widetilde{Y}_\infty \to (\mathbb{R} \times \mathbb{R} \times \mathcal{S}^1 \times \{t_n\})_n \subset \widetilde{Y}_n$. We claim that this map is a homeomorphism. To this end, let $g^n \colon (\mathbb{R}\times\mathbb{R}\times\mathcal{S}^1\times\{t_n\})_n \subset \widetilde{Y}_n \to (\mathbb{R}\times\mathbb{R}\times\mathcal{S}^1\times\{\tilde{t}\})_\infty \subset \widetilde{Y}_\infty$ be given by

$$g^n((r_1, r_2, z_3, t_n)) = (f^n_1((r_1, r_2, z_3, t_n)), \dots, f^n_{n-1}((r_1, r_2, z_3, t_n)),$$

$$(r_1, r_2, z_3, t_n), g^n_{n+1}((r_1, r_2, z_3, t_n)), \dots, g^n_{n+m}((r_1, r_2, z_3, t_n)), \dots).$$

Since it is easy to see that $g^n \circ \tilde{f}_n|_{(\mathbb{R}\times\mathbb{R}\times\mathcal{S}^1\times\{\tilde{t}\})_\infty} = 1_{(\mathbb{R}\times\mathbb{R}\times\mathcal{S}^1\times\{\tilde{t}\})_\infty}$ and $\tilde{f}_n|_{(\mathbb{R}\times\mathbb{R}\times\mathcal{S}^1\times\{\tilde{t}\})_\infty} \circ g^n = 1_{(\mathbb{R}\times\mathbb{R}\times\mathcal{S}^1\times\{t_n\})_n}$, $\tilde{f}_n|_{(\mathbb{R}\times\mathbb{R}\times\mathcal{S}^1\times\{\tilde{t}\})_\infty}$ is a homeomorphism.

Let $\widetilde{X}_n = q_n^{-1}(X_n)$, then, $\widetilde{X}_n$ is a locally compact, metric space with 2^{n-1} components, each of which is locally homeomorphic to $\mathbb{M}$. Let $\widetilde{\mathbb{MR}} = \varprojlim\{\widetilde{X}_n, \tilde{f}^{n+1}_n\}$, and let $\mathbb{K}$ be a component of $\widetilde{\mathbb{MR}}$.

$\mathbb{K}$ is contained in a component of $\widetilde{Y}_\infty$. Let $(\mathbb{R} \times \mathbb{R} \times \mathcal{S}^1 \times \{\tilde{t}\})_\infty$ be such a component. Since $\tilde{f}_1|_{(\mathbb{R}\times\mathbb{R}\times\mathcal{S}^1\times\{\tilde{t}\})_\infty}$ is a homeomorphism, then $\tilde{f}_1|_{\mathbb{K}}$ is a homeomorphism onto $\tilde{f}_1(\mathbb{K}) = \widetilde{X}_1$. Since $\widetilde{X}_1$ is locally homeomorphic to $\mathbb{M}$, $\mathbb{K}$ is locally homeomorphic to $\mathbb{M}$. Therefore, $\mathbb{K}$ is locally connected.

Q.E.D.

REFERENCES

[1] J. M. Aarts and P. Van Emde Boas, Continua as Remainders in Compact Extensions, Nieuw Archief voor Wiskunde 3 (1967), 34–37.

[2] J. M. Aarts and L. G. Oversteegen, The Product Structure of Homogeneous Spaces, Indag. Math., 1 (1990), 1–5.

[3] L. V. Ahlfors, *Conformal Invariants, Topics in Geometric Function Theory*, Series in Higher Mathematics, McGraw–Hill Book Co., New York, 1973.

[4] R. D. Anderson, A Characterization of the Universal Curve and a Proof of its Homogeneity, Ann. Math., 67 (1958), 313–324.

[5] R. D. Anderson, One–dimensional Continuous Curves and a Homogeneity Theorem, Ann. Math., 68 (1958), 1–16.

[6] D. P. Bellamy and L. Lum, The Cyclic Connectivity of Homogeneous Arcwise Connected Continua, Trans. Amer. Math. Soc., 266 (1981), 389–396.

[7] J. H. Case, Another 1–dimensional Homogeneous Continuum Which Contains an Arc, Pacific J. Math., 11(1961), 455–469.

[8] C. O. Christenson and W. L. Voxman, *Aspects of Topology*, Monographs and Textbooks in Pure and Applied Math., Vol. 39, Marcel Dekker, Inc., New York, Basel, 1977.

[9] R. Fenn, What is the Geometry of a Surface?, The Amer. Math. Monthly, 90 (1983), 87–98.

[10] J. Grispolakis and E. D. Tymchatyn, On Confluent Mappings and Essential Mappings–A Survey, Rocky Mountain J. Math., 11 (1981), 131–153.

[11] C. L. Hagopian, A Characterization of Solenoids, Pacific J. Math., 68 (1977), 425–435.

[12] J. G. Hocking and G. S. Young, *Topology*, Dover Publications, Inc., New York, 1988.

[13] S. T. Hu, *Theory of Retracts*, Wayne State University Press, Detroit, 1965.

[14] J. Krasinkiewicz, On One–point Union of Two Circles, Houston J. Math., 2 (1976), 91–95.

[15] J. Krasinkiewicz, On Two Theorems of Dyer, Colloq. Math., 50 (1986), 201–208.

[16] K. Kuratowski, *Topology*, Vol. II, Academic Press, New York, N. Y., 1968.

[17] E. L. Lima, *Grupo Fundamental e Espaços de Recobrimento*, Instituto de Matemática Pura e Aplicada, CNPq, (Projeto Euclides), 1993. (Portuguese)

[18] J. C. Macías, *El Teorema de Descomposición Terminal de Rogers*, Tesis de Maestría, Facultad de Ciencias Físico Matemáticas, B. U. A. P., 1999. (Spanish)

[19] S. Macías, Covering Spaces of Homogeneous Continua, Topology Appl., (1994), 157–177.

[20] S. Macías, Homogeneous Continua for Which the Set Function T is Continuous, to appear in Houston Journal of Mathematics.

[21] T. Maćkowiak and E. D. Tymchatyn, Continuous Mappings on Continua II, Dissertationes Math., 225 (1984), 1–57.

[22] M. C. McCord, Inverse Limit Sequences with Covering Maps, Trans. Amer. Math. Soc., 114 (1965), 197–209.

[23] P. Minc and J. T. Rogers, Jr., Some New Examples of Homogeneous Curves, Topology Proc., 10 (1985), 347–356.

[24] J. Munkres, *Topology*, second edition, Prentice Hall, Upper Saddle River, NJ, 2000.

[25] S. B. Nadler, Jr., *Dimension Theory: An Introduction with Exercises*, Aportaciones Matemáticas, Serie Textos # 18, Sociedad Matemática Mexicana, 2002.

[26] J. T. Rogers, Jr., Homogeneous Separating Plane Continua are Decomposable, Michigan J. Math., 28 (1981), 317–321.

[27] J. T. Rogers, Jr., Decompositions of Homogeneous Continua, Pacific J. Math., 99 (1982), 137–144.

[28] J. T. Rogers, Jr., Homogeneous Hereditarily Indecomposable Continua Are Tree–like, Houston J. Math., 8, (1982), 421–428.

[29] J. T. Rogers, Jr., An Aposyndetic Homogeneous Curve That is not Locally Connected, Houston J. Math., 9 (1983), 433–440.

[30] J. T. Rogers, Jr., Cell–like Decompositions of Homogeneous Continua, Proc. Amer. Math. Soc., 87 (1983), 375–377.

[31] J. T. Rogers, Jr., Homogeneous Curves That Contain Arcs, Topology Appl., 21 (1985), 95–101.

[32] J. T. Rogers, Jr., Orbits of Higher–dimensional Hereditarily Indecomposable Continua, Proc. Amer. Math. Soc., 95 (1985), 483–486.

[33] J. T. Rogers, Jr., Hyperbolic Ends And Continua, Michigan Math. J., 34 (1987), 337–347.

[34] J. T. Rogers, Jr., Decompositions of Continua Over the Hyperbolic Plane, Trans. Amer. Math. Soc., 310 (1988), 277–291.

[35] J. T. Rogers, Jr., Higher Dimensional Aposyndetic Decompositions, Proc. Amer. Math. Soc., 131 (2003), 3285–3288.

[36] P. Scott, The Geometries of 3–manifolds, Bull. London Math. Soc., 15 (1983), 401–487.

[37] G. T. Whyburn, Topological Characterization of the Sierpiński Curve, Fund. Math., 45 (1958), 320–324.

Chapter 6

n–FOLD HYPERSPACES

We give a brief overview of n–fold hyperspaces. Other hyperspaces have been studied extensively in [40] and [21]. Throughout this chapter, σ denotes the union map and $\mathcal{H}$ denotes the Hausdorff metric.

6.1   General Properties

We present general properties of n–fold hyperspaces.

6.1.1. Lemma. *Let $n \in \mathbb{N}$, and let X be a continuum. If $\mathcal{A}$ is a connected subset of $\mathcal{C}_n(X)$, then $\bigcup \mathcal{A} = \bigcup \{A \mid A \in \mathcal{A}\}$ has at most n components.*

Proof. Suppose the result is not true. Then there exists a connected subset $\mathcal{A}$ of $\mathcal{C}_n(X)$ such that $\bigcup \mathcal{A}$ has at least $n + 1$ components. Thus, we can find $n + 1$ pairwise separated subsets, $C_1, \ldots,$

287

C_{n+1}, of X such that $\bigcup \mathcal{A} = \bigcup\limits_{j=1}^{n+1} C_j$. Let

$$\mathcal{B} = \left\{ A \in \mathcal{A} \mid A \subset \bigcup_{j=1}^{n} C_j \right\}$$

and

$$\mathcal{D} = \{ A \in \mathcal{A} \mid A \cap C_{n+1} \neq \emptyset \}.$$

Then $\mathcal{B}$ and $\mathcal{D}$ are separated subsets of $\mathcal{C}_n(X)$ and $\mathcal{A} = \mathcal{B} \cup \mathcal{D}$, a contradiction. Therefore, $\bigcup \mathcal{A}$ has at most n components.

Q.E.D.

6.1.2. Corollary. *Let $n \in \mathbb{N}$. If X is a continuum and $\mathcal{A} \in \mathcal{C}\left(\mathcal{C}_n(X)\right)$, then $\sigma\left(\mathcal{A}\right) \in \mathcal{C}_n(X)$.*

Proof. By Lemma 1.8.11, $\sigma\left(\mathcal{A}\right)$ is a closed subset of X. By Lemma 6.1.1, $\sigma\left(\mathcal{A}\right)$ has at most n components. Therefore, $\sigma\left(\mathcal{A}\right) \in \mathcal{C}_n(X)$.

Q.E.D.

6.1.3. Notation. Let X be a continuum. To simplify notation, we write: $\langle U_1, \dots, U_m \rangle_n$, to denote the intersection of the open set $\langle U_1, \dots, U_m \rangle$, of the Vietoris Topology, with $\mathcal{C}_n(X)$.

6.1.4. Theorem. *Let $n \in \mathbb{N}$. Then the continuum X is locally connected if and only if $\mathcal{C}_n(X)$ is locally connected.*

Proof. If X is locally connected, then Wojdysławski has shown that $\mathcal{C}_n(X)$ is an absolute retract (Théorème II$_m$ of [47]). Hence, it is locally connected ((2.6) (p. 101) of [5]).

Suppose $\mathcal{C}_n(X)$ is locally connected. Let x be a point in X and let U be an open subset of X containing x. Since $\mathcal{C}_n(X)$ is locally connected, there exists a connected open subset $\mathcal{V}$ of $\mathcal{C}_n(X)$ such that $\{x\} \in \mathcal{V} \subset Cl(\mathcal{V}) \subset \langle U \rangle_n$. Let $\langle V_1, \dots, V_k \rangle_n$ be a basic open set of $\mathcal{C}_n(X)$ such that $\{x\} \in \langle V_1, \dots, V_k \rangle_n \subset \mathcal{V}$. Let $V = \bigcap\limits_{j=1}^{k} V_j$.

Now, if $y \in V$, then $y \in \sigma(Cl(\mathcal{V}))$. Since $\{x\} \in Cl(\mathcal{V})$, we have that $\sigma(Cl(\mathcal{V})) \in \mathcal{C}(X)$ ((1.49) of [40]). Thus, both x and y belong to $\sigma(Cl(\mathcal{V})) \subset U$. Hence, X is connected im kleinen at x. Since x was an arbitrary point of X, we have that X is locally connected (Theorem 1.7.12).

Q.E.D.

6.1.5. Definition. A *free arc* in a metric space X is an arc α such that $\alpha \setminus \{\text{end points}\}$ is an open subset of X.

A proof of the following Theorem may be found in Theorem 7.1 of [29].

6.1.6. Theorem. *If X is a locally connected continuum without free arcs, then $\mathcal{C}_n(X)$ is homeomorphic to the Hilbert cube for each $n \in \mathbb{N}$.*

6.1.7. Theorem. *Let X be a continuum. If $n \in \mathbb{N}$, then $\mathcal{C}_n(X)$ is nowhere dense in $\mathcal{C}_{n+1}(X)$. In particular, $\mathcal{C}_n(X)$ is nowhere dense in 2^X.*

Proof. Let $n \in \mathbb{N}$. Suppose $Int_{\mathcal{C}_{n+1}(X)}(\mathcal{C}_n(X)) \neq \emptyset$. Recall that $\mathcal{F}(X)$ is dense in 2^X (see the proof of Corollary 1.8.9). Hence, X does not belong to $Int_{\mathcal{C}_{n+1}(X)}(\mathcal{C}_n(X))$. Let $A \in Int_{\mathcal{C}_{n+1}(X)}(\mathcal{C}_n(X) \setminus \{X\})$, and suppose that $A_1, \ldots, A_k$ are the components of A. Thus, $k \leq n$. Then there exists $\varepsilon > 0$ such that $\mathcal{V}_{\varepsilon}^{\mathcal{H}}(A) \cap \mathcal{C}_{n+1}(X) \subset Int_{\mathcal{C}_{n+1}(X)}(\mathcal{C}_n(X))$. Without loss of generality, we assume that ε is small enough that $\mathcal{V}_{\varepsilon}^{d}(A_j) \cap \mathcal{V}_{\varepsilon}^{d}(A_\ell) = \emptyset$ if and only if $j \neq \ell$ and $j, \ell \in \{1, \ldots, k\}$. Suppose that $n - k = m$ and let $x_1, \ldots, x_{m+1} \in \mathcal{V}_{\varepsilon}^{d}(A_1) \setminus A_1$ be $m+1$ distinct points. Let $B = A \cup \{x_1, \ldots, x_{m+1}\}$. Then $B \in \mathcal{C}_{n+1}(X) \setminus \mathcal{C}_n(X)$ and $\mathcal{H}(A, B) < \varepsilon$. Therefore $Int_{\mathcal{C}_{n+1}(X)}(\mathcal{C}_n(X)) = \emptyset$.

Similarly, $Int_{2^X}(\mathcal{C}_n(X)) = \emptyset$.

Q.E.D.

6.1.8. Lemma. *Let X be a continuum, with metric d. If $n \in \mathbb{N}$, then there exist n pairwise disjoint nondegenerate subcontinua of X.*

Proof. Let $n \in \mathbb{N}$, and let $x_1, \ldots, x_n$ be n distinct points of X. Let $\eta = \min\{d(x_j, x_k) \mid j, k \in \{1, \ldots, n\}$ and $j \neq k\}$. Then $\eta > 0$ and $\mathcal{V}_\eta^d(x_j) \cap \mathcal{V}_\eta^d(x_k) = \emptyset$ for each $j, k \in \{1, \ldots, n\}$ and $j \neq k$. By Corollary 1.7.23, there exists a subcontinuum A_j of X such that $\{x_j\} \subsetneq A_j \subset \mathcal{V}_\eta^d(x_j)$ for every $j \in \{1, \ldots, n\}$. Therefore, $A_1, \ldots, A_n$ are n pairwise disjoint nondegenerate subcontinua of X.

$\hspace{9cm}$ **Q.E.D.**

6.1.9. Theorem. *Let X be a continuum. If $n \in \mathbb{N}$, then $\mathcal{C}_n(X)$ contains an n–cell.*

Proof. Let $A_1, \ldots, A_n$ be n pairwise disjoint nondegenerate subcontinua of X (Lemma 6.1.8). For each $j \in \{1, \ldots, n\}$, let $a_j \in A_j$, and let $\alpha_j \colon [0, 1] \to \mathcal{C}(X)$ be an order arc such that $\alpha_j(0) = \{a_j\}$ and $\alpha_j(1) = A_j$ (Theorem 1.8.20). Then the map $\xi \colon [0, 1]^n \to \mathcal{C}_n(X)$ given by $\xi((t_1, \ldots, t_n)) = \alpha_1(t_1) \cup \cdots \cup \alpha_n(t_n)$ is an embedding of $[0, 1]^n$ in $\mathcal{C}_n(X)$.

$\hspace{9cm}$ **Q.E.D.**

If the continuum X contains decomposable continua, more can be said.

6.1.10. Theorem. *Let X be a continuum and let $n \in \mathbb{N}$. If X contains k pairwise disjoint decomposable subcontinua $(k \leq n)$, then $\mathcal{C}_n(X)$ contains a $(k + n)$–cell.*

Proof. First suppose $k < n$. Let $M_1, \ldots, M_k$ be k pairwise disjoint decomposable subcontinua of X. Suppose that $M_j = A_j \cup B_j$, where A_j and B_j are continua, for each $j \in \{1, \ldots, k\}$. By the proof of (1.145) of [40], we may assume that for each $j \in \{1, \ldots, k\}$, $A_j \cap B_j$ is connected, $A_j \setminus (A_j \cap B_j) \neq \emptyset$, $B_j \setminus (A_j \cap B_j) \neq \emptyset$, and $[A_j \setminus (A_j \cap B_j)] \cap [B_j \setminus (A_j \cap B_j)] = \emptyset$. Let $C_{k+1}, \ldots, C_n$ be $n - k$ pairwise disjoint nondegenerate subcontinua of X such that $M_j \cap C_\ell = \emptyset$ for every $j \in \{1, \ldots, k\}$ and every $\ell \in \{k+1, \ldots, n\}$. For each $j \in \{1, \ldots, k\}$, let $\alpha_j \colon [0, 1] \to \mathcal{C}(A_j)$ and $\beta_j \colon [0, 1] \to \mathcal{C}(B_j)$ be order arcs such that $\alpha_j(0) = A_j \cap B_j$, $\alpha_j(1) = A_j$, $\beta_j(0) = A_j \cap B_j$, and $\beta_j(1) = B_j$ (Theorem 1.8.20). For each $\ell \in \{k+1, \dots, n\}$, let $x_\ell \in C_\ell$. Let $\gamma_\ell \colon [0,1] \to \mathcal{C}(C_\ell)$ be an order arc such that $\gamma_\ell(0) = \{x_\ell\}$ and $\gamma_\ell(1) = C_\ell$, $\ell \in \{k+1, \dots, n\}$. Since $[0,1]^{k+n}$ is homeomorphic to $[0,1]^{2k} \times [0,1]^{n-k}$, we need an embedding of $[0,1]^{2k} \times [0,1]^{n-k}$ into $\mathcal{C}_n(X)$. The map $\xi \colon [0,1]^{2k} \times [0,1]^{n-k} \to \mathcal{C}_n(X)$ given by

$$\xi((t_1, \dots, t_{2k}), (t_1, \dots, t_{n-k})) =$$

$$\left(\bigcup_{j=1}^{k} (\alpha_j(t_{2j-1}) \cup \beta_j(t_{2j})) \right) \cup \left(\bigcup_{\ell=1}^{n-k} \gamma_{k+\ell}(t_\ell) \right)$$

is an embedding of $[0,1]^{k+n}$ in $\mathcal{C}_n(X)$. When $k = n$, repeat the argument without using the γ's.

Q.E.D.

As a consequence of Theorem 6.1.10, we have the following two Corollaries:

6.1.11. Corollary. *If X is a continuum which is not hereditarily indecomposable, then $\dim(\mathcal{C}_n(X)) \geq n+1$.*

6.1.12. Corollary. *Let X be a continuum and let $n \in \mathbb{N}$. If X contains n pairwise disjoint decomposable subcontinua, then $\mathcal{C}_n(X)$ contains a $2n$–cell.*

6.1.13. Corollary. *If X is a continuum containing an arc, then $\mathcal{C}_n(X)$ contains a $2n$–cell for each $n \in \mathbb{N}$.*

6.1.14. Theorem. *Let X be a continuum. If $n \in \mathbb{N}$, then the following are equivalent:*

(1) 2^X is contractible;

(2) $\mathcal{C}_n(X)$ is contractible;

(3) $\mathcal{C}(X)$ is contractible.

Proof. Suppose 2^X is contractible. Then there exists a map $H'\colon 2^X \times [0,1] \to 2^X$ such that for each $A \in 2^X$, $H'((A,0)) = A$ and $H'((A,1)) = X$. Let $H\colon 2^X \times [0,1] \to 2^X$ be the *segment homotopy associated* with H' defined by

$$H((A,t)) = \bigcup\{H'(A,s) \mid 0 \le s \le t\}.$$

Then H is continuous ((16.3) of [40]). Observe that for each $A \in 2^X$, $H((A,0)) = A$, $H((A,1)) = X$, and $H(\{A\} \times [0,1])$ is an order arc from A to X. Note that if $A \in \mathcal{C}_n(X)$ and $B \in H(\{A\}\times[0,1])$, then $B \in \mathcal{C}_n(X)$ (Theorem 1.8.20). Therefore, $G = H|_{\mathcal{C}_n(X)\times[0,1]}\colon \mathcal{C}_n(X)\times [0,1] \to \mathcal{C}_n(X)$ is a continuous function such that for each $A \in \mathcal{C}_n(X)$, $G((A,0)) = A$ and $G((A,1)) = X$. Hence, $\mathcal{C}_n(X)$ is contractible. A similar argument shows that if $\mathcal{C}_n(X)$ is contractible, then $\mathcal{C}(X)$ is contractible. The other implication is contained in the proof of (16.7) of [40].

$$\text{Q.E.D.}$$

6.1.15. Definition. A continuum X is said to have the *property of Kelley* provided that given any $\varepsilon > 0$, there exists $\delta > 0$ such that if $a, b \in X$, $d(a,b) < \delta$, and $a \in A \in \mathcal{C}(X)$, then there exists $B \in \mathcal{C}(X)$ such that $b \in B$ and $\mathcal{H}(A,B) < \varepsilon$. This number δ is called a *Kelley number* for the given ε.

6.1.16. Corollary. *If X is a continuum having the property of Kelley, then $\mathcal{C}_n(X)$ is contractible for each $n \in \mathbb{N}$.*

Proof. If X is a continuum having the property of Kelley, then 2^X and $\mathcal{C}(X)$ are contractible ((16.15) of [40]). Hence, the result follows from Theorem 6.1.14.

$$\text{Q.E.D.}$$

6.1.17. Lemma. *Let $n \in \mathbb{N}$, and let X be a continuum having the property of Kelley. If $\mathcal{W}$ is a subcontinuum of $\mathcal{C}_n(X)$ having nonempty interior, then $\sigma\left(\mathcal{W}\right) \in \mathcal{C}_n(X)$ and $Int_X(\sigma\left(\mathcal{W}\right)) \neq \emptyset$.*

Proof. By Corollary 6.1.2, $\sigma(\mathcal{W}) \in \mathcal{C}_n(X)$. Let $A \in Int_{\mathcal{C}_n(X)}(\mathcal{W})$. Let $A_1, \ldots, A_k$ be the components of A. Then $k \leq n$. Let $\varepsilon > 0$ be given such that $\mathcal{V}_\varepsilon^{\mathcal{H}}(A) \cap \mathcal{C}_n(X) \subset \mathcal{W}$, and such that $\mathcal{V}_\varepsilon^d(A_j) \cap \mathcal{V}_\varepsilon^d(A_\ell) = \emptyset$ when $j \neq \ell$. Let $\delta > 0$ be a Kelley number for the given ε. Clearly, $A \subset \sigma(\mathcal{W})$. For each $j \in \{1, \ldots, k\}$, let $a_j \in A_j$. For every $j \in \{1, \ldots, k\}$, let $x_j \in \mathcal{V}_\delta^d(a_j)$. Since X has the property of Kelley, there exists a subcontinuum B_j of X such that $x_j \in B_j$ and $\mathcal{H}(A_j, B_j) < \varepsilon$ for each $j \in \{1, \ldots, k\}$. Let $B = \bigcup_{j=1}^{k} B_j$; then

$B \in \mathcal{C}_n(X)$. Note that $\mathcal{V}_\varepsilon^d(A) = \bigcup_{j=1}^{k} \mathcal{V}_\varepsilon^d(A_j)$, $\mathcal{V}_\varepsilon^d(B) = \bigcup_{j=1}^{k} \mathcal{V}_\varepsilon^d(B_j)$, and $\mathcal{H}(A_j, B_j) < \varepsilon$ for each $j \in \{1, \ldots, k\}$; hence, $\mathcal{H}(A, B) < \varepsilon$.

Thus, $B \in \mathcal{W}$, and $B \subset \sigma(\mathcal{W})$. Therefore, $\bigcup_{j=1}^{k} \mathcal{V}_\delta^d(a_j) \subset \sigma(\mathcal{W})$, and $Int_X(\sigma(\mathcal{W})) \neq \emptyset$.

Q.E.D.

6.1.18. Theorem. *Let $n \in \mathbb{N}$. If X is an indecomposable continuum having the property of Kelley, then X is the only point at which $\mathcal{C}_n(X)$ is locally connected.*

Proof. Using order arcs (Theorem 1.8.20), it is easy to see that for each $\varepsilon > 0$, $\mathcal{V}_\varepsilon^{\mathcal{H}}(X) \cap \mathcal{C}_n(X)$ is arcwise connected. Suppose that $\mathcal{C}_n(X)$ is locally connected at the point $A \neq X$. Since $\mathcal{C}_n(X)$ is locally connected at A, there exists a subcontinuum $\mathcal{W}$ of $\mathcal{C}_n(X)$ such that $A \in Int_{\mathcal{C}_n(X)}(\mathcal{W})$ and $\sigma(\mathcal{W}) \neq X$. By Lemma 6.1.17, $\sigma(\mathcal{W}) \in \mathcal{C}_n(X)$ and $Int_X(\sigma(\mathcal{W})) \neq \emptyset$. Since $\sigma(\mathcal{W})$ has finitely many components and $Int_X(\sigma(\mathcal{W})) \neq \emptyset$, it follows that at least one of the components of $\sigma(\mathcal{W})$ has nonempty interior, which is impossible because $\sigma(\mathcal{W}) \neq X$ and X is an indecomposable continuum (Corollary 1.7.21).

Q.E.D.

As a consequence of the previous Theorem, we have the following result:

6.1.19. Theorem. *Let $n \in \mathbb{N}$. If X is a hereditarily indecomposable continuum, then X is the only point at which $C_n(X)$ is locally connected.*

Proof. Since hereditarily indecomposable continua have the property of Kelley ((16.27) of [40]), we have that the result follows from Theorem 6.1.18.

Q.E.D.

The following result is easy to establish.

6.1.20. Lemma. *Let $n \in \mathbb{N}$, $n \geq 2$, and let X be a continuum. Let A be a point of $C_n(X) \setminus C_{n-1}(X)$, and suppose $A_1, \ldots, A_n$ are the components of A. Let $\varepsilon > 0$ be such that $\mathcal{V}_{2\varepsilon}^d(A_j) \cap \mathcal{V}_{2\varepsilon}^d(A_k) = \emptyset$ if and only if $j \neq k$ and $j, k \in \{1, \ldots, n\}$. Let B be a point of $C_n(X) \setminus C_{n-1}(X)$, and suppose $B_1, \ldots, B_n$ are the components of B. Then $\mathcal{H}(A, B) < \varepsilon$ if and only if $\mathcal{H}^2(\{A_1, \ldots, A_n\}, \{B_1, \ldots, B_n\}) < \varepsilon$, where $\mathcal{H}^2$ is the Hausdorff metric on $\mathcal{F}_n(\mathcal{C}(X))$ induced by $\mathcal{H}$.*

As a consequence of Lemma 6.1.20 , we have the following result.

6.1.21. Theorem. *Let $n \in \mathbb{N}$, $n \geq 2$. If X is a continuum, then the function $f_n \colon C_n(X) \setminus C_{n-1}(X) \to \mathcal{F}_n(\mathcal{C}(X))$ given by*

$$f_n(A) = \{K \mid K \text{ is a component of } A\}$$

is an embedding.

The next Corollary says that given a continuum X, if $\mathcal{C}(X)$ is of finite dimension, then all the n–fold hyperspaces of X are of finite dimension too.

6.1.22. Corollary. *If X is a continuum such that $\mathcal{C}(X)$ is of finite dimension then for each $n \geq 2$, $\dim(C_n(X)) \leq n(\dim(\mathcal{C}(X)))$.*

Proof. Let $n \geq 2$. Then $\dim(\mathcal{F}_n(\mathcal{C}(X))) \leq n(\dim(\mathcal{C}(X)))$ (Lemma 3.1 of [9]). Also, $\dim(C_n(X)) \leq \dim(\mathcal{F}_n(\mathcal{C}(X)))$ (Theorem 6.1.21 and 7.3 of [42]). Therefore, $\dim(C_n(X)) \leq n(\dim(\mathcal{C}(X)))$.

Q.E.D.

6.1.23. Corollary. *If X is a continuum such that all its nondegenerate proper subcontinua are arcs, then $\dim(\mathcal{C}_n(X)) = 2n$ for every $n \in \mathbb{N}$.*

Proof. Since the hyperspace of subcontinua of an arc is a 2–cell (5.1 of [21]), $\dim(\mathcal{C}(Y)) < 3$ for every proper subcontinuum Y of X. Hence, by Theorem 2.2 of [43], $\dim(\mathcal{C}(X)) < 3$. Also, by 22.18 of [21], $\dim(\mathcal{C}(X)) \geq 2$. Thus, $\dim(\mathcal{C}(X)) = 2$. Therefore, by Corollaries 6.1.13 and 6.1.22, $\dim(\mathcal{C}_n(X)) = 2n$ for every $n \in \mathbb{N}$.

Q.E.D.

6.2 Unicoherence

We show that $\mathcal{C}_n(X)$ has trivial shape and it is unicoherent for every $n \in \mathbb{N}$.

6.2.1. Definition. A compactum X has *trivial shape* if X is homeomorphic to $\varprojlim\{X_n, f_n^{n+1}\}$, where each X_n is an absolute neighborhood retract, and for each $n \in \mathbb{N}$, there exists $m \geq n$ such that f_n^m is homotopic to a constant map.

6.2.2. Theorem. *If X is a continuum, then $\mathcal{C}_n(X)$ has trivial shape for each $n \in \mathbb{N}$.*

Proof. Let $n \in \mathbb{N}$. Note that by Theorem 2.1.52, X is homeomorphic to $\varprojlim\{X_m, f_m^{m+1}\}$, where each X_m is a polyhedron. Hence, by Théorème II$_m$ of [47], $\mathcal{C}_n(X_m)$ is an absolute retract. Also, by Theorem 2.3.4, $\mathcal{C}_n(X)$ is homeomorphic to $\varprojlim\{\mathcal{C}_n(X_m), \mathcal{C}_n(f_m^{m+1})\}$. Since absolute retracts are contractible (1.6.7 of [37]), by Theorem 1.3.12, all the maps $\mathcal{C}_n(f_k^m)$ are homotopic to a constant map. Therefore, $\mathcal{C}_n(X)$ has trivial shape.

Q.E.D.

6.2.3. Remark. Given a continuum X, $\check{H}^1(X;\mathbb{Z})$ denotes the first Čech cohomology group of X with integer coefficients. It is known that $\check{H}^1(X;\mathbb{Z}) = \{0\}$ if and only if each map from X into $\mathcal{S}^1$ is homotopic to a constant map (8.1 of [10]). This implies, by 19.7 of [21], that if $\check{H}^1(X;\mathbb{Z}) = \{0\}$, then X is unicoherent.

6.2.4. Theorem. *If X is a continuum, then for every $n \in \mathbb{N}$, each map from $\mathcal{C}_n(X)$ to the unit circle, $\mathcal{S}^1$, is homotopic to a constant map. In particular, we have that $\mathcal{C}_n(X)$ is unicoherent.*

Proof. By Theorem 2.1.52, X is homeomorphic to the inverse limit of an inverse sequence of compact and connected polyhedra, $\{X_m, f_m^{m+1}\}$. Hence, $\mathcal{C}_n(X_m)$ is a continuum which is an absolute retract (Théorème II$_m$of [47]) for each $n \in \mathbb{N}$.

By 2.4 (p. 86) of [5] and 8.1 of [10], each map from $\mathcal{C}_n(X_m)$ into $\mathcal{S}^1$ is homotopic to a constant map. Then we obtain that $\check{H}^1(\mathcal{C}_n(X_m);\mathbb{Z}) = \{0\}$ (8.1 of [10]). By the continuity theorem for Čech cohomology (Theorem 7–7 of [44]), $\check{H}^1(\mathcal{C}_n(X),\mathbb{Z}) = \{0\}$. Thus, every map from $\mathcal{C}_n(X)$ into $\mathcal{S}^1$ is homotopic to a constant map (8.1 of [10]). Therefore, by Remark 6.2.3, we have that $\mathcal{C}_n(X)$ is unicoherent for each $n \in \mathbb{N}$.

$$\text{Q.E.D.}$$

6.3   Aposyndesis

We prove that for every $n \in \mathbb{N}$, $\mathcal{C}_n(X)$ is colocally connected and finitely aposyndetic.

6.3.1. Theorem. *Let X be a continuum, with metric d, and let $n \in \mathbb{N}$ be given. Then $\mathcal{C}_n(X)$ is colocally connected.*

Proof. Let A be a point of $\mathcal{C}_n(X)$. We consider two cases.

First assume that $A \in \mathcal{C}_n(X) \setminus \mathcal{F}_n(X)$. Let $\varepsilon > 0$ be given such that $(\mathcal{V}_\varepsilon^{\mathcal{H}}(A) \cap \mathcal{C}_n(X)) \cap \mathcal{F}_n(X) = \emptyset$. To see that $\mathcal{C}_n(X) \setminus$

$(\mathcal{V}_\varepsilon^{\mathcal{H}}(A) \cap \mathcal{C}_n(X))$ is connected, let $B \in \mathcal{C}_n(X) \setminus (\mathcal{V}_\varepsilon^{\mathcal{H}}(A) \cap \mathcal{C}_n(X))$, and let $B_1, \ldots, B_\ell$ be the components of B. Then $\ell \leq n$. Since $B \in \mathcal{C}_n(X) \setminus (\mathcal{V}_\varepsilon^{\mathcal{H}}(A) \cap \mathcal{C}_n(X))$, we have that $\mathcal{H}(A, B) \geq \varepsilon$. Hence, either $A \not\subset \mathcal{V}_\varepsilon^d(B)$ or $B \not\subset \mathcal{V}_\varepsilon^d(A)$. If $A \not\subset \mathcal{V}_\varepsilon^d(B)$, then there exists a point a in A such that $a \notin \mathcal{V}_\varepsilon^d(B)$. Thus, for every $b \in B$, $d(a, b) \geq \varepsilon$. For each $j \in \{1, \ldots, \ell\}$, let $b_j \in B_j$. Let $\alpha \colon [0, 1] \to \mathcal{C}_n(X)$ be an order arc from $\{b_1, \ldots, b_\ell\}$ to B (Theorem 1.8.20). Since for each $t \in [0, 1]$, $\alpha(t) \subset B$, and for every $b \in B$, $d(a, b) \geq \varepsilon$, we have that $\alpha(t) \notin \mathcal{V}_\varepsilon^{\mathcal{H}}(A) \cap \mathcal{C}_n(X)$ for any $t \in [0, 1]$. Thus, $\alpha([0, 1]) \cup \mathcal{F}_n(X) \subset \mathcal{C}_n(X) \setminus (\mathcal{V}_\varepsilon^{\mathcal{H}}(A) \cap \mathcal{C}_n(X))$.

If $B \not\subset \mathcal{V}_\varepsilon^d(A)$, then there exists a point b in B such that $b \notin \mathcal{V}_\varepsilon^d(A)$. Thus, for every point a of A, $d(b, a) \geq \varepsilon$. Without loss of generality, we assume that $b \in B_1$. For each $j \in \{2, \ldots, \ell\}$, let b_j be any point of B_j. Let $\beta \colon [0, 1] \to \mathcal{C}_n(X)$ be an order arc from $\{b, b_2, \ldots, b_\ell\}$ to B. Since for each $t \in [0, 1]$, $b \in \beta(t) \subset B$ and $d(b, a) \geq \varepsilon$ for any $a \in A$, we have that $\beta(t) \notin \mathcal{V}_\varepsilon^{\mathcal{H}}(A) \cap \mathcal{C}_n(X)$ for any $t \in [0, 1]$. Thus, $\beta([0, 1]) \cup \mathcal{F}_n(X) \subset \mathcal{C}_n(X) \setminus (\mathcal{V}_\varepsilon^{\mathcal{H}}(A) \cap \mathcal{C}_n(X))$.

Therefore, if $A \in \mathcal{C}_n(X) \setminus \mathcal{F}_n(X)$ and $\varepsilon > 0$ is given such that $(\mathcal{V}_\varepsilon^{\mathcal{H}}(A) \cap \mathcal{C}_n(X)) \cap \mathcal{F}_n(X) = \emptyset$, then each element B of $\mathcal{C}_n(X) \setminus (\mathcal{V}_\varepsilon^{\mathcal{H}}(A) \cap \mathcal{C}_n(X))$ can be joined with $\mathcal{F}_n(X)$ by an order arc completely contained in $\mathcal{C}_n(X) \setminus (\mathcal{V}_\varepsilon^{\mathcal{H}}(A) \cap \mathcal{C}_n(X))$.

Now, assume that $A \in \mathcal{F}_n(X)$. Suppose $A = \{a_1, \ldots, a_k\}$ and let $\varepsilon > 0$ be given such that $\mathcal{V}_\varepsilon^d(a_j) \cap \mathcal{V}_\varepsilon^d(a_m) = \emptyset$ if and only if $j \neq m$ and $j, m \in \{1, \ldots, k\}$. Let $B \in \mathcal{C}_n(X) \setminus (\mathcal{V}_\varepsilon^{\mathcal{H}}(A) \cap \mathcal{C}_n(X))$ and let $B_1, \ldots, B_\ell$ be the components of B. Then $\ell \leq n$. Hence, $\mathcal{H}(A, B) \geq \varepsilon$. Thus, we have that either $A \not\subset \mathcal{V}_\varepsilon^d(B)$ or $B \not\subset \mathcal{V}_\varepsilon^d(A)$.

If $A \not\subset \mathcal{V}_\varepsilon^d(B)$, then there exists a point a in A such that for each point b of B, $d(a, b) \geq \varepsilon$. Without loss of generality, we assume that $a = a_1$. Hence, $B \cap \mathcal{V}_\varepsilon^d(a_1) = \emptyset$. Let $\alpha \colon [0, 1] \to \mathcal{C}_n(X)$ be an order arc from B to X. We claim that for each $t \in [0, 1]$, $\mathcal{H}(\alpha(t), A) \geq \varepsilon$. To show this, suppose it is not true. Then there exists a point t_0 in $[0, 1]$ such that $\mathcal{H}(\alpha(t_0), A) < \varepsilon$. Thus, we have that $A \subset \mathcal{V}_\varepsilon^d(\alpha(t_0))$ and $\alpha(t_0) \subset \mathcal{V}_\varepsilon^d(A)$. Since $A \subset \mathcal{V}_\varepsilon^d(\alpha(t_0))$, we have that for each $j \in \{1, \ldots, k\}$, $\alpha(t_0) \cap \mathcal{V}_\varepsilon^d(a_j) \neq \emptyset$. On the other hand, since $\alpha(t_0) \subset \mathcal{V}_\varepsilon^d(A) = \bigcup_{j=1}^{k} \mathcal{V}_\varepsilon^d(a_j)$, and these balls are pairwise disjoint, we have that each component of $\alpha(t_0)$ is contained in one such ball. In particular, there is a component of $\alpha(t_0)$ contained in $\mathcal{V}^d_\varepsilon(a_1)$. Hence, $B \cap \mathcal{V}^d_\varepsilon(a_1) \neq \emptyset$, a contradiction. Therefore, $\alpha([0,1]) \subset \mathcal{C}_n(X) \setminus (\mathcal{V}^{\mathcal{H}}_\varepsilon(A) \cap \mathcal{C}_n(X))$.

If $B \not\subset \mathcal{V}^d_\varepsilon(A)$, then given an order arc $\beta\colon [0,1] \to \mathcal{C}_n(X)$ from B to X, we have that $\beta([0,1]) \subset \mathcal{C}_n(X) \setminus (\mathcal{V}^{\mathcal{H}}_\varepsilon(A) \cap \mathcal{C}_n(X))$.

Therefore if $A \in \mathcal{F}_n(X)$, where $A = \{a_1, \dots, a_k\}$, and $\varepsilon > 0$ is given such that $\mathcal{V}^d_\varepsilon(a_j) \cap \mathcal{V}^d_\varepsilon(a_m) = \emptyset$ if and only if $j \neq m$ and $j, m \in \{1, \dots, k\}$, then each element $B \in \mathcal{C}_n(X) \setminus (\mathcal{V}^{\mathcal{H}}_\varepsilon(A) \cap \mathcal{C}_n(X))$ can be joined with $\{X\}$ by an order arc contained in $\mathcal{C}_n(X) \setminus (\mathcal{V}^{\mathcal{H}}_\varepsilon(A) \cap \mathcal{C}_n(X))$.

Therefore $\mathcal{C}_n(X)$ is colocally connected.

Q.E.D.

6.3.2. Definition. A continuum X is said to be *finitely aposyndetic* provided that for each finite subset K of X, $\mathcal{T}(K) = K$.

6.3.3. Corollary. *If X is a continuum and $n \in \mathbb{N}$, then $\mathcal{C}_n(X)$ is aposyndetic and finitely aposyndetic.*

Proof. Clearly, any colocally connected continuum is aposyndetic. Given a continuum X and a positive integer n, since $\mathcal{C}_n(X)$ is unicoherent (Theorem 6.2.4) and aposyndetic, we have that $\mathcal{C}_n(X)$ is finitely aposyndetic (Corollary 1 of [3]).

Q.E.D.

6.4   Arcwise Accessibility

We present results concerning arcwise accessibility of points of the n–fold symmetric product from the n–fold hyperspace of a given continuum.

6.4.1. Definition. Let X be a continuum. Let Σ_1 and Σ_2 be two arcwise connected closed subsets of 2^X, such that $\Sigma_2 \subset \Sigma_1$. A member A of Σ_2 is said to be *arcwise accessible from* $\Sigma_1 \setminus \Sigma_2$ *beginning with* K if and only if there is an arc $\alpha \colon [0,1] \to \Sigma_1$ such that $\alpha(0) = K$, $\alpha(1) = A$ and $\alpha(t) \in \Sigma_1 \setminus \Sigma_2$ for all $t < 1$.

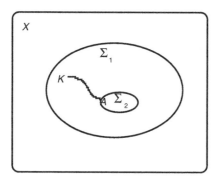

We begin observing that in the proof of (2.2) of [39], Nadler actually showed the following:

6.4.2. Theorem. *Let X be a continuum. If A is a nondegenerate subcontinuum of X and q is any point of A, then there exist a point $p \in A \setminus \{q\}$ and an order arc $\alpha \colon [0,1] \to \mathcal{C}(A)$ such that $\alpha(0) = \{q\}$, $\alpha(1) = A$ and $p \notin \alpha(t)$ for any $t < 1$.*

6.4.3. Theorem. *Let $n \in \mathbb{N}$, $n \geq 2$. Let X be a continuum and let A be an element of $\mathcal{C}_n(X)$ having exactly n components and at least one of them is nondegenerate. Then A is arcwise accessible from $\mathcal{C}_{n+1}(X) \setminus \mathcal{C}_n(X)$, beginning with an element in $\mathcal{F}_{n+1}(X) \setminus \mathcal{F}_n(X)$.*

Proof. Suppose $A_1, \dots, A_n$ are the components of A and that A_n is not degenerate. For each $j \in \{1, \dots, n\}$, let $q_j \in A_j$. By Theorem 6.4.2, there exist $q_{n+1} \in A_n$ and an order arc $\alpha_n \colon [0,1] \to \mathcal{C}(A_n)$ such that $\alpha_n(0) = \{q_n\}$, $\alpha_n(1) = A_n$ and $q_{n+1} \notin \alpha_n(t)$ for any $t < 1$. For each $j \in \{1, \dots, n-1\}$, let $\alpha_j \colon [0,1] \to \mathcal{C}(A_j)$ be an order arc such that $\alpha_j(0) = \{q_j\}$, $\alpha_j(1) = A_j$ (Theorem 1.8.20).

Let $\gamma \colon [0,1] \to \mathcal{C}_{n+1}(X)$ be given by $\gamma(t) = \alpha_1(t) \cup \dots \cup \alpha_n(t) \cup \{q_{n+1}\}$. Since for each $j \in \{1, \dots, n\}$, α_j is continuous and the union is also continuous (Lemma 1.8.11), then γ is continuous. Also, we have that $\gamma(0) = \{q_1, \dots, q_{n+1}\}$, $\gamma(1) = A$, and $\gamma(t) \in \mathcal{C}_{n+1}(X) \setminus$

$\mathcal{C}_n(X)$ for each $t < 1$. Therefore, A is arcwise accessible from $\mathcal{C}_{n+1}(X) \setminus \mathcal{C}_n(X)$, beginning with an element in $\mathcal{F}_{n+1}(X) \setminus \mathcal{F}_n(X)$.

Q.E.D.

6.4.4. Corollary. *Let $n \in \mathbb{N}$, $n \geq 3$. Let X be a continuum. If x is a point of X such that $\{x\}$ is arcwise accessible from $\mathcal{C}_2(X) \setminus \mathcal{C}(X)$ with an arc $\alpha \colon [0,1] \to \mathcal{C}_2(X)$ such that $\alpha([0,1]) \cap (\mathcal{C}_2(X) \setminus \mathcal{F}_2(X)) \neq \emptyset$, then $\{x\}$ is arcwise accessible from $\mathcal{C}_n(X) \setminus \mathcal{C}(X)$.*

6.4.5. Theorem. *Let $n \in \mathbb{N}$, $n \geq 3$. Let X be a continuum, and let a be a point of X such that $\{a\}$ is arcwise accessible from $2^X \setminus \mathcal{C}(X)$. Then each element A of $\mathcal{F}_n(X) \setminus \mathcal{F}_{n-1}(X)$ containing a is arcwise accessible from $2^X \setminus \mathcal{C}_n(X)$.*

Proof. Let $\{a_1, \dots, a_n\} \in \mathcal{F}_n(X) \setminus \mathcal{F}_{n-1}(X)$ containing a. Without loss of generality, we assume that $a = a_n$. Let U be an open set of X such that $a_n \in U$, and for each $j \in \{1, \dots, n-1\}$, $a_j \notin U$. Since $\{a_n\}$ is arcwise accessible from $2^X \setminus \mathcal{C}(X)$, there exists an arc $\alpha \colon [0,1] \to 2^X$ such that $\alpha(1) = \{a_n\}$ and $\alpha(t) \in 2^X \setminus \mathcal{C}(X)$ for each $t < 1$. By continuity, there exists $t_1 \in [0,1)$ such that $\alpha(t) \in U$ for each $t \geq t_1$. Let $\beta \colon [0,1] \to 2^X$ be given by $\beta(s) = \{a_1, \dots, a_{n-1}\} \cup \alpha((1-s)t_1 + s)$. Since α is continuous and the union function is also continuous (Lemma 1.8.11), we have that β is continuous. By construction, we also have that $\beta(1) = \{a_1, \dots, a_n\}$, and $\beta(t) \in 2^X \setminus \mathcal{C}_n(X)$ for each $t < 1$. Therefore, $\{a_1, \dots, a_n\}$ is arcwise accessible from $2^X \setminus \mathcal{C}_n(X)$.

Q.E.D.

From the proof of Theorem 6.4.5 we have the following:

6.4.6. Corollary. *Let $n, m \in \mathbb{N}$, $n \geq 3$. Let X be a continuum and let a be a point of X such that $\{a\}$ is arcwise accessible from $\mathcal{C}_m(X) \setminus \mathcal{C}(X)$. Then each element A of $\mathcal{F}_n(X) \setminus \mathcal{F}_{n-1}(X)$ such that $a \in A$ is arcwise accessible from $\mathcal{C}_{m+n-1}(X) \setminus \mathcal{C}_n(X)$.*

6.4.7. Theorem. *Let* $n \in \mathbb{N}$, $n \geq 2$. *Let* X *be a continuum with metric* d. *Suppose that for each* $j \in \{1, \dots, n\}$, a_j *is a point of* X *such that* $\{a_j\}$ *is not arcwise accessible from* $2^X \setminus \mathcal{C}(X)$. *Then* $\{a_1, \dots, a_n\}$ *is not arcwise accessible from* $2^X \setminus \mathcal{C}_n(X)$.

Proof. Suppose that $\{a_1, \dots, a_n\}$ is arcwise accessible from $2^X \setminus \mathcal{C}_n(X)$. Then there exists an arc $\alpha \colon [0, 1] \to 2^X$ such that $\alpha(1) = \{a_1, \dots, a_n\}$ and $\alpha(t) \in 2^X \setminus \mathcal{C}_n(X)$ for each $t < 1$. Let $U_1, \dots, U_n$ be open sets of X such that for each $j \in \{1, \dots, n\}$, $a_j \in U_j$, and $Cl(U_j) \cap Cl(U_k) = \emptyset$ if and only if $j, k \in \{1, \dots, n\}$ and $j \neq k$. By continuity, there exists $t_1 \in [0, 1)$ such that $\alpha(t) \in \langle U_1, \dots, U_n \rangle$ for each $t \geq t_1$.

For each $j \in \{1, \dots, n\}$, let $\beta_j \colon [0, 1] \to 2^X$ be given by $\beta_j(s) = \alpha((1 - s)t_1 + s) \cap U_j$. Since the family $\{U_1, \dots, U_n\}$ of open sets has pairwise disjoint closures, β_j is well defined for each $j \in \{1, \dots, n\}$. To see that each β_j is continuous, let $j \in \{1, \dots, n\}$ and $\varepsilon > 0$ such that $d(Cl(U_k), Cl(U_\ell)) > \varepsilon$, for each $k, \ell \in \{1, \dots, n\}$ and $k \neq \ell$. Let $s_0 \in [0, 1]$. By continuity, there exists $\delta > 0$ such that if $|s_0 - s_1| < \delta$, then $\mathcal{H}(\alpha((1 - s_0)t_1 + s_0), \alpha((1 - s_1)t_1 + s_1)) < \varepsilon$. Let $x \in \alpha((1 - s_0)t_1 + s_0) \cap U_j$. Since $\mathcal{H}(\alpha((1 - s_0)t_1 + s_0), \alpha((1 - s_1)t_1 + s_1)) < \varepsilon$, there exists $y \in \alpha((1 - s_1)t_1 + s_1)$ such that $d(x, y) < \varepsilon$. Since $y \in \bigcup_{k=1}^{n} U_k$, there exists $\ell \in \{1, \dots, n\}$ such that $y \in U_\ell$. Since $d(x, y) < \varepsilon$ and $d(Cl(U_k), Cl(U_\ell)) > \varepsilon$ for each $k \in \{1, \dots, n\}$ and $k \neq \ell$, we have that $\ell = j$ and $y \in U_j$. Therefore, $y \in \alpha((1 - s_1)t_1 + s_1) \cap U_j$, and $\alpha((1 - s_0)t_1 + s_0) \cap U_j \subset \mathcal{V}_\varepsilon^d(\alpha((1 - s_1)t_1 + s_1) \cap U_j)$. Similarly $\alpha((1 - s_1)t_1 + s_1) \cap U_j \subset \mathcal{V}_\varepsilon^d(\alpha((1 - s_0)t_1 + s_0) \cap U_j)$. Thus, $\mathcal{H}(\alpha((1 - s_0)t_1 + s_0) \cap U_j, \alpha((1 - s_1)t_1 + s_1) \cap U_j) = \mathcal{H}(\beta_j(s_0), \beta_j(s_1)) < \varepsilon$. Hence, β_j is continuous.

Note that for each $j \in \{1, \dots, n\}$, $\beta_j(1) = \{a_j\}$. Since, for each $j \in \{1, \dots, n\}$, $\{a_j\}$ is not arcwise accessible from $2^X \setminus \mathcal{C}(X)$, we have that, for each $j \in \{1, \dots, n\}$, there exists $s_j \in [0, 1)$ such that $\beta_j(s) \in \mathcal{C}(X)$ for every $s \geq s_j$. Let $s_* = \max\{s_1, \dots, s_n\}$. Then for each $j \in \{1, \dots, n\}$, $\beta_j(s) \in \mathcal{C}(X)$ for every $s \geq s_*$.

Let $s \geq s_*$. Then $\bigcup_{j=1}^{n} \beta_j(s) = \bigcup_{j=1}^{n} \alpha((1 - s)t_1 + s) \cap U_j = \alpha((1 - s)t_1 + s) \cap \bigcup_{j=1}^{n} U_j = \alpha((1 - s)t_1 + s)$. Also, if $s \geq s_*$, then

$\bigcup\limits_{j=1}^{n} \beta_j(s) \in \mathcal{C}_n(X)$. Thus, for each $s \geq s_*$, $\alpha((1-s)t_1 + s) \in \mathcal{C}_n(X)$, a contradiction. Therefore, $\{a_1, \ldots, a_n\}$ is not arcwise accessible from $2^X \setminus \mathcal{C}_n(X)$.

Q.E.D.

6.4.8. Corollary. *Let $n \in \mathbb{N}$, $n \geq 2$. If X is a hereditarily indecomposable continuum, then $\{a_1, \ldots, a_n\}$ is not arcwise accessible from $2^X \setminus \mathcal{C}_n(X)$ for any $\{a_1, \ldots, a_n\} \in \mathcal{F}_n(X)$.*

Proof. Since X is hereditarily indecomposable, no singleton is arcwise accessible from $2^X \setminus \mathcal{C}(X)$ ((3.4) of [39]). Thus, we have the result follows from Theorem 6.4.7.

Q.E.D.

6.5   Points that Arcwise Disconnect

We consider when a point arcwise disconnects $\mathcal{C}_n(X)$. We also study the arc components of $\mathcal{C}_n(X) \setminus \{X\}$, where X is an indecomposable continuum.

6.5.1. Lemma. *Let X be a continuum and let $n \in \mathbb{N}$ be given. If A is a proper subcontinuum of X, then $\mathcal{C}_n(X) \setminus \mathcal{C}_n(A)$ is arcwise connected.*

Proof. Each element of $\mathcal{C}_n(X) \setminus \mathcal{C}_n(A)$ can be joined with X with an order arc contained in $\mathcal{C}_n(X) \setminus \mathcal{C}_n(A)$ (Theorem 1.8.20).

Q.E.D.

The proof of the following Theorem is similar to the one given in (11.3) of [40].

6.5.2. Theorem. *Let X be a continuum and let $n \in \mathbb{N}$ be given. If $A \in \mathcal{C}_n(X)$ is such that $\mathcal{C}_n(X) \setminus \{A\}$ is not arcwise connected, then A is connected.*

The following Theorem characterizes indecomposable continua.

6.5.3. Theorem. *A nondegenerate continuum X is indecomposable if and only if for each $n \in \mathbb{N}$, $\mathcal{C}_n(X) \setminus \{X\}$ is not arcwise connected.*

Proof. Suppose X is decomposable. We show that for each $n \in \mathbb{N}$, $\mathcal{C}_n(X) \setminus \{X\}$ is arcwise connected. The proof is done by induction. The result is known for $n = 1$ ((1.51) of [40]). Let $n = 2$, and let A and B be two elements of $\mathcal{C}_2(X) \setminus \{X\}$. If both A and B are connected then there exists an arc in $\mathcal{C}(X) \setminus \{X\}$ joining A and B ((1.51) of [40]). Suppose A has two components, say A_1 and A_2. Since X is decomposable, there exist two proper subcontinua H and K of X such that $X = K \cup H$. We show there exists an arc in $\mathcal{C}_2(X) \setminus \{X\}$ joining A and an element of $\mathcal{C}(X)$. We have to consider several cases.

If either $A \subset H$ or $A \subset K$, then there exists an order arc joining A with H or K (Theorem 1.8.20), and we are done.

If $A_2 = K$, then let $x \in H \cap K$. Let $\alpha \colon [0,1] \to \mathcal{C}_2(X)$ be an order arc joining $A_1 \cup \{x\}$ and $A_1 \cup K = A$. Let $\beta \colon [0,1] \to \mathcal{C}_2(X)$ be an order arc joining $A_1 \cup \{x\}$ and H. Then $\alpha([0,1]) \cup \beta([0,1])$ contains an arc having A and H as its end points, contained in $\mathcal{C}_2(X) \setminus \{X\}$.

If $A_1 \subset H$, $A_2 \cap (H \cap K) \neq \emptyset$, and $A_2 \neq K$, then $H \cup A_2$ is a proper subcontinuum of X, and we can take an order arc joining A to $H \cup A_2$.

If $A_1 \cap (H \cap K) \neq \emptyset$ and $A_2 \cap (H \cap K) \neq \emptyset$, then let $a_j \in A_j \cap H$ for $j \in \{1, 2\}$. Let $\alpha \colon [0,1] \to \mathcal{C}_2(X)$ be an order arc joining $\{a_1, a_2\}$ and A, and let $\beta \colon [0,1] \to \mathcal{C}_2(X)$ be an order arc having $\{a_1, a_2\}$ and H as its end points. Then $\alpha([0,1]) \cup \beta([0,1])$ is contained in $\mathcal{C}_2(X) \setminus \{X\}$ and contains an arc joining A and H.

If $A_1 \subset H \setminus K$ and $A_2 \subset K \setminus H$, then let $x \in H \cap K$. Let $\alpha \colon [0,1] \to \mathcal{C}_2(X)$ be an order arc from A to $H \cup A_2$. Let $\beta[0,1] \to \mathcal{C}_2(X)$ be an order arc from $\{x\} \cup A_2$ to $H \cup A_2$. Let $\gamma \colon [0,1] \to \mathcal{C}_2(X)$ be an order arc from $\{x\} \cup A_2$ to K. Then $\alpha([0,1]) \cup \beta([0,1]) \cup \gamma([0,1])$ contains an arc joining A and K contained in $\mathcal{C}_2(X) \setminus \{X\}$. The rest of the cases are similar to the ones treated.

Now, let $n \geq 3$ and suppose $\mathcal{C}_n(X) \setminus \{X\}$ is arcwise connected. We show that $\mathcal{C}_{n+1}(X) \setminus \{X\}$ is arcwise connected. Let A and B be two points in $\mathcal{C}_{n+1}(X) \setminus \{X\}$. If both A and B belong to $\mathcal{C}_n(X)$, by induction hypothesis we can find an arc in $\mathcal{C}_n(X) \setminus \{X\} \subset \mathcal{C}_{n+1}(X) \setminus \{X\}$ having A and B as its end points. Hence, assume that A has $n+1$ components. Let $A_1, \ldots, A_{n+1}$ be the components of A. Since X is decomposable, there exist two proper subcontinua H and K of it such that $X = H \cup K$. Since $n \geq 3$, at least two components of A intersect either H or K; suppose that two components of A intersect H. Without loss of generality, we assume that $A_n \cap H \neq \emptyset$ and $A_{n+1} \cap H \neq \emptyset$. For each $j \in \{1, \ldots, n-1\}$, let $a_j \in A_j$. Take $a_n \in A_n \cap H$ and $a_{n+1} \in A_{n+1} \cap H$. Let $\alpha \colon [0,1] \to \mathcal{C}_{n+1}(X)$ be an order arc from $\{a_1, \ldots, a_{n+1}\}$ to A. Let $\beta \colon [0,1] \to \mathcal{C}_2(X)$ be an order arc from $\{a_n, a_{n+1}\}$ to H. Let $\gamma \colon [0,1] \to \mathcal{C}_{n+1}(X)$ be given by $\gamma(t) = \{a_1, \ldots, a_{n-1}\} \cup \beta(t)$. Then γ is continuous, $\gamma(0) = \{a_1, \ldots, a_{n-1}\} \cup \beta(0) = \{a_1 \ldots, a_{n+1}\}$ and $\gamma(1) = \{a_1, \ldots, a_{n-1}\} \cup \beta(1) = \{a_1, \ldots, a_{n-1}\} \cup H \in \mathcal{C}_n(X) \setminus \{X\}$. Hence, $\alpha([0,1]) \cup \gamma([0,1])$ is contained in $\mathcal{C}_{n+1}(X) \setminus \{X\}$ and contains an arc having A and $\gamma(1)$ as its end points. Similarly, if B has $n+1$ components, we can find an arc in $\mathcal{C}_{n+1}(X) \setminus \{X\}$ having B and an element of $\mathcal{C}_n(X)$ as its end points. Thus by induction hypothesis, we are done.

The proof of the reverse implication is similar to the one given in (11.4) of [40].

Q.E.D.

The next Theorem tells us more about what type of subcontinua may arcwise disconnect the hyperspaces.

6.5.4. Theorem. *Let X be a continuum, and let E be a nondegenerate proper subcontinuum of X. Consider the following statements:*

(1) E is a terminal subcontinuum of X.

(2) $2^X \setminus \{E\}$ is not arcwise connected.

(3) For each $n \in \mathbb{N}$, $\mathcal{C}_n(X) \setminus \{E\}$ is not arcwise connected.

(4) $\mathcal{C}(X) \setminus \{E\}$ is not arcwise connected.

Then (1) implies (2), (3) and (4). Furthermore, if E is decomposable then all four statements are equivalent.

Proof. The proof of (1) implies (2) and (4) are given in (11.5) of [40]. The proof of (1) implies (3) is similar to the one given in (11.5) of [40].

Now suppose E is a decomposable nondegenerate subcontinuum of X. The equivalence between (1), (2) and (4) is given in (11.5) of [40]. Suppose E is not terminal, then $E \neq X$. We show that for each $n \in \mathbb{N}$, $\mathcal{C}_n(X) \setminus \{E\}$ is arcwise connected. This is done by induction.

For $n = 1$ the result is known. Suppose $\mathcal{C}_n(X) \setminus \{E\}$ is arcwise connected. To show that $\mathcal{C}_{n+1}(X) \setminus \{E\}$ is arcwise connected, let A and B be two points in $\mathcal{C}_{n+1}(X) \setminus \{E\}$. If both A and B belong to $\mathcal{C}_{n+1}(X) \setminus \mathcal{C}_{n+1}(E)$, then there exists an arc having A and B as its end points and contained in $\mathcal{C}_{n+1}(X) \setminus \{E\}$ (Lemma 6.5.1). If both A and B belong to $\mathcal{C}_{n+1}(E)$, then by Theorem 6.5.3, there exists an arc joining A and B and contained in $\mathcal{C}_{n+1}(E) \setminus \{E\} \subset \mathcal{C}_{n+1}(X) \setminus \{E\}$. Thus, suppose, without loss of generality, that $A \in \mathcal{C}_{n+1}(E) \setminus \{E\}$ and $B \in \mathcal{C}_{n+1}(X) \setminus \mathcal{C}_n(E)$.

If A has less than $n + 1$ components then, by induction hypothesis, there exists an arc in $\mathcal{C}_n(X) \setminus \{E\}$ joining A and X. So, assume A has exactly $n + 1$ components. Since E is decomposable, by Theorem 6.5.3, there exists an arc joining A and an element A' of $\mathcal{C}_n(E) \setminus \{E\}$. By induction hypothesis, there exists an arc in $\mathcal{C}_n(X) \setminus \{E\}$ joining A' and X. Hence, there exists an arc in $\mathcal{C}_{n+1}(X) \setminus \{E\}$ joining A and X. Since $B \in \mathcal{C}_{n+1}(X) \setminus \mathcal{C}_{n+1}(E)$, by Lemma 6.5.1, there exits an arc having B and X as its end points.

Therefore, there exists an arc in $\mathcal{C}_{n+1}(X) \setminus \{E\}$ joining A and B.

Q.E.D.

6.5.5. Theorem. *If X is a continuum, then for any $E \in 2^X$, the following are equivalent:*

(1) $2^X \setminus \{E\}$ is not arcwise connected.

(2) $\mathcal{C}(X) \setminus \{E\}$ is not arcwise connected.

(3) For each $n \in \mathbb{N}$, $\mathcal{C}_n(X) \setminus \{E\}$ is not arcwise connected.

Proof. The proof of the equivalence of (1) and (2) is given in (11.8) of [40]. Clearly (3) implies (2). We show that (2) implies (3).

Let $E \in 2^X$ and suppose $\mathcal{C}(X) \setminus \{E\}$ is not arcwise connected. Then $E \in \mathcal{C}(X)$ ((11.3) of [40]). Let $n \geq 2$. If $E = X$, then, by Theorem 6.5.3, X is indecomposable and $\mathcal{C}_n(X) \setminus \{E\}$ is not arcwise connected.

Suppose $E \neq X$. If E is decomposable, then, by Theorem 6.5.4, E is a terminal subcontinuum of X and $\mathcal{C}_n(X) \setminus \{E\}$ is not arcwise connected.

Thus, assume E is indecomposable. Let $B \in \mathcal{C}(E) \subset \mathcal{C}_n(X)$ and let $A \in \mathcal{C}_n(X) \setminus \mathcal{C}_n(E)$. Let $\alpha \colon [0,1] \to \mathcal{C}_n(X)$ be an arc such that $\alpha(0) = B$ and $\alpha(1) = A$. Let $\beta \colon [0,1] \to \mathcal{C}_n(X)$ be given by $\beta(t) = \sigma(\alpha([0,t]))$. Then β is an order arc from B to $\sigma(\alpha([0,1]))$. Hence, $\beta([0,1]) \subset \mathcal{C}(X)$ ((1.11) of [40]). In particular, $\sigma(\alpha([0,1]))$ is connected. Since $\beta(0) = B \subset E$, $\beta(1) = \sigma(\alpha([0,1]))$, $(\sigma(\alpha([0,1]))) \cap (\mathcal{C}(X) \setminus \mathcal{C}(E)) \neq \emptyset$, and $\mathcal{C}(X) \setminus \{E\}$ is not arcwise connected, we have that there exists $t_0 \in [0,1]$ such that $\beta(t_0) = E$. Let $t_1 = \min\{t \in [0,1] \mid \beta(t) = E\}$. Then $t_1 > 0$ and $\beta(t_1) = E$. Since for each $t < t_1$, $\beta(t)$ is a proper subcontinuum of E and E is indecomposable, we have that $\beta(t)$ is nowhere dense in E (Corollary 1.7.21). Note that, for every $t < t_1$, $E = \beta(t) \cup (\sigma(\alpha([t,t_1])))$, and $\sigma(\alpha([t,t_1]))$ is compact. Hence, $E = \sigma(\alpha([t,t_1]))$. By continuity, we have that $E = \alpha(t_1)$.

Therefore, $\mathcal{C}_n(X) \setminus \{E\}$ is not arcwise connected.

Q.E.D.

6.5.6. Theorem. *If E, A and B are subcontinua of the continuum X and $n \in \mathbb{N}$, then the following are equivalent:*

(1) If Γ is an arc in $\mathcal{C}(X)$ such that $A, B \in \Gamma$, then $E \in \Gamma$.

(2) If Γ is an arc in $\mathcal{C}_n(X)$ such that $A, B \in \Gamma$, then $E \in \Gamma$.

(3) If Γ is an arc in 2^X such that $A, B \in \Gamma$, then $E \in \Gamma$.

Proof. Clearly (3) implies (2) and (2) implies (1). The proof of the implication from (1) to (3) is in (11.13) of [40].

Q.E.D.

6.5.7. Corollary. *Let E be a subcontinuum of the continuum X. If Λ is an arc component of $2^X \setminus \{E\}$ and $\Lambda \cap \mathcal{C}_n(X) \neq \emptyset$, for some $n \in \mathbb{N}$, then $\Lambda \cap \mathcal{C}_n(X)$ is an arc component of $\mathcal{C}_n(X) \setminus \{E\}$.*

6.5.8. Theorem. *If X is a continuum, then the following are equivalent:*

(1) X is hereditarily indecomposable.

(2) For each nondegenerate subcontinuum E of X, $2^X \setminus \{E\}$ is not arcwise connected.

(3) For each nondegenerate subcontinuum E of X, $C_n(X) \setminus \{E\}$ is not arcwise connected, for each $n \in \mathbb{N}$.

(4) For each nondegenerate subcontinuum E of X, $C(X) \setminus \{E\}$ is not arcwise connected.

Proof. By Theorem 6.5.5, we have that (2), (3) and (4) are equivalent. The proof of the equivalence between (4) and (1) is given in (11.15) of [40].

$$\textbf{Q.E.D.}$$

6.5.9. Definition. Let X be a continuum, and let $p \in X$. The *composant of p in X* is the union of all proper subcontinua of X containing p.

6.5.10. Notation. Given a continuum X, a positive integer n and a composant κ of X, let $C_n(\kappa)$ denote the set $C_n(\kappa) = \{A \in C_n(X) \mid A \subset \kappa\}$.

In the next two Theorems we describe the arc components of $C_n(X) \setminus \{X\}$, where X is an indecomposable continuum.

6.5.11. Theorem. *Let n be an integer greater than one. If κ is a composant of an indecomposable continuum X, then $C_n(\kappa)$ is an arc component of $C_n(X) \setminus \{X\}$.*

Proof. By Theorem 6.5.3, $C_n(X) \setminus \{X\}$ is not arcwise connected. First, observe that $C_n(\kappa)$ is arcwise connected. To see this, let $A \in C_n(\kappa)$. Then, since A has finitely many components and X is indecomposable, it is easy to show that there exists a proper subcontinuum B of X containing A. Thus, there exists an order arc from A to B (Theorem 1.8.20). Since $\mathcal{C}(\kappa)$ is arcwise connected ((1.52.1) of [40]), we have that $\mathcal{C}_n(\kappa)$ is arcwise connected.

Let $\mathcal{A}$ be the arc component of $\mathcal{C}_n(X) \setminus \{X\}$ containing $\mathcal{C}_n(\kappa)$, and suppose there exists $B \in \mathcal{A} \setminus \mathcal{C}_n(\kappa)$. Let $A \in \mathcal{C}(\kappa)$. Since A and B belong to $\mathcal{A}$, there is an arc $\alpha \colon [0,1] \to \mathcal{A}$ such that $\alpha(0) = A$ and $\alpha(1) = B$. Let $\beta \colon [0,1] \to \mathcal{C}_n(X)$ be given by $\beta(t) = \sigma(\alpha([0,t]))$. Then β is well defined (Corollary 6.1.2), β is an order arc and $\beta(0) = \alpha(0) = A$. Hence, $\beta(t) \in \mathcal{C}(X)$ for each $t \in [0,1]$ ((1.11) of [40]), and $B \subset \beta(1)$. Since B is not contained in κ and $B \subset \beta(1)$, we have that $\beta(1)$ is a subcontinuum of X intersecting two different composants of it. Thus, $\beta(1) = X$. Let $t_0 = \min\{t \in [0,1] \mid \beta(t) = X\}$. Then $\beta(t_0) = X$ and $t_0 > 0$. Observe that if $0 \le t < t_0$, then $\beta(t)$ is a nowhere dense subset of X (Corollary 1.7.21). Note that for each $0 < t < t_0$, $X = \beta(t_0) = \beta(t) \cup (\sigma(\alpha([t,t_0])))$. Since $\beta(t)$ is nowhere dense in X, we have that $X = \sigma(\alpha([t,t_0]))$ for each $t < t_0$. By continuity, $\alpha(t_0) = X$, a contradiction. Therefore, $\mathcal{C}_n(\kappa) = \mathcal{A}$.

Q.E.D.

6.5.12. Theorem. *Let n be an integer greater than one, and let X be an indecomposable continuum. If $\mathcal{A}$ is an arc component of $\mathcal{C}_n(X) \setminus \{X\}$, which is not of the form $\mathcal{C}_n(\kappa)$, where κ is a composant of X, then there exist finitely many composants $\kappa_1, \ldots, \kappa_\ell$ of X and ℓ positive integers $m_1, \ldots, m_\ell$ such that there exists a one–to–one map from $\displaystyle\prod_{j=1}^{\ell} \mathcal{C}_{m_j}(\kappa_j)$ (with the "max" metric ρ_1) onto $\mathcal{A}$.*

Proof. Let A_0 be a point of $\mathcal{A}$, and let $\kappa_1, \ldots, \kappa_\ell$ be the composants of X which intersect A_0. Since $\mathcal{A}$ is not of the form $\mathcal{C}_n(\kappa)$, $\ell \ge 2$.

First, we show that every element of $\mathcal{A}$ intersects each κ_j for each $j \in \{1, \ldots, \ell\}$. To see this, suppose that there is a point B of $\mathcal{A}$ such that $B \cap \kappa_j = \emptyset$, for some $j \in \{1, \ldots, \ell\}$. Since A_0 and B belong to $\mathcal{A}$, there exists an arc $\alpha \colon [0,1] \to \mathcal{A}$ such that $\alpha(0) = A_0$ and $\alpha(1) = B$. Let $\beta \colon [0,1] \to \mathcal{C}_n(X)$ be given by $\beta(t) = \sigma(\alpha([0,t]))$. Then β is well defined (Corollary 6.1.2), β is an order arc, $\beta(0) = \alpha(0) = A_0$, and $\beta(1) = \sigma(\alpha([0,1]))$ is an element of $\mathcal{C}_n(X)$ intersecting κ_j and containing B. On the other hand, the map $\gamma \colon [0,1] \to \mathcal{C}_n(X)$ given by $\gamma(t) = \sigma(\alpha([1-t,1]))$ is also well defined, γ is an order arc, $\gamma(0) = \alpha(1) = B$ and $\gamma(1) = \sigma(\alpha([0,1])) = \beta(1)$. Thus, there exists an order arc in $\mathcal{C}_n(X)$ from B to $\beta(1)$, $B \cap \kappa_j = \emptyset$ and $\beta(1) \cap \kappa_j \neq \emptyset$; this contradicts Theorem 1.8.20 if $\sigma(\alpha([0,1]))$ is a proper subset of X. Otherwise, a similar argument to the one given in Theorem 6.5.11 shows that there exists $t_0 \in [0,1]$ such that $\alpha(t_0) = X$, which is also a contradiction. Therefore, every element of $\mathcal{A}$ intersects each κ_j, $j \in \{1,\ldots,\ell\}$. A similar argument proves that if $A \in \mathcal{A}$ and $A \cap \kappa \neq \emptyset$, for some composant κ of X, then $\kappa \in \{\kappa_1,\ldots,\kappa_\ell\}$.

For each $j \in \{1,\ldots,\ell\}$, let

$$m_j = \max\{\text{number of components of } A \text{ contained in } \kappa_j \mid A \in \mathcal{A}\}.$$

Let us observe that $\sum_{j=1}^{\ell} m_j = n$. Let $f \colon \prod_{j=1}^{\ell} \mathcal{C}_{m_j}(\kappa_j) \to \mathcal{A}$ be given by

$$f((A_1,\ldots,A_\ell)) = \bigcup_{j=1}^{\ell} A_j.$$

Let us see first that f is well defined. Clearly, $f((A_1,\ldots,A_\ell)) \in \mathcal{C}_n(X)$. On the other hand, for each $j \in \{1,\ldots,\ell\}$, A_j and $A_0 \cap \kappa_j$ both belong to $\mathcal{C}_{m_j}(\kappa_j)$. Since $\mathcal{C}_{m_j}(\kappa_j)$ is arcwise connected, there is an arc $\alpha_j \colon [0,1] \to \mathcal{C}_{m_j}(\kappa_j)$ such that $\alpha_j(0) = A_0 \cap \kappa_j$ and $\alpha_j(1) = A_j$. Hence, $\alpha \colon [0,1] \to \mathcal{C}_n(X) \setminus \{X\}$ given by $\alpha(t) = \bigcup_{j=1}^{\ell} \alpha_j(t)$ is a path joining A_0 and $\bigcup_{j=1}^{\ell} A_j$. Therefore, $\bigcup_{j=1}^{\ell} A_j \in \mathcal{A}$.

If A is a point of $\mathcal{A}$, then $(A \cap \kappa_1,\ldots,A \cap \kappa_\ell)$ is an element of $\prod_{j=1}^{\ell} \mathcal{C}_{m_j}(\kappa_j)$ such that $f((A \cap \kappa_1,\ldots,A \cap \kappa_\ell)) = \bigcup_{j=1}^{\ell} (A \cap \kappa_j) = A$. Thus, f is surjective.

Let $(A_1,\ldots,A_\ell)$ and $(B_1,\ldots,B_\ell)$ be two different points of $\prod_{j=1}^{\ell} \mathcal{C}_{m_j}(\kappa_j)$. Then $A_{j_0} \neq B_{j_0}$ for some $j_0 \in \{1,\ldots,\ell\}$. Hence, $\bigcup_{j=1}^{\ell} A_j \neq \bigcup_{j=1}^{\ell} B_j$, being a disjoint union. Therefore, f is one–to–one.

To see that f is continuous, let $\varepsilon > 0$ be given. Let $(A_1, \ldots, A_\ell)$ and $(B_1, \ldots, B_\ell)$ be two points of $\prod_{j=1}^{\ell} \mathcal{C}_{m_j}(\kappa_j)$ such that

$$\rho_1\left((A_1, \ldots, A_\ell), (B_1, \ldots, B_\ell)\right) < \frac{\varepsilon}{2}.$$

Then for each $j \in \{1, \ldots, \ell\}$, $\mathcal{H}(A_j, B_j) < \frac{\varepsilon}{2}$. Hence, for every $j \in \{1, \ldots, \ell\}$, $A_j \subset \mathcal{V}_{\frac{\varepsilon}{2}}^d(B_j) \subset \mathcal{V}_{\frac{\varepsilon}{2}}^d\left(\bigcup_{r=1}^{\ell} B_r\right)$ and $B_j \subset \mathcal{V}_{\frac{\varepsilon}{2}}^d(A_j) \subset \mathcal{V}_{\frac{\varepsilon}{2}}^d\left(\bigcup_{r=1}^{\ell} A_r\right)$. Thus, we have that $\bigcup_{j=1}^{\ell} A_j \subset \mathcal{V}_{\frac{\varepsilon}{2}}^d\left(\bigcup_{j=1}^{\ell} B_j\right)$ and $\bigcup_{j=1}^{\ell} B_j \subset \mathcal{V}_{\frac{\varepsilon}{2}}^d\left(\bigcup_{j=1}^{\ell} A_j\right)$. This implies that

$$\mathcal{H}\left(\bigcup_{j=1}^{\ell} A_j, \bigcup_{j=1}^{\ell} B_j\right) \leq \frac{\varepsilon}{2} < \varepsilon.$$

Therefore, f is continuous.

Q.E.D.

6.5.13. Theorem. *Let n be an integer greater than one. Let X be an indecomposable continuum. If $\mathcal{A}$ is an arc component of $\mathcal{C}_n(X)\setminus\{X\}$, which is not of the form $\mathcal{C}_n(\kappa)$, where κ is a composant of X, then for any arc $\alpha\colon [0,1] \to \mathcal{A}$, $\sigma(\alpha([0,1]))$ is not connected.*

Proof. Let $\alpha\colon [0,1] \to \mathcal{A}$ be an arc and suppose $\sigma(\alpha([0,1]))$ is connected. Observe that $\alpha(0)$ is a nonconnected subset of $\sigma(\alpha([0,1]))$ intersecting at least two composants of X. Hence, $\sigma(\alpha([0,1])) = X$. An argument similar to the one given in the proof of Theorem 6.5.11 shows that there exists a point t_0 in $[0,1]$ such that $\alpha(t_0) = X$, which is not possible.

Therefore, $\sigma(\alpha([0,1]))$ is not connected.

Q.E.D.

We end this section showing that hereditarily indecomposable continua have unique n–fold hyperspaces.

6.5.14. Theorem. *Let $n, m \in \mathbb{N}$, $m \geq 2$. Let X be a hereditarily indecomposable continuum. If Y is a continuum such that $\mathcal{C}_m(Y)$ is homeomorphic to $\mathcal{C}_n(X)$, then Y is homeomorphic to X.*

Proof. Let $h: \mathcal{C}_n(X) \twoheadrightarrow \mathcal{C}_m(Y)$ be a homeomorphism. We consider first the case $n = 1$. Since X is hereditarily indecomposable, $\mathcal{C}(X)$ is uniquely arcwise connected ((1.61) of [40]). Hence, $\mathcal{C}_m(Y)$ is uniquely arcwise connected. Then, by Theorem 6.1.9, $m = 1$. A contradiction.

Suppose now that $n \geq 2$. Since X is hereditarily indecomposable, X is the only point at which $\mathcal{C}_n(X)$ is locally connected (Theorem 6.1.19). Hence, $h(X) = Y$.

Now, we show that $h(\mathcal{C}(X)) \subset \mathcal{C}(Y)$. To see this, let $A \in \mathcal{C}(X) \setminus \mathcal{F}_n(X)$, then $\mathcal{C}_n(X) \setminus \{A\}$ is not arcwise connected (Theorem 6.5.8). Thus, $\mathcal{C}_m(Y) \setminus \{h(A)\}$ is not arcwise connected. Hence, $h(A) \in \mathcal{C}(Y)$ (Theorem 6.5.2). Since $\mathcal{C}(Y)$ is closed in $\mathcal{C}_m(Y)$ and $\mathcal{F}_1(X) \subset Cl_{\mathcal{C}_n(X)}(\mathcal{C}(X) \setminus \mathcal{F}_1(X))$, we have that $h(\mathcal{C}(X)) \subset \mathcal{C}(Y)$.

Next, we prove that $h(\mathcal{F}_1(X)) \subset \mathcal{F}_1(Y)$. To this end, suppose there exists a point $\{x\}$ in $\mathcal{F}_1(X)$ such that $h(\{x\}) \in \mathcal{C}(Y) \setminus \mathcal{F}_1(Y)$. Then there exists an order arc $\alpha: [0, 1] \to \mathcal{C}_2(Y)$ such that $\alpha(0) \in \mathcal{F}_2(Y)$, $\alpha(1) = \{x\}$, and $\alpha([0, 1)) \subset \mathcal{C}_2(Y) \setminus \mathcal{C}(Y)$ (see (2.2) of [39]). Hence, $h^{-1} \circ \alpha: [0, 1] \to \mathcal{C}_n(X)$ is an arc such that $h^{-1} \circ \alpha(1) = \{x\}$ and $h^{-1} \circ \alpha([0, 1)) \subset \mathcal{C}_n(X) \setminus \mathcal{C}(X)$. This contradicts the fact that in 2^X, such arcs do not exist (i.e., singletons are not arcwise accessible from $2^X \setminus \mathcal{C}(X)$) ((3.4) of [39]). Therefore, $h(\mathcal{F}_1(X)) \subset \mathcal{F}_1(Y)$.

Let $Y' \in \mathcal{C}(Y)$ be such that $\mathcal{F}_1(Y') = h(\mathcal{F}_1(X))$. Note that $\mathcal{C}(Y') \subset \mathcal{C}(Y)$ and $h^{-1}(\mathcal{C}(Y'))$ is an arcwise connected subcontinuum of $\mathcal{C}_n(X)$. Thus, $\mathcal{C}(X) \cap h^{-1}(\mathcal{C}(Y'))$ is arcwise connected ((5.2) of [39]), and $\mathcal{F}_1(X) \subset \mathcal{C}(X) \cap h^{-1}(\mathcal{C}(Y'))$. We claim that $h^{-1}(Y') \in \mathcal{C}(X)$. Suppose, to the contrary, that $h^{-1}(Y') \in \mathcal{C}_n(X) \setminus \mathcal{C}(X)$. Let x_1 and x_2 be two points in different composants of X. Let $\beta_1, \beta_2: [0, 1] \to h^{-1}(\mathcal{C}(Y'))$ be two arcs such that $\beta_j(0) = \{x_j\}$ and $\beta_j(1) = h^{-1}(Y')$, $j \in \{1, 2\}$. By (3.4) of [39], we have that $\beta_j([0, 1]) \cap \mathcal{C}(X) \neq \{\{x_j\}\}$, $j \in \{1, 2\}$. Hence, $(\beta_1([0, 1]) \cup \beta_2([0, 1])) \cap \mathcal{C}(X)$ contains an arc from $\{x_1\}$ to $\{x_2\}$ ((5.2) of [39]). Since X is indecomposable and x_1 and x_2 are in different composants of X, we have that $X = \sigma([(\beta_1([0, 1]) \cup \beta_2([0, 1])) \cap \mathcal{C}(X)])$ ((1.51) of [40]). Then $X \in (\beta_1([0, 1]) \cup \beta_2([0, 1])) \cap \mathcal{C}(X)$ ((1.50) of [40]). Thus, $X \in h^{-1}(\mathcal{C}(Y'))$. It follows that $Y = h(X) \in \mathcal{C}(Y')$ and $Y = Y'$, a contradiction. Therefore, $h^{-1}(Y') \in \mathcal{C}(X)$.

We assert that $h^{-1}(Y') = X$. To see this, let x_1' and x_2' be two points in different composants of X. Since $\mathcal{C}(X) \cap h^{-1}(\mathcal{C}(Y'))$ is arcwise connected, there exist two arcs $\alpha_1, \alpha_2 \colon [0,1] \to \mathcal{C}(X) \cap h^{-1}(\mathcal{C}(Y'))$ such that $\alpha_j(0) = \{x_j'\}$ and $\alpha_j(1) = h^{-1}(Y')$, $j \in \{1,2\}$. Then $\alpha_1([0,1]) \cup \alpha_2([0,1])$ contains an arc from $\{x_1'\}$ to $\{x_2'\}$. Since X is indecomposable, we have that $X = \sigma(\alpha_1([0,1]) \cap \alpha_2([0,1]))$ ((1.51) of [40]). Thus, $X \in \alpha_1([0,1]) \cup \alpha_2([0,1])$ ((1.50) of [40]). Then $h(X) \in \mathcal{C}(Y')$. Hence, since $h(X) = Y$, $Y \in \mathcal{C}(Y')$. Hence, $Y' = Y$. Thus, by definition of Y', $h(\mathcal{F}_1(X)) = \mathcal{F}_1(Y)$.

Therefore, Y is homeomorphic to X.

Q.E.D.

6.5.15. Corollary. *Let $n \in \mathbb{N}$, and let X be a hereditarily indecomposable continuum. If Y is a continuum such that $\mathcal{C}_n(Y)$ is homeomorphic to $\mathcal{C}_n(X)$, then X and Y are homeomorphic.*

Proof. By Theorem 6.5.14, we only have to consider the case when $n = 1$. Since X is hereditarily indecomposable, $\mathcal{C}(X)$ is uniquely arcwise connected ((1.61) of [40]). Hence, $\mathcal{C}(Y)$ is uniquely arcwise connected. Thus, by (1.61) of [40], Y is hereditarily indecomposable. Therefore, Y is homeomorphic to X ((0.60) of [40]).

Q.E.D.

6.6 $\mathcal{C}_n^*$–smoothness

We study the continuity of taking n–fold hyperspaces.

6.6.1. Definition. Let $n \in \mathbb{N}$. A continuum X is $\mathcal{C}_n^*$–*smooth at* $A \in \mathcal{C}_n(X)$, provided that for any sequence $\{A_k\}_{k=1}^\infty$ of elements of $\mathcal{C}_n(X)$ converging to A, the sequence $\{\mathcal{C}_n(A_k)\}_{k=1}^\infty$ of hyperspaces converges to $\mathcal{C}_n(A)$; i.e., the map $\mathcal{C}_n^* \colon \mathcal{C}_n(X) \to 2^{2^X}$ given by $\mathcal{C}_n^*(A) = \mathcal{C}_n(A)$ is continuous at A. A continuum X is $\mathcal{C}_n^*$–*smooth* if it is $\mathcal{C}_n^*$–smooth at each element of $\mathcal{C}_n(X)$; i.e., $\mathcal{C}_n^*$ is continuous.

6.6.2. Remark. For $n = 1$, C_n^*–smoothness is called C^*–*smoothness* instead of C_1^*–smoothness. C^*–smoothness was introduced in Chapter XV of [40].

6.6.3. Theorem. *If X is a C^*–smooth homogeneous continuum, then X is indecomposable. Moreover, if X is a C^*–smooth homogeneous plane continuum, then X is hereditarily indecomposable.*

Proof. Let X be a C^*–smooth homogeneous continuum. Then X is hereditarily unicoherent ((3.4) of [13]). By Theorem 1 of [22], X is indecomposable.

If X is a C^*–smooth homogeneous plane continuum, we have that X is indecomposable. Since any indecomposable homogeneous plane continuum is hereditarily indecomposable (Theorem 1 of [15]), X is hereditarily indecomposable.

$$\textbf{Q.E.D.}$$

The following result is easy to prove.

6.6.4. Lemma. *Let $n \in \mathbb{N}$. Let X be a continuum and let $\{A_k\}_{k=1}^{\infty}$ be a sequence in $C_n(X)$ converging to A. If $\lim_{k \to \infty} C_n(A_k)$ exists, then $\lim_{k \to \infty} C_n(A_k) \subset C_n(A)$.*

6.6.5. Theorem. *Let X be a continuum. If A is a subcontinuum of X, then the following are equivalent:*

(1) X is C^*–smooth at A;

(2) $C_n^*|_{C(X)}$ is continuous at A for all $n \in \mathbb{N}$;

(3) $C_n^*|_{C(X)}$ is continuous at A for some $n \in \mathbb{N}$.

Proof. Assume X is C^*–smooth at A, and let $n \geq 2$. We prove $C_n^*|_{C(X)}$ is continuous at A. Let $\{A_k\}_{k=1}^{\infty}$ be a sequence of subcontinua of X converging to A. Let B be any element of $C_n(A)$. Let $B_1, \ldots, B_\ell$ ($\ell \leq n$) be the components of B. Hence, each B_j is a subcontinuum of A, $j \in \{1, \ldots, \ell\}$. Since X is C^*–smooth at A, there exist subcontinua $B_k^1, \ldots, B_k^\ell$ of A_k for each $k \in \mathbb{N}$ such that

$\lim_{k\to\infty} B_k^j = B_j$ for each $j \in \{1,\dots,\ell\}$. Hence, $B_k = \bigcup_{j=1}^{\ell} B_k^j$ is an element of $\mathcal{C}_n(A_k)$ for each $k \in \mathbb{N}$, and $\lim_{k\to\infty} B_k = B$ (Lemma 1.8.11). Therefore, $\mathcal{C}_n(A) \subset \lim_{k\to\infty} \mathcal{C}_n(A_k)$. By Lemma 6.6.4, we may conclude that $\lim_{k\to\infty} \mathcal{C}_n(A_k) = \mathcal{C}_n(A)$. Therefore, $\mathcal{C}_n^*|_{\mathcal{C}(X)}$ is continuous at A.

Assume $\mathcal{C}_n^*|_{\mathcal{C}(X)}$ is continuous at A for some $n \geq 1$ and some element A of $\mathcal{C}(X)$. We prove X is $\mathcal{C}^*$–smooth. Let $\{A_k\}_{k=1}^{\infty}$ be a sequence of subcontinua of X converging to A. Let B be a non-degenerate proper subcontinuum of A. Let $x_1,\dots,x_{n-1}$ be $n-1$ distinct points in $A\backslash B$. Let $D = B \cup \{x_1,\dots,x_{n-1}\}$. Since $\mathcal{C}_n^*|_{\mathcal{C}(X)}$ is continuous at A, there exists $D_k \in \mathcal{C}_n(A_k)$ for each $k \in \mathbb{N}$ such that the sequence $\{D_k\}_{k=1}^{\infty}$ converges to D. Since D has n components, we assume without loss of generality that D_k also has n components for any $k \in \mathbb{N}$. Since n is the maximum number of components we allow, there exists a component D_k^1 of D_k such that $\{D_k^1\}_{k=1}^{\infty}$ converges to B. Therefore, $\mathcal{C}(A) \subset \lim_{k\to\infty} \mathcal{C}(A_k)$. By Lemma 6.6.4, we conclude that $\lim_{k\to\infty} \mathcal{C}(A_k) = \mathcal{C}(A)$.

The fact that (2) implies (3) is obvious.

Q.E.D.

6.6.6. Definition. A continuum X is *absolutely $\mathcal{C}^*$–smooth*, provided that for any continuum Z in which X can be embedded and for each sequence $\{A_k\}_{k=1}^{\infty}$ of elements of $\mathcal{C}(Z)$ converging to X, the sequence $\{\mathcal{C}(A_k)\}_{k=1}^{\infty}$ of hyperspaces converges to $\mathcal{C}(X)$.

With a proof similar to the one given for Theorem 6.6.5, we have the following result:

6.6.7. Theorem. *Let X be a continuum. Then the following statements are equivalent:*

(1) X is absolutely $\mathcal{C}^*$–smooth;

(2) for any continuum Z in which X is embedded, $\mathcal{C}_n^*|_{\mathcal{C}(Z)}$ is continuous at X for all $n \in \mathbb{N}$;

(3) for any continuum Z in which X is embedded, $\mathcal{C}_n^*|_{\mathcal{C}(Z)}$ is continuous at X for some $n \in \mathbb{N}$.

6.6.8. Lemma. *Let $n \in \mathbb{N}$. Let X be a continuum, with metric d, let A be an indecomposable subcontinuum of X and let $\{B_m\}_{m=1}^{\infty}$ be a sequence of elements of $\mathcal{C}_n(X)$ converging to A. Then there exists a subsequence $\{B_{m_k}\}_{k=1}^{\infty}$ of $\{B_m\}_{m=1}^{\infty}$ such that for each k, there exists a component D_k of B_{m_k} such that the sequence $\{D_k\}_{k=1}^{\infty}$ of continua converges to A.*

Proof. Since A is an indecomposable continuum, A has uncountably many mutually disjoint composants (11.15 and 11.17 of [41]). Let $a_1, \ldots, a_{n+1}$ be $n+1$ points in $n+1$ distinct composants of A. We may assume that $\mathcal{V}_{\frac{1}{\ell}}^d(a_i) \cap \mathcal{V}_{\frac{1}{\ell}}^d(a_j) = \emptyset$ if $i \neq j$ for each positive integer ℓ.

Since $\{B_m\}_{m=1}^{\infty}$ converges to A, for each ℓ, there exists an integer m_ℓ such that $\mathcal{H}(A, B_{m_\ell}) < \dfrac{1}{\ell}$. Thus, $B_{m_\ell} \cap \mathcal{V}_{\frac{1}{\ell}}^d(a_j) \neq \emptyset$ for each $j \in \{1, \ldots, n+1\}$. Since B_{m_ℓ} has at most n components, we have that at least one of the components of B_{m_ℓ} intersects two of the balls $\mathcal{V}_{\frac{1}{\ell}}^d(a_j)$, $j \in \{1, \ldots, n+1\}$.

Since we only have $n+1$ balls, there exist $j_0, j_1 \in \{1, \ldots, n+1\}$ such that for infinitely many indices k, B_{m_k} has a component D_k such that $D_k \cap \mathcal{V}_{\frac{1}{k}}^d(a_{j_0}) \neq \emptyset$ and $D_k \cap \mathcal{V}_{\frac{1}{k}}^d(a_{j_1}) \neq \emptyset$ for each k. Since $\mathcal{C}(X)$ is compact (Theorem 1.8.5), we assume without loss of generality that the sequence $\{D_k\}_{k=1}^{\infty}$ converges to a subcontinuum D of A. Since a_{j_0} and a_{j_1} belong to D and they are in different composants of A, we conclude that $D = A$.

$$\text{Q.E.D.}$$

6.6.9. Remark. The converse of Lemma 6.6.8 is false, as can be seen from the argument of Example 3.4 of [34].

6.6.10. Lemma. *Let X be a decomposable continuum, with metric d, and let A and B be nondegenerate proper subcontinua of X such that $X = A \cup B$. Assume that there exist two order arcs $\alpha, \beta \colon [0,1] \to \mathcal{C}(X)$ with the following properties: $\alpha(0) \in \mathcal{F}_1(A)$, $\alpha(1) = A$, $\beta(0) \in \mathcal{F}_1(B)$, $\beta(1) = B$ and $(A \cap B) \cap (\alpha(t) \cup \beta(t)) = \emptyset$ for each $t \in [0,1)$. Then X is not $\mathcal{C}_n^*$–smooth at X for any $n \geq 2$.*

Proof. Suppose X is $\mathcal{C}_n^*$–smooth at X. Let $\{t_m\}_{m=1}^\infty$ be an increasing sequence of numbers in $[0,1)$ converging to 1. For each $m \in \mathbb{N}$, let $D_m = \alpha(t_m) \cup \beta(t_m)$. For each $m \in \mathbb{N}$, $(A \cap B) \cap (\alpha(t_m) \cup \beta(t_m)) = \emptyset$, hence, $D_m \in \mathcal{C}_2(X) \setminus \mathcal{C}(X)$.

Let R be a component of $A \cap B$. Let H and K be proper subcontinua of A and B, respectively, such that they properly contain R (Corollary 1.7.23). Let $x_1, \ldots, x_{n-1}$ be $n-1$ distinct points of $X \setminus (H \cup K)$. Let $L = \{x_1, \ldots, x_{n-1}\} \cup (H \cup K)$. Let $\varepsilon > 0$ be such that the following hold:

$$\mathcal{V}_{2\varepsilon}^d(x_i) \cap \mathcal{V}_{2\varepsilon}^d(x_j) = \emptyset \text{ if and only if } i \neq j,$$

$$\{x_1, \ldots, x_{n-1}\} \cap \mathcal{V}_{2\varepsilon}^d(H \cup K) = \emptyset,$$

$$\bigcup_{j=1}^{n-1} \mathcal{V}_{2\varepsilon}^d(x_j) \cap (H \cup K) = \emptyset,$$

$$H \setminus \mathcal{V}_{2\varepsilon}^d(K) \neq \emptyset, \text{ and } K \setminus \mathcal{V}_{2\varepsilon}^d(H) \neq \emptyset.$$

Since X is $\mathcal{C}_n^*$–smooth at X, there exists $m_0 \in \mathbb{N}$ such that if $m \geq m_0$, then there exists $E_m \in \mathcal{C}_n(D_m)$ such that $\mathcal{H}(E_m, L) < \varepsilon$. Let $m' \geq m_0$. Then $E_{m'} \subset \mathcal{V}_\varepsilon^d(L) = \left(\bigcup_{j=1}^{n-1} \mathcal{V}_\varepsilon^d(x_j) \right) \cup \mathcal{V}_\varepsilon^d(H \cup K)$, $E_{m'} \cap \mathcal{V}_\varepsilon^d(x_j) \neq \emptyset$ for each $j \in \{1, \ldots, n-1\}$, and $E_{m'} \cap \mathcal{V}_\varepsilon^d(H \cup K) \neq \emptyset$. Hence, $E_{m'}$ has exactly n components. Let $G_1, \ldots, G_n$ be the components of $E_{m'}$. Since the ε–balls about each $x_1, \ldots, x_{n-1}$ and $H \cup K$ are pairwise disjoint we assume, without loss of generality, that $G_j \subset \mathcal{V}_\varepsilon^d(x_j)$ for each $j \in \{1, \ldots, n-1\}$ and $G_n \subset \mathcal{V}_\varepsilon^d(H \cup K)$. Since G_n is a subcontinuum of $D_{m'}$, G_n is contained either in $\alpha(t_{m'})$ or in $\beta(s_{m'})$. Suppose that G_n is contained in $\alpha(t_{m'})$. Let $x \in K \setminus \mathcal{V}_{2\varepsilon}^d(H)$. Then for each point z of $E_{m'}$, $d(y, z) \geq \varepsilon$. This is a contradiction; therefore, X is not $\mathcal{C}_n^*$–smooth at X.

Q.E.D.

The following result characterizes the class of continua for which the map $\mathcal{C}_n^*$ is continuous for $n \geq 2$.

6.6.11. Theorem. *A continuum X is $\mathcal{C}_n^*$–smooth for some $n \geq 2$ if and only if X is hereditarily indecomposable.*

Proof. Suppose X is hereditarily indecomposable. Then X is C_n^*–smooth by Lemma 6.6.8 and (1.207.8) of [40].

Suppose that X is C_n^*–smooth for some integer $n \geq 2$. Then condition (3) of Theorem 6.6.7 is satisfied. Hence, X is C^*–smooth by Theorem 6.6.7. Since X is C^*–smooth, X is hereditarily unicoherent ((3.4) of [13]).

Suppose X is decomposable. Then there exist two proper subcontinua A and B of X such that $X = A \cup B$.

Let $a \in A \setminus B$ and $b \in B \setminus A$. Let $\alpha, \beta \colon [0,1] \to \mathcal{C}(X)$ be order arcs such that $\alpha(0) = \{a\}$, $\alpha(1) = A$, $\beta(0) = \{b\}$ and $\beta(1) = B$ (Theorem 1.8.20). Let t_0 and s_0 be points of $[0,1]$ such that $\alpha(t_0) \cap \beta(s_0) \neq \emptyset$ and such that for each $t < t_0$ and each $s < s_0$, $\alpha(t) \cap \beta(s) = \emptyset$. Note that $t_0 > 0$ and $s_0 > 0$. Let $\{t_k\}_{k=1}^\infty$ and $\{s_k\}_{k=1}^\infty$ be increasing sequences in $[0,1]$ converging to t_0 and s_0, respectively.

Let $Y = \alpha(t_0) \cup \beta(s_0)$. Then Y is a subcontinuum of X. Then, by Lemma 6.6.10, X is not C_n^*–smooth at Y, a contradiction. Therefore, X is indecomposable.

A similar argument shows that each subcontinuum of X is indecomposable.

Q.E.D.

We now present some results about the points at which a continuum X is C_n^*–smooth.

6.6.12. Theorem. *Let X be a continuum and let A be an element of $C_n(X)$ for some $n \geq 2$. If X is C_n^*–smooth at A, then X is C^*–smooth at each component of A.*

Proof. Let A be an element of $C_n(X)$ and suppose X is C_n^*–smooth at A. Observe that if A is connected, then X is C^*–smooth at A by Theorem 6.6.5.

Suppose A has at least two components. Let $A_1, \ldots, A_k$ be the components of A. We show that X is C^*–smooth at A_1. Let $\{K_m\}_{m=1}^\infty$ be a sequence of subcontinua of X converging to A_1. Without loss of generality, we assume that $K_m \cap \left(\bigcup_{j=2}^k A_j \right) = \emptyset$ for each $m \in \mathbb{N}$. Let L be a subcontinuum of A_1.

Let $\alpha\colon [0,1] \to \mathcal{C}(X)$ be an order arc such that $\alpha(0) \in \mathcal{F}_1(A_2)$ and $\alpha(1) = A_2$ (Theorem 1.8.20). Let $\{t_m\}_{m=1}^{\infty}$ be an increasing sequence of numbers in $[0,1)$ converging to 1. For each $m \in \mathbb{N}$, let $p_m^{(1)}, \ldots, p_m^{(n-k)}$ be $n-k$ distinct points in $A_2 \setminus \alpha(t_m)$.

For each $m \in \mathbb{N}$, let

$$F_m = K_m \cup \alpha(t_m) \cup \left(\bigcup_{j=3}^{k} A_j\right) \cup \{p_m^{(1)}, \ldots, p_m^{(n-k)}\}.$$

Then $\lim_{m\to\infty} F_m = A$. Since X is $\mathcal{C}_n^*$–smooth at A, for every $m \in \mathbb{N}$, there exists an element D_m of $\mathcal{C}_n(F_m)$ such that $\lim_{m\to\infty} D_m = L \cup$

$$\alpha(t_1) \cup \left(\bigcup_{j=3}^{k} A_j\right) \cup \{p_1^{(1)}, \ldots, p_1^{(n-k)}\}.$$

For each $m \in \mathbb{N}$, let $L_m = D_m \cap K_m$. Then L_m is a subcontinuum of K_m and $\lim_{m\to\infty} L_m = L$. Therefore, X is $\mathcal{C}^*$–smooth at A_1. Similarly, X is $\mathcal{C}^*$–smooth at the other components of A.

Q.E.D.

6.6.13. Lemma. *Let C be a closed subset of a space Z. Let $A = Cl(Z \setminus C)$ and let $B = Cl(Z \setminus A)$. Then $A = Cl(Z \setminus B)$.*

Proof. Since A is closed in Z, $Cl(Int(A)) \subset A$; thus, since

$$A = Cl(Z\backslash C) = Cl(Int(Z\backslash C)) \subset Cl(Int(Cl(Z\backslash C))) = Cl(Int(A)),$$

we have that $A = Cl(Int(A))$. Therefore, since $Int(A) = Z\backslash Cl(Z\backslash A) = Z \setminus B$, $A = Cl(Z \setminus B)$.

Q.E.D.

6.6.14. Theorem. *If X is an irreducible continuum such that X is $\mathcal{C}_n^*$–smooth at X for some $n \geq 2$, then X is indecomposable.*

Proof. Assume that a and b are points about which X is irreducible. Suppose X is decomposable. Let C be a nondegenerate proper subcontinuum of X, with nonempty interior, containing b. Let $A = Cl(X \setminus C)$ and $B = Cl(X \setminus A)$.

Then A and B are subcontinua of X (Theorem 1.7.26) containing a and b, respectively. Note that $A = Cl(X\backslash B)$, by Lemma 6.6.13.

Since $A \cap B = Bd(A) = Bd(B)$ and since $B = Cl(X \setminus A)$ (and $A = Cl(X \setminus B)$), A (and B, respectively) is irreducible between a (b, respectively) and any point of $A \cap B$ (11.42 of [41]).

Let $\alpha, \beta \colon [0,1] \to \mathcal{C}(X)$ be order arcs such that $\alpha(0) = \{a\}$, $\alpha(1) = A$, $\beta(0) = \{b\}$ and $\beta(1) = B$ (Theorem 1.8.20).

Notice that for any $t \in [0,1)$, $(A \cap B) \cap \alpha(t) = \emptyset$ and $(A \cap B) \cap \beta(t) = \emptyset$. By Lemma 6.6.10, X is not $\mathcal{C}_n^*$–smooth at X, a contradiction. Therefore, X is indecomposable.

$$\textbf{Q.E.D.}$$

6.6.15. Theorem. *Let X be a continuum and let A be an element of $\mathcal{C}_n(X)$ with exactly n components, $n \geq 2$. Then X is $\mathcal{C}_n^*$–smooth at A if and only if X is $\mathcal{C}^*$–smooth at each component of A.*

Proof. Let $A \in \mathcal{C}_n(X) \setminus \mathcal{C}_{n-1}(X)$. If X is $\mathcal{C}_n^*$–smooth at A, then X is $\mathcal{C}^*$–smooth at each component of A by Theorem 6.6.12.

Let A be an element of $\mathcal{C}_n(X)$ with n components $A_1, \ldots, A_n$. Suppose X is $\mathcal{C}^*$–smooth at each A_j for each $j \in \{1, \ldots, n\}$.

Let $\{B_k\}_{k=1}^{\infty}$ be a sequence of elements of $\mathcal{C}_n(X)$ converging to A. Since A has n components, without loss of generality, we assume that B_k has n components, $B_k^1, \ldots, B_k^n$, for each $k \in \mathbb{N}$. In fact, we may suppose that $\lim_{k \to \infty} B_k^j = A_j$ for each $j \in \{1, \ldots, n\}$.

Let C be an element of $\mathcal{C}_n(A)$. Let $A_{j_1}, \ldots, A_{j_\ell}$ be the components of A intersecting C, i.e., $C = \bigcup_{i=1}^{\ell}(A_{j_i} \cap C)$. Let $C_{j_i} = A_{j_i} \cap C$ for each $i \in \{1, \ldots, \ell\}$. Since X is $\mathcal{C}^*$–smooth at A_{j_i}, there exists a subcontinuum $D_k^{j_i}$ of $B_k^{j_i}$ for each $i \in \{1, \ldots, \ell\}$ such that $\lim_{k \to \infty} D_k^{j_i} = C_{j_i}$. For $k \in \mathbb{N}$, let $D_k = \bigcup_{i=1}^{\ell} D_k^{j_i}$. Hence, $D_k \in \mathcal{C}_n(B_k)$ and $\lim_{k \to \infty} D_k = C$.

Therefore, X is $\mathcal{C}_n^*$–smooth at A by Lemma 6.6.4.

$$\textbf{Q.E.D.}$$

6.6.16. Theorem. *Let X be a continuum. If A is an element of $\mathcal{C}_n(X)$ for some $n \geq 2$ such that all the components of A are indecomposable and X is $\mathcal{C}^*$–smooth at each component of A, then X is $\mathcal{C}_n^*$–smooth at A.*

Proof. Let A be an element of $\mathcal{C}_n(X)$. Let $A_1, \ldots, A_\ell$ ($\ell \leq n$) be the components of A. Suppose A_j is an indecomposable continuum and X is $\mathcal{C}^*$–smooth at A_j for each $j \in \{1, \ldots, \ell\}$.

Let $\{B_k\}_{k=1}^\infty$ be a sequence of elements of $\mathcal{C}_n(X)$ converging to A. Let C be an element of $\mathcal{C}_n(A)$. Let $A_{j_1}, \ldots, A_{j_s}$ be the components of A intersecting C, i.e., $C = \bigcup_{i=1}^s (A_{j_i} \cap C)$. Let $C_{j_i} = A_{j_i} \cap C$ for each $i \in \{1, \ldots, s\}$.

In what follows, $k \in \mathbb{N}$ and $i \in \{1, \ldots, s\}$. Since each A_{j_i} is indecomposable, by a similar argument to the one given in Lemma 6.6.8, there are components $B_k^{j_i}$ of B_k such that $\lim_{k \to \infty} B_k^{j_i} = A_{j_i}$. Since X is $\mathcal{C}^*$–smooth at each A_{j_i}, there are subcontinua $D_k^{j_i}$ of $B_k^{j_i}$ such that $\lim_{k \to \infty} D_k^{j_i} = C_{j_i}$. Let $D_k = \bigcup_{i=1}^s D_k^{j_i}$. Then $D_k \in \mathcal{C}_n(B_k)$ and $\lim_{k \to \infty} D_k = C$. Therefore, X is $\mathcal{C}_n^*$–smooth at A by Lemma 6.6.4.

Q.E.D.

6.6.17. Corollary. *Let X be a continuum and let $n \geq 2$. If A is an element of $\mathcal{C}_n(X)$ such that all the components of A are hereditarily indecomposable, then X is $\mathcal{C}_n^*$–smooth at A.*

Proof. The corollary follows from Theorem 6.6.16 and the fact that hereditarily indecomposable continua are absolutely $\mathcal{C}^*$–smooth continua (by (14.14.1) of [40] and 3.2 of [14]).

Q.E.D.

6.7 Retractions

We present results about retractions between the hyperspaces of locally connected continua.

6.7.1. Lemma. *Let $n \in \mathbb{N}$. If X is a continuum containing an open set with uncountably many components, then $\mathcal{C}_n(X)$ contains an open set with uncountably many components.*

Proof. Let U be an open subset of X with uncountably many components. Let $\Gamma = \langle U \rangle_n$. Then Γ is an open subset of $\mathcal{C}_n(X)$, and $\bigcup \Gamma \subset U$. Since for each $x \in U$, $\{x\} \in \Gamma$, we have that $\bigcup \Gamma = U$. By Lemma 6.1.1, if Λ is a component of Γ, then $\bigcup \Lambda$ has at most n components. Since U has uncountably many components, $\bigcup \Gamma = U$ and for each component Λ of Γ, $\bigcup \Lambda$ has at most n components, we have that Γ has uncountably many components.

$$\mathbf{Q.E.D.}$$

We begin considering maps between some of the hyperspaces.

6.7.2. Theorem. *Let $n \in \mathbb{N}$. Then for any continuum X, there exists a map of 2^X onto $\mathcal{C}_n(X)$.*

Proof. If X is locally connected, then 2^X and $\mathcal{C}_n(X)$ are locally connected continua (Théorème II and Théorème II_m of [47]). Hence, the result follows from 8.19 of [41].

Suppose that X is not locally connected. Then there exists a map f of 2^X onto the cone over the Cantor set ((1.39) of [40]). Also, there exists a map g of the cone over the Cantor set onto $\mathcal{C}_n(X)$ (Remark (p. 29) and Theorem 2.7 of [23]). Thus $g \circ f$ is a map of 2^X onto $\mathcal{C}_n(X)$.

$$\mathbf{Q.E.D.}$$

6.7.3. Theorem. *Let $n \in \mathbb{N}$, $n \geq 2$. If X is a continuum containing an open set with uncountably many components, then there exists a map of $\mathcal{C}_n(X)$ onto $\mathcal{C}(X)$.*

Proof. Since X contains an open subset with uncountably many components, $\mathcal{C}_n(X)$ contains an open subset with uncountably many components (Lemma 6.7.1). Thus, there exists a map f of $\mathcal{C}_n(X)$ onto the cone over the Cantor set (Theorem II of [2]). Also, there exists a map g of the cone over the Cantor set onto $\mathcal{C}(X)$ (Theorem 2.7 of [23]). Hence, $g \circ f$ is a map of $\mathcal{C}_n(X)$ onto $\mathcal{C}(X)$.

$$\mathbf{Q.E.D.}$$

6.7.4. Definition. Let Z be a metric space. By a *deformation* we mean a map $H\colon Z \times [0,1] \to Z$ such that for each $z \in Z$, $H((z,1)) = z$. Let $A = \{H((z,0)) \mid z \in Z\}$. If the map $h\colon Z \to A$ given by $h(z) = H((z,0))$ is a retraction from Z onto A, then H is a *deformation retraction from Z onto A*. If H is a deformation retraction from Z onto A such that for each $z \in A$ and each $t \in [0,1]$, $H((z,t)) = z$, then H is a *strong deformation retraction from Z onto A*. The set A is called *deformation retract* (*strong deformation retract*, respectively).

The next Lemma provides a sufficient condition for a continuum X for the nonexistence of a deformation of the m–fold hyperspace of X onto a subset of the n–fold hyperspace of X, $m > n$.

6.7.5. Lemma. *Let $n, m \in \mathbb{N}$ and let X be a nonlocally connected continuum, with metric d. Let p be a point of X at which X is not connected im kleinen, and let $\mathcal{A} \subset \mathcal{C}_n(X)$ such that $\{p\} \in \mathcal{A}$. If $m > n$, then there does not exist a deformation, H, of $\mathcal{C}_m(X)$ onto $\mathcal{A}$ such that $H((\{p\}, t)) = \{p\}$ for each $t \in [0,1]$.*

Proof. Since X is not connected im kleinen at p, there exists a neighborhood U of p and a sequence $\{K_j\}_{j=1}^{\infty}$ of components of $Cl_X(U)$ converging to a continuum $K \subset Cl_X(U)$ such that $p \in K$ and such that for each $j \in \mathbb{N}$, $K_j \cap K = \emptyset$ ((12.1) (p. 18) of [45]). Let $\{p_j\}_{j=1}^{\infty}$ be a sequence in $Cl_X(U)$ converging to p and such that $p_j \in K_j$ for each $j \in \mathbb{N}$.

Suppose there exists a deformation $H\colon \mathcal{C}_m(X) \times [0,1] \twoheadrightarrow \mathcal{C}_m(X)$ from $\mathcal{C}_m(X)$ onto $\mathcal{A}$ such that for each $t \in [0,1]$, $H((\{p\}, t)) = \{p\}$. Let H' be the segment homotopy associated with H ((16.3) of [40]), given by $H'((A,t)) = \sigma(\{H((A,s)) \mid 0 \le s \le t\})$. Observe that for each $t \in [0,1]$, $H'((\{p\}, t)) = \{p\}$ and for each $A \in \mathcal{C}_m(X)$, $H'((A,0)) = H((A,0)) \in \mathcal{C}_n(X)$. Let $\varepsilon > 0$ such that $\mathcal{V}_\varepsilon^d(p) \subset U$. Then there exists $\delta > 0$ such that if $A \in \mathcal{C}_m(X)$ and $\mathcal{H}(A, \{p\}) < \delta$, then for each $t \in [0,1]$, $\mathcal{H}(H'((A,t)), H'((\{p\}, t))) = \mathcal{H}(H'((A,t)), \{p\}) < \varepsilon$. For each $j \in \mathbb{N}$, let $A_j = \{p\} \cup \{p_{j+\ell}\}_{\ell=1}^{m-1}$. Hence, there exists $j_0 \in \mathbb{N}$ such that if $j \ge j_0$, then $\mathcal{H}(A_j, \{p\}) < \delta$. Choose $j \ge j_0$. Then $\{H'(A_j, t) \mid t \in [0,1]\}$ is a connected subset of $\mathcal{V}_\varepsilon^{\mathcal{H}}(\{p\})$ such that $H'((A_j, 0)) \in \mathcal{C}_n(X)$. Thus $\sigma(\{H'((A_j, t)) \mid t \in$

$[0, 1]\})$ is a closed subset of X having at most n components (by a similar argument to the one given in Lemma 6.1.1). Since $A_j \subset \sigma(\{H'((A_j, t)) \mid t \in [0, 1]\}) \subset \mathcal{V}_\varepsilon^d(p) \subset U$ and since $A_j \cap K \neq \emptyset$ and $A_j \cap K_{j+\ell} \neq \emptyset$ for each $\ell \in \{1, \ldots, m-1\}$, we obtain a contradiction. We conclude that there does not exist a deformation H of $\mathcal{C}_m(X)$ onto $\mathcal{A}$ such that for each $t \in [0, 1]$, $H((\{p\}, t)) = \{p\}$.

Q.E.D.

6.7.6. Corollary. *Let $m, n \in \mathbb{N}$. If X is a nonlocally connected continuum and $m > n$, then there does not exist a strong deformation retraction of $\mathcal{C}_m(X)$ onto $\mathcal{F}_n(X)$.*

6.7.7. Theorem. *Let $n \in \mathbb{N}$, $n \geq 2$, and let X be a continuum. If $\mathcal{F}_1(X)$ is a deformation retract of 2^X, then $\mathcal{F}_1(X)$ is a deformation retract of $\mathcal{C}_n(X)$.*

Proof. Let $H \colon 2^X \times [0, 1] \twoheadrightarrow 2^X$ be a deformation retraction of 2^X onto $\mathcal{F}_1(X)$. Let $r(A) = H((A, 0))$, for each $A \in 2^X$.

Define $G \colon \mathcal{C}_n(X) \times [0, 1] \twoheadrightarrow \mathcal{C}_n(X)$ by

$$G((A, t)) = \begin{cases} \sigma(\{H((A, s)) \mid 0 \leq s \leq 2t\}) & \text{if } t \in \left[0, \dfrac{1}{2}\right]; \\[2ex] \sigma(\{H((A, s)) \mid 2t - 1 \leq s \leq 1\}) & \text{if } t \in \left[\dfrac{1}{2}, 1\right]. \end{cases}$$

Note that if $t = \dfrac{1}{2}$, then both definitions of G give the value $\sigma(\{H((A, s)) \mid 0 \leq s \leq 1\})$. Let $t \in \left[0, \dfrac{1}{2}\right]$. Since H is continuous, $\{H((A, s)) \mid 0 \leq s \leq 2t\}$ is a closed connected subset of 2^X that contains the element $H((A, 0)) = A$ and $A \in \mathcal{C}_n(X)$. By similar argument to the one given in Lemma 6.1.1, $G((A, t)) \in \mathcal{C}_n(X)$ for each $A \in \mathcal{C}_n(X)$. Similarly, since $H((A, 1)) \in \mathcal{F}_1(X)$, $G((A, t)) \in \mathcal{C}_n(X)$ for each $t \in \left[\dfrac{1}{2}, 1\right]$ and each $A \in \mathcal{C}_n(X)$. Hence, G is well defined. Since H and σ (Lemma 1.8.11) are continuous, it follows that G is continuous also.

Note that $G((A,0)) = r(A)$ and $G((A,1)) = A$. Therefore, $\mathcal{F}_1(X)$ is a deformation retract of $\mathcal{C}_n(X)$.

Q.E.D.

The proof of the following Theorem is similar to the one given in Theorem 6.7.7.

6.7.8. Theorem. *Let $n \in \mathbb{N}$, $n \geq 2$, and let X be a continuum. If $\mathcal{F}_1(X)$ is a deformation retract of $\mathcal{C}_n(X)$, then $\mathcal{F}_1(X)$ is a deformation retract of $\mathcal{C}(X)$.*

6.7.9. Definition. A metric ρ for a continuum X is said to be *convex* provided that given two points x and y of X there exists a point z in X such that $\rho(x, z) = \dfrac{\rho(x, y)}{2} = \rho(z, y)$.

The following result was proved by R. H. Bing [4] and E. E. Moise [38], independently.

6.7.10. Theorem. *Every locally connected continuum admits a convex metric.*

6.7.11. Definition. If X is a locally connected continuum with a convex metric ρ, let $K_\rho \colon [0, \infty) \times 2^X \to 2^X$ be given by

$$K_\rho((t, A)) = \{x \in X \mid \rho(x, y) \leq t \text{ for some } y \in A\}.$$

6.7.12. Remark. If X is a locally connected continuum with a convex metric ρ, then given two points x and y of X, there exists an arc $\gamma \colon [0, \rho(x, y)] \to X$ such that γ is an isometry ((0.65.3)(a) of [40]). This implies that the set $K_\rho((t, A))$ has at most as many components as A for each $t \in [0, \infty)$. Also, observe that if $t \geq \operatorname{diam}(X)$, then $K_\rho((t, A)) = X$ for each $A \in 2^X$.

6.7.13. Theorem. *Let $n \in \mathbb{N}$ and let X be a locally connected continuum with a convex metric ρ. Then the function $\alpha_\rho^n \colon 2^X \to \mathbb{R}$ given by*

$$\alpha_\rho^n(A) = \inf\{t \geq 0 \mid K_\rho((t, A)) \in \mathcal{C}_n(X)\}$$

satisfies the following:

(a) $K_\rho((\alpha_\rho^n(A), A))$ belongs to $\mathcal{C}_n(X)$ for each $A \in 2^X$.

(b) If A and B belong to 2^X, $t \geq 0$, $K_\rho((t, A)) \in \mathcal{C}_n(X)$ and $\mathcal{H}_\rho(A, B) \leq \eta$, then $K_\rho((t + \eta, B)) \in \mathcal{C}_n(X)$.

(c) α_ρ^n is continuous.

Proof. Since X is locally connected with a convex metric, K_ρ is continuous ((0.65.3)(f) of [40]). To see $K_\rho((\alpha_\rho^n(A), A))$ belongs to $\mathcal{C}_n(X)$ for each $A \in 2^X$, observe that, by definition of α_ρ^n, there exists a decreasing sequence, $\{t_j\}_{j=1}^\infty$, of real numbers converging to $\alpha_\rho^n(A)$ such that $K_\rho((t_j, A))$ belongs to $\mathcal{C}_n(X)$ for each $j \in \mathbb{N}$. By the continuity of K_ρ, we have that $\lim_{j \to \infty} K_\rho((t_j, A)) = K_\rho((\alpha_\rho^n(A), A))$. Since $\mathcal{C}_n(X)$ is compact (Theorem 1.8.5), we have that $K_\rho((\alpha_\rho^n(A), A)) \in \mathcal{C}_n(X)$.

Now, suppose A and B belong to 2^X, $t \geq 0$, $K_\rho((t, A)) \in \mathcal{C}_n(X)$ and $\mathcal{H}_\rho(A, B) \leq \eta$. Observe first that $A \subset K_\rho((t, A)) \subset K_\rho((t + \eta, B))$. Let $x \in K_\rho((t + \eta, B))$, then there exists $b \in B$ such that $\rho(x, b) \leq t + \eta$. Since $b \in B$ and $\mathcal{H}_\rho(A, B) \leq \eta$, there exists $a \in A$ such that $\rho(a, b) \leq \eta$. Since ρ is a convex metric, there exist arcs $\gamma_1 \colon [0, \rho(x, b)] \to X$ and $\gamma_2 \colon [0, \rho(b, a)] \to X$ such that $\gamma_1(0) = x$, $\gamma_1(\rho(x, b)) = b$, $\gamma_2(0) = b$, $\gamma_2(\rho(b, a)) = a$, and both γ_1 and γ_2 are isometries ((0.65.3)(a) of [40]). Let $G_x = \gamma_1([0, \rho(x, b)]) \cup \gamma_2([0, \rho(b, a)])$. Then G_x is a connected subset of $K_\rho((t + \eta, B))$ containing x and intersecting $K_\rho((t, A))$. Hence, $K_\rho((t + \eta, B))$ has at most as many components as $K_\rho((t, A))$. Therefore, $K_\rho((t + \eta, B)) \in \mathcal{C}_n(X)$.

To show α_ρ^n is continuous, let $\eta > 0$. Let A and B be two elements of 2^X such that $\mathcal{H}_\rho(A, B) \leq \eta$. By (a), $K_\rho((\alpha_\rho^n(A), A))$ belongs to $\mathcal{C}_n(X)$. Hence, by (b), $K_\rho((\alpha_\rho^n(A) + \eta, B))$ also belongs to $\mathcal{C}_n(X)$. By definition of $\alpha_\rho^n(B)$, we have that $\alpha_\rho^n(B) \leq \alpha_\rho^n(A) + \eta$. Interchanging the roles of A and B, in the above argument, we obtain $\alpha_\rho^n(A) \leq \alpha_\rho^n(B) + \eta$. Hence, $|\alpha_\rho^n(A) - \alpha_\rho^n(B)| \leq \eta$. Therefore, α_ρ^n is continuous.

Q.E.D.

The next five results give conditions for the existence of retractions between the hyperspaces of locally connected continua.

6.7.14. Theorem. *Let $n, m \in \mathbb{N}$, $m > n$, and let X be a continuum. Then $\mathcal{C}_n(X)$ is a strong deformation retract of $\mathcal{C}_m(X)$ if and only if X is locally connected.*

Proof. If X is not locally connected, then there is not a strong deformation retraction from $\mathcal{C}_m(X)$ onto $\mathcal{C}_n(X)$ (Lemma 6.7.5).

If X is a locally connected continuum, then the map $H \colon \mathcal{C}_m(X) \times [0,1] \twoheadrightarrow \mathcal{C}_m(X)$ given by

$$H((A,t)) = K_\rho \left(((1-t)\alpha_\rho^n(A), A) \right)$$

is a strong deformation retraction from $\mathcal{C}_m(X)$ onto $\mathcal{C}_n(X)$ (Remark 6.7.12 and Theorem 6.7.13).

Q.E.D.

The proof of the following Theorem is similar to the one given for Theorem 6.7.14.

6.7.15. Theorem. *Let $n \in \mathbb{N}$ and let X be a continuum. Then $\mathcal{C}_n(X)$ is a strong deformation retract of 2^X if and only if X is locally connected.*

6.7.16. Lemma. *Let $n \in \mathbb{N}$. If X is a locally connected continuum, then $\mathcal{F}_1(X)$ is a retract of $\mathcal{C}_n(X)$ if and only if X is an absolute retract.*

Proof. Since X is a locally connected continuum, $\mathcal{C}_n(X)$ is an absolute retract (Théorème II$_m$ of [47]). If $\mathcal{F}_1(X)$ is a retract of $\mathcal{C}_n(X)$, then $\mathcal{F}_1(X)$ is an absolute retract (see (2.2) (p. 86) of [5]). Since X is homeomorphic to $\mathcal{F}_1(X)$, X is an absolute retract.

If X is an absolute retract, then $\mathcal{F}_1(X)$ is a retract of $\mathcal{C}_n(X)$ (1.5.2 of [37]).

Q.E.D.

6.7.17. Theorem. *Let $n \in \mathbb{N}$. If X is a locally connected continuum, then the following are equivalent:*

(1) X is an absolute retract.

(2) $\mathcal{F}_1(X)$ is a retract of $\mathcal{C}_n(X)$.

(3) $\mathcal{F}_1(X)$ is a deformation retract of $\mathcal{C}_n(X)$.

(4) $\mathcal{F}_1(X)$ is a strong deformation retract of $\mathcal{C}_n(X)$.

Proof. Clearly (4) implies (3), and (3) implies (2). Note that (2) implies (1) by Lemma 6.7.16. Suppose now, X is an absolute retract. Then $\mathcal{C}_n(X)$ is an absolute retract (Théorème II$_m$ of [47]). Hence, there exists a retraction $r_n \colon 2^X \to \mathcal{C}_n(X)$. By 2.5 of [12], there exists a strong deformation retraction $G \colon 2^X \times [0, 1] \to 2^X$, from 2^X onto $\mathcal{F}_1(X)$. Thus, $r_n \circ G|_{\mathcal{C}_n(X) \times [0,1]}$ is a strong deformation retraction from $\mathcal{C}_n(X)$ onto $\mathcal{F}_1(X)$. Therefore, (1) implies (4).

Q.E.D.

6.7.18. Theorem. *Let $n \in \mathbb{N}$. If X is a locally connected continuum, then the following hold:*

(1) $\mathcal{C}_n(X)$ is a retract of 2^X.

(2) $\mathcal{C}_n(X)$ is a deformation retract of 2^X.

(3) $\mathcal{C}_n(X)$ is a strong deformation retract of 2^X.

(4) $\mathcal{C}(X)$ is a retract of $\mathcal{C}_n(X)$.

(5) $\mathcal{C}(X)$ is a deformation retract of $\mathcal{C}_n(X)$.

(6) $\mathcal{C}(X)$ is a strong deformation retract of $\mathcal{C}_n(X)$.

Proof. By Théorème II$_m$ of [47], $\mathcal{C}_n(X)$ is an absolute retract. Hence, (1) and (4) hold. Since X is locally connected, X has the property of Kelley. Thus, $\mathcal{C}_n(X)$ is contractible (Corollary 6.1.16) and 2^X is contractible too (Theorem 6.1.14). This implies, by 32E.4 of [46], the equivalence of (1) and (2) and of (4) and (5). By Theorems 6.7.15 and 6.7.14, (3) and (6) are equivalent to the fact that X is locally connected.

Q.E.D.

6.8   Graphs

We present results about the n–fold hyperspaces of graphs.

6.8.1. Definition. A *graph* is a continuum which can be written as the union of finitely many arcs, any two of which are either disjoint or intersect only in one or both of their end points.

6.8.2. Definition. Let X be a graph and let A be a subset of X. Let β be a cardinal number. We say that A *is of order less than or equal to* β *in* X, written $\text{ord}(A, X) \leq \beta$, provided that for each open subset U of X containing A, there exists an open subset V of X such that $A \subset V \subset U$ and $Bd_X(V)$ has cardinality less than or equal to β. We say that A *is of order* β *in* X, written $\text{ord}(A, X) = \beta$, provided that $\text{ord}(A, X) \leq \beta$ and $\text{ord}(A, X) \not\leq \alpha$ for any cardinal number $\alpha < \beta$. A point x of a graph X is *a ramification point of* X if and only if $\text{ord}(\{x\}, X) \geq 3$. A point a of a graph X is a *end point of* X if and only if $\text{ord}(\{a\}, X) = 1$.

The next Theorem characterizes graphs as the class of locally connected continua X having finite dimensional hyperspaces $\mathcal{C}_n(X)$.

6.8.3. Theorem. *A locally connected continuum X is a graph if and only if for each $n \in \mathbb{N}$, $\mathcal{C}_n(X)$ is of finite dimension.*

Proof. Suppose X is a graph. Then $\dim(\mathcal{C}(X)) < \infty$ (Lemma 5.2 of [23]). Hence, given $n \in \mathbb{N}$, by Corollary 6.1.22, we have that $\dim(\mathcal{C}_n(X)) \leq n(\dim(\mathcal{C}(X)) < \infty$.

Suppose X is a locally connected continuum such that for each $n \in \mathbb{N}$, $\dim(\mathcal{C}_n(X)) < \infty$. Then we have that $\dim(\mathcal{C}(X)) < \infty$. Thus, X is a graph (Lemma 5.2 of [23]).

Q.E.D.

6.8.4. Theorem. *Let $n \in \mathbb{N}$, $n \geq 2$. Then $\mathcal{C}_n([0,1]) \setminus \mathcal{C}_{n-1}([0,1])$ is embeddable in $\mathbb{R}^{2n}$.*

Proof. Given $A \in \mathcal{C}_n([0,1]) \setminus \mathcal{C}_{n-1}([0,1])$, without loss of generality, we assume that $[a_1, a_1'], \ldots, [a_n, a_n']$ are the components of A and $a_1 \leq a_1' < a_2 \leq a_2' < \ldots < a_n \leq a_n'$. Define $\xi \colon \mathcal{C}_n([0,1]) \setminus \mathcal{C}_{n-1}([0,1]) \to \mathbb{R}^{2n}$ by $\xi(A) = (a_1, a_1', \ldots, a_n, a_n')$. Clearly, ξ is a one–to–one function. To see its continuity, let $\varepsilon > 0$ and let $B = \bigcup_{j=1}^{n} [b_j, b_j']$ be a point of $\mathcal{C}_n([0,1]) \setminus \mathcal{C}_{n-1}([0,1])$ such that $\mathcal{H}(A, B) < \varepsilon$. Then for each $j \in \{1, \ldots, n\}$, $|a_j - b_j| < \varepsilon$ and $|a_j' - b_j'| < \varepsilon$. Hence, $D(\xi(A), \xi(B)) = \max\{|a_j - b_j|, |a_j' - b_j'| \mid j \in \{1, \ldots, n\}\} < \varepsilon$. Thus, ξ is continuous. Note that $\xi\left(\mathcal{V}_\varepsilon^{\mathcal{H}}(A) \cap \mathcal{C}_n([0,1]) \setminus \mathcal{C}_{n-1}([0,1])\right) = \mathcal{V}_\varepsilon^{\mathbb{R}^{2n}}(\xi(A)) \cap \xi\left(\mathcal{C}_n([0,1]) \setminus \mathcal{C}_{n-1}([0,1])\right)$. Therefore, ξ is an embedding.

$$\textbf{Q.E.D.}$$

6.8.5. Theorem. *Let $n \in \mathbb{N}$, and let $\mathcal{E}_n = \{(x_1, \ldots, x_{2n}) \in \mathbb{R}^{2n} \mid 0 \leq x_1 \leq \ldots \leq x_{2n} \leq 1\}$. Then the function $f \colon \mathcal{E}_n \to \mathcal{C}_n([0,1])$ given by*

$$f((x_1, \ldots, x_{2n})) = \bigcup_{j=1}^{n} [x_{2j-1}, x_{2j}]$$

is continuous and surjective. Furthermore, $\mathcal{E}_n / \mathcal{G}_f$ is homeomorphic to $\mathcal{C}_n([0,1])$.

Proof. Let $\varepsilon > 0$ be given. If $(x_1, \ldots, x_{2n})$ and $(x_1', \ldots, x_{2n}')$ belong to $\mathcal{E}_n$ and $D((x_1, \ldots, x_{2n}), (x_1', \ldots, x_{2n}')) = \max\{|x_j - x_j'| \mid j \in \{1, \ldots, 2n\}\} < \varepsilon$, then we have that

$$\mathcal{H}(f((x_1, \ldots, x_{2n})), f((x_1', \ldots, x_{2n}'))) \leq$$

$$\max\{|x_j - x_j'| \mid j \in \{1, \ldots, 2n\}\} < \varepsilon.$$

Thus, f is continuous.

If $A = \bigcup_{j=1}^{k} [x_j, y_j]$, where $k \leq n$, then $(x_1, y_1, \ldots, x_k, y_k, y_k, \ldots, y_k)$ belongs to $\mathcal{E}_n$ and $f((x_1, y_1, \ldots, x_k, y_k, y_k, \ldots, y_k)) = A$. Then f is surjective. The fact that $\mathcal{E}_n/\mathcal{G}_f$ is homeomorphic to $\mathcal{C}_n([0,1])$ follows from Theorem 1.2.10.

Q.E.D.

6.8.6. Notation. Let $\mathcal{S}^1$ denote the unit circle in the plane. If $\theta \in \mathcal{S}^1$ and $t \in [0,1)$, then let $A(\theta, t)$ be the arc (possibly degenerate) contained in $\mathcal{S}^1$ having θ as mid point and length $2\pi t$. For $t = 1$, $A(\theta, 1)$ denotes $\mathcal{S}^1$.

The following Lemma is easily established.

6.8.7. Lemma. *Let $\varepsilon > 0$ be given. If θ and φ belong to $\mathcal{S}^1$, t and s belong to $[0,1]$ and satisfy that $||\theta - \varphi|| < \varepsilon$ and $|t - s| < \dfrac{\varepsilon}{2\pi}$, then $\mathcal{H}(A(\theta, t), A(\varphi, s)) < \varepsilon$.*

6.8.8. Notation. Let $n \in \mathbb{N}$. If $n > 1$, let $\mathbb{T}^n = \underbrace{\mathcal{S}^1 \times \cdots \times \mathcal{S}^1}_{n \text{ times}}$. If $n = 1$, then $\mathbb{T}^1 = \mathcal{S}^1$.

6.8.9. Theorem. *Let $n \in \mathbb{N}$. If $f \colon \mathbb{T}^n \times [0,1]^n \to \mathcal{C}_n(\mathcal{S}^1)$ is a function given by*

$$f((\theta_1, \ldots, \theta_n, t_1, \ldots, t_n)) = \bigcup_{j=1}^{n} A(\theta_j, t_j),$$

then f is continuous and surjective. Furthermore, $(\mathbb{T}^n \times [0,1]^n)/\mathcal{G}_f$ is homeomorphic to $\mathcal{C}_n(\mathcal{S}^1)$.

Proof. Let $\varepsilon > 0$ be given. If $(\theta_1, \ldots, \theta_n, t_1, \ldots, t_n), (\varphi_1, \ldots, \varphi_n, s_1, \ldots, s_n) \in \mathbb{T}^n \times [0,1]^n$ and $D((\theta_1, \ldots, \theta_n, t_1, \ldots, t_n), (\varphi_1, \ldots, \varphi_n, s_1, \ldots, s_n)) < \dfrac{\varepsilon}{2\pi}$, then $|\theta_j - \varphi_j| < \dfrac{\varepsilon}{2\pi} < \varepsilon$ and $|t_j - s_j| < \dfrac{\varepsilon}{2\pi}$ for each $j \in \{1, \ldots, n\}$. By Lemma 6.8.7, it follows that $\bigcup_{j=1}^{n} A(\theta_j, t_j) \subset$

$$\mathcal{V}_\varepsilon^d \left(\bigcup_{j=1}^n A(\varphi_j, s_j) \right) \text{ and } \bigcup_{j=1}^n A(\varphi_j, s_j) \subset \mathcal{V}_\varepsilon^d \left(\bigcup_{j=1}^n A(\theta_j, t_j) \right). \text{ Thus,}$$

$$\mathcal{H} \left(\bigcup_{j=1}^n A(\theta_j, t_j), \bigcup_{j=1}^n A(\varphi_j, s_j) \right) < \varepsilon. \text{ Therefore, } f \text{ is continuous.}$$

Let $A \in \mathcal{C}_n(\mathcal{S}^1)$, and suppose $A_1, \ldots, A_k$ be the components of A, with $k \leq n$. For each $j \in \{1, \ldots, k\}$, let θ_j be the midpoint of A_j and let t_j be the length of A_j. Then $\left(\theta_1, \ldots, \theta_k, \underbrace{\theta_k, \ldots, \theta_k}_{n-k \text{ times}}, \dfrac{t_1}{2\pi}, \ldots, \right.$

$\left. \dfrac{t_k}{2\pi}, \underbrace{\dfrac{t_k}{2\pi}, \ldots, \dfrac{t_k}{2\pi}}_{n-k \text{ times}} \right)$ is an element of $\mathbb{T}^n \times [0,1]^n$ whose image under f is A. Thus, f is surjective.

The fact that $(\mathbb{T}^n \times [0,1]^n)/\mathcal{G}_f$ is homeomorphic to $\mathcal{C}_n(\mathcal{S}^1)$ follows from Theorem 1.2.10.

Q.E.D.

As a consequence of Corollary 6.1.23 we have the following Theorem:

6.8.10. Theorem. $\dim(\mathcal{C}_n([0,1])) = \dim(\mathcal{C}_n(\mathcal{S}^1)) = 2n$ *for every* $n \in \mathbb{N}$.

The proof of the following Theorem is due to R. Schori:

6.8.11. Theorem. $\mathcal{C}_2([0,1])$ *is homeomorphic to* $[0,1]^4$.

Proof. Let $D^1 = \{A \in \mathcal{C}_2([0,1]) \mid 1 \in A\}$ and $D_0^1 = \{A \in \mathcal{C}_2([0,1]) \mid \{0,1\} \subset A\}$. We divide the proof in three steps.

Step 1. $\mathcal{C}_2([0,1])$ *is homeomorphic to* $K(D^1)$ (*the cone over* D^1).

Let $f \colon K(D^1) \twoheadrightarrow \mathcal{C}_2([0,1])$ be given by $f((A,t)) = (1-t)A = \{(1-t)a \mid a \in A\}$. Since $f((A,1)) = \{0\}$ for every $A \in \mathcal{C}_2([0,1])$, f is a well defined map. Let $A, B \in D^1$ and let $s, t \in [0,1]$ such that $f((A,t)) = f((B,s))$. Since $1 \in A \cap B$, $1 - t = \max(1 - t)A$ and $1 - s = \max(1 - s)B$. This implies that $t = s$. If $t = s = 1$, then $(A,1)$ and $(B,1)$ both represent the vertex of $K(D^1)$.

Hence, we assume that $t < 1$. Since $(1-t)A = (1-t)B$, $A = B$. Therefore, f is one–to–one. Now, let $A \in \mathcal{C}_2([0,1]) \setminus \{\{0\}\}$. Take $t = 1 - \max A$. Then $\max \left(\dfrac{1}{1-t} \right) A = 1$. Thus, $\left(\dfrac{1}{1-t} \right) A \in D^1$ and $f\left(\left(\left(\dfrac{1}{1-t} \right) A, t \right) \right) = A$. Hence, f is surjective. Therefore, $\mathcal{C}_2([0,1])$ is homeomorphic to $K(D^1)$.

Step 2. D^1 *is homeomorphic to* $K(D_0^1)$.

Let $g\colon K(D_0^1) \twoheadrightarrow D^1$ be given by $g((A,t)) = t + (1-t)A = \{t + (1-t)a \mid a \in A\}$. Proceeding as in Step 1, it can be seen that g is a homeomorphism.

Step 3. D_0^1 *is homeomorphic to* $[0,1]^2$.

Let $T = \{(a,b) \in \mathbb{R}^2 \mid 0 \le a \le b \le 1\}$, and let $S = T/\Delta$, where $\Delta = \{(a,b) \in T \mid a = b\}$. Note that S is homeomorphic to $[0,1]^2$. Let $h\colon T \twoheadrightarrow D_0^1$ be given by $h((a,b)) = [0,a] \cup [b,1]$. Then h is continuous. Note that $h((a,b)) = h((c,d))$ if and only if $a = c$ and $b = d$. Hence, by Theorem 1.2.10, S and D_0^1 are homeomorphic.

Q.E.D.

Recall that for any continuum X and any $n \in \mathbb{N}$, $\mathcal{C}_n(X)$ is unicoherent (Theorem 6.2.4). The next Theorem, whose proof is contained in Lemma 2.3 of [17], says that unicoherence of these hyperspaces may be destroyed by removing a point.

6.8.12. Theorem. $\mathcal{C}_2(\mathcal{S}^1) \setminus \{\mathcal{S}^1\}$ *is not unicoherent.*

The proof of the following Theorem is the content of [20].

6.8.13. Theorem. $\mathcal{C}_2(\mathcal{S}^1)$ *is homeomorphic to the cone over the solid torus.*

The proof of the following Theorem is the content of [17] and [19].

6.8.14. Theorem. *Let X be a graph and let Y be a continuum such that $\mathcal{C}_n(Y)$ is homeomorphic to $\mathcal{C}_n(X)$. Then:*

(1) if $n = 1$ and X is neither an arc nor a simple closed curve, then Y is homeomorphic to X,

(2) if $n \geq 2$, then Y is homeomorphic to X.

6.9 Cones

Recall that given a space Z, $K(Z)$ denotes the cone over Z.

Theorem 6.8.11 gives an example of a continuum whose 2–fold hyperspace is a cone. Corollary 6.9.4 extends this example. Note that Theorem 6.8.13 also gives such an example.

We need the following definitions.

6.9.1. Definition. A *dendroid* is an arcwise connected and hereditarily unicoherent continuum. A *fan* is a dendroid with exactly one ramification point (i.e., with only one point which is the common part of three otherwise disjoint arcs) [8]. The unique ramification point of a fan F is called the *top of F*; τ always denotes the top of a fan. By an *end point of a fan F* we mean an end point in the classical sense, that is, a point e of F which is a nonseparating point of any arc in F that contains e; $E(F)$ denotes the set of all end points of a fan F. A *leg of a fan F* is the unique arc in F from τ to some end point of F. Given two points x and y of a fan F, xy denotes the unique arc in F joining x and y. Given an $m \in \mathbb{N}$, an *m–od* is a fan for which $E(F)$ has exactly m elements.

6.9.2. Definition. A fan F is said to be *smooth* provided that whenever $\{x_i\}_{i=1}^{\infty}$ is a sequence in F converging to a point x of F, then the sequence of arcs $\{\tau x_i\}_{i=1}^{\infty}$ converges to the arc τx.

Given a fan F, let $\mathcal{G}(F)$ denote either of the hyperspaces 2^F, $\mathcal{C}_n(F)$, or $\mathcal{F}_n(F)$, for $n \in \mathbb{N}$.

6.9.3. Theorem. *If F is a fan, with metric d, which is homeomorphic to the cone over a compact metric space, then $\mathcal{G}(F)$ is homeomorphic to the cone over a continuum.*

Proof. Let F be a fan which is a cone. Then F is smooth. We assume by (Corollary 4 (p. 90) of [11] and Theorem 9 (p. 27) of [8]) that F is embedded in $\mathbb{R}^2$, $\tau = (0,0)$ is the top of F and the legs of F are convex arcs of length one (4.2 of [35]). Given two points a and b of $\mathbb{R}^2$, $[a,b]$ denotes the convex arc in $\mathbb{R}^2$ whose end points are a and b, and $\|a\|$ denotes the norm of a in $\mathbb{R}^2$. Given an element A of $\mathcal{G}(F)$ and $r \geq 0$, $rA = \{ra \mid a \in A\}$. Note that for $r = 0$, $rA = \{(0,0)\} = \{\tau\}$.

Let $E(F) = \{e_\lambda\}_{\lambda \in \Lambda}$. Then $F = K(E(F))$ by 4.2 of [35]. Note that this equality implies that $E(F)$ is closed in F. Hence, $E(F)$ is a compactum.

Let $\mathcal{B} = \bigcup\{\{A \in \mathcal{G}(F) \mid e_\lambda \in A\} \mid \lambda \in \Lambda\}$.

Let $\varphi \colon \mathcal{B} \times I \to \mathcal{G}(F)$ be given by

$$\varphi((A,t)) = (1-t)A.$$

Clearly, φ is well defined. Observe that if $t \in [0,1)$ and $A \in \mathcal{B}$, then $\{\tau\} \in \varphi((A,t))$ if and only if $\tau \in A$. We show that φ is continuous. Let $\varepsilon > 0$ be given and let $\delta = \dfrac{\varepsilon}{2}$. Let $A, B \in \mathcal{G}(F)$ and let $t, s \in [0,1]$ such that $\mathcal{H}(A,B) < \delta$ and $|t - s| < \delta$. Let $a \in A$. Then there exists $b \in B$ such that $\|a - b\| < \delta$. Note that

$$\|(1-t)a - (1-s)b\| \leq \|(1-t)a - (1-t)b\| + \|(1-t)b - (1-s)b\| \leq$$

$$(1-t)\|a - b\| + |s - t|\|b\| \leq \|a - b\| + |s - t| < 2\delta = \varepsilon.$$

Thus, $\varphi((A,t)) \subset \mathcal{V}^d_\varepsilon(\varphi((B,s)))$. Similarly, we have that $\varphi((B,s)) \subset \mathcal{V}^d_\varepsilon(\varphi((A,t)))$. Therefore, $\mathcal{H}(\varphi((A,t)), \varphi((B,s))) < \varepsilon$ and φ is continuous.

We show that φ is one–to–one on $\mathcal{B} \times [0,1)$. Let $t, s \in [0,1)$, and let $A, B \in \mathcal{B}$. Suppose that $\varphi((A,t)) = \varphi((B,s))$. Since $A, B \in \mathcal{B}$, there exist $e_\lambda, e_{\lambda'} \in E(F)$ such that $e_\lambda \in A$ and $e_{\lambda'} \in B$.

Case (1). $\tau \notin A$. Then $\tau \notin B$. Let $[a, e_\lambda]$ be the component of A containing e_λ. Let $[b, e_{\lambda'}]$ be the component of B containing $e_{\lambda'}$. Since $\varphi((A, t)) = \varphi((B, s))$, there exist $[b_{\lambda'}, c_{\lambda'}] \subset A$ and $[b_\lambda, c_\lambda] \subset B$ such that

$$[(1 - t)a, (1 - t)e_\lambda] = [(1 - s)b_\lambda, (1 - s)c_\lambda]$$

and

$$[(1 - s)b, (1 - s)e_{\lambda'}] = [(1 - t)b_{\lambda'}, (1 - t)c_{\lambda'}].$$

From the first equality we obtain that $(1 - t)e_\lambda = (1 - s)c_\lambda$, which implies that $1 - t = (1 - s)\|c_\lambda\| \leq 1 - s$. From the second inequality we obtain that $(1 - s)e_{\lambda'} = (1 - t)c_{\lambda'}$, which implies that $1 - s = (1 - t)\|c_{\lambda'}\| \leq 1 - t$. Therefore, $t = s$. Hence, $A = B$.

Case (2). $\tau \in A$. Then $\tau \in B$. Let us observe that either $[\tau, e_\lambda] \subset A$ or there exists $a \in A$ such that $[a, e_\lambda] \subset A$. In either case, as in Case (1), we conclude that $(1 - t)e_\lambda = (1 - s)c_\lambda$, for some $c_\lambda \in E(B)$, and that $(1 - s)e_{\lambda'} = (1 - t)c_{\lambda'}$, for some $c_{\lambda'} \in E(A)$. These two equalities imply that $t = s$. Hence, $A = B$.

We show that φ is surjective. Let $B \in \mathcal{G}(F)$. If $B = \{\tau\}$, then $\varphi((A, 1)) = \{\tau\}$ for any $A \in \mathcal{B}$. Thus, assume $B \neq \{\tau\}$. If $B \cap E(F) \neq \emptyset$, then $\varphi((B, 0)) = B$.

Suppose $B \cap E(F) = \emptyset$. Let $t = \inf\{\|b - e_\lambda\| \mid b \in B$ and $e_\lambda \in E(F)\}$. Since $B \neq \{\tau\}$, $t \neq 1$. Then there exists $\lambda_0 \in \Lambda$ such that $\|b_{\lambda_0} - e_{\lambda_0}\| = t$, where $b_{\lambda_0} \in B \cap [\tau, e_{\lambda_0}]$.

Let $A = \dfrac{1}{1 - t}B$. Note that for λ_0, $\dfrac{1}{1 - t}b_{\lambda_0} \in A \cap [\tau, e_{\lambda_0}]$. Since $\dfrac{1}{1 - t}b_{\lambda_0} = e_{\lambda_0}$, we have that $e_{\lambda_0} \in A$. Hence, $A \in \mathcal{B}$, and $\varphi((A, t)) = \dfrac{1 - t}{1 - t}B = B$.

By Theorem 1.2.10, the hyperspace $\mathcal{G}(F)$ is homeomorphic to $K(\mathcal{B})$. Since no point of $\mathcal{G}(F)$ arcwise disconnects 2^F ((11.5) of [40]), we have that $\mathcal{B}$ is a continuum.

Q.E.D.

Note that a similar proof to the one given for Theorem 6.9.3 shows:

6.9.4. Corollary. *If $\mathcal{G}([0,1]) \in \{2^{[0,1]}, \mathcal{C}_n([0,1]), \mathcal{F}_n([0,1])\}$, then $\mathcal{G}([0,1])$ is homeomorphic to the cone over a continuum.*

Since, clearly, a simple m–od is a fan homeomorphic to the cone over a finite set, we have the following:

6.9.5. Corollary. *Let m and n be positive integers. If F is a simple m–od, and $\mathcal{G}(F) \in \{\mathcal{F}_n(F), \mathcal{C}_n(F)\}$, then $\mathcal{G}(F)$ is homeomorphic to the cone over a finite–dimensional continuum.*

The next Theorem shows that for $n \geq 2$, no n–fold hyperspace of a finite–dimensional continuum is homeomorphic to its cone.

6.9.6. Theorem. *Let X be a finite–dimensional continuum. Then for each integer $n \geq 2$, $\mathcal{C}_n(X)$ is not homeomorphic to $K(X)$.*

Proof. Let $n \geq 2$ and suppose $\mathcal{C}_n(X)$ is homeomorphic to $K(X)$. Since X is of finite dimension, $K(X)$ is of finite dimension too. In fact $\dim(K(X)) = \dim(X) + 1$ ((8.0) of [40]). Since $\mathcal{C}(X) \subset \mathcal{C}_n(X)$ and $\mathcal{C}_n(X)$ is homeomorphic to $K(X)$, we have that $\dim(\mathcal{C}(X)) < \infty$. Hence, by the dimension theorem, $\dim(X) = 1$ (Theorem 2.1 of [26]). Thus, $\dim(K(X)) = 2$, and $\dim(\mathcal{C}_n(X)) = 2$. Therefore, since $\mathcal{C}_n(X)$ contains an n–cell (Theorem 6.1.9), $n = 2$. We consider two cases.

Case (1). *X contains a proper decomposable subcontinuum.* Then, by Theorem 6.1.10, $\mathcal{C}_2(X)$ contains a 3–cell, a contradiction to the fact that $\dim(\mathcal{C}_2(X)) = 2$.

Case (2). *All proper subcontinua of X are indecomposable.* Then, by Lemma 1.7.24, X is hereditarily indecomposable. Hence, $K(X)$ is uniquely arcwise connected. On the other hand, since $\mathcal{C}_2(X)$ contains 2–cells (Theorem 6.1.9), $\mathcal{C}_2(X)$ is not uniquely arcwise connected.

Therefore, $\mathcal{C}_n(X)$ is not homeomorphic to $K(X)$.

Q.E.D.

6.9.7. Lemma. *Let X be a continuum such that $\mathcal{C}(X)$ is finite dimensional. If A is a nondegenerate indecomposable proper subcontinuum of X, then at most a finite number of composants of A have the property that some subcontinuum of X contains a point of $X \setminus A$ and a point of the composant but does not contain A. Also, $\mathcal{C}_n(X) \setminus \{A\}$ has uncountably many arc components.*

Proof. The first part follows by the proof of $(*)$ (p. 312) of [40]. By the proof of (v) (p. 312) of [40], $\mathcal{C}(X) \setminus \{A\}$ has uncountably many arc components. Hence, $\mathcal{C}_n(X) \setminus \{A\}$ is not arcwise connected (Theorem 6.5.5) and has uncountably many arc components by 11.15 of [41] and Theorem 6.5.11.

Q.E.D.

6.9.8. Theorem. *Let X be a continuum and let $n \geq 2$ be an integer. If Z is a finite–dimensional continuum such that $K(Z)$ is homeomorphic to $\mathcal{C}_n(X)$, then $\dim(X) = 1$ and X contains at most one nondegenerate indecomposable continuum. Hence, X is not hereditarily indecomposable.*

Proof. Let $n \geq 2$. Let $h : \mathcal{C}_n(X) \to K(Z)$ be a homeomorphism. The proof of the fact that $\dim(X) = 1$ is similar to the one given in Theorem 6.9.6.

Suppose X contains two nondegenerate indecomposable continua, A and B. Then $\mathcal{C}_n(X) \setminus \{A\}$ and $\mathcal{C}_n(X) \setminus \{B\}$ have infinitely many arc components (Lemma 6.9.7). Hence, $K(Z) \setminus \{h(A)\}$ and $K(Z) \setminus \{h(B)\}$ both have infinitely many arc components. On the other hand, for each $p \in K(Z) \setminus \{v_Z\}$, $K(Z) \setminus \{p\}$ has at most two arc components. Therefore, X contains at most one nondegenerate indecomposable subcontinuum.

Q.E.D.

6.9.9. Theorem. *Let X be a continuum and let $n \geq 2$. Then every 2–cell in $\mathcal{C}_n(X)$ is nowhere dense.*

Proof. First, suppose $n \geq 3$. Let $\mathcal{U}$ be any nonempty open set in $\mathcal{C}_n(X)$. Then there exists $A \in \mathcal{U}$ such that A has exactly n components, $A_1, \ldots, A_n$ (Theorem 6.1.7). By Corollary 1.7.23, for each $j \in \{1, \dots, n\}$, there is a subcontinuum B_j of X such that B_j contains A_j properly, $B_j \cap B_\ell = \emptyset$ if $j \neq \ell$ and $B = \bigcup_{j=1}^{n} B_j \in \mathcal{U}$.

For each $j \in \{1, \dots, n\}$, let $\alpha_j \colon [0, 1] \to \mathcal{C}(X)$ be an order arc such that $\alpha_j(0) = A_j$ and $\alpha_j(1) = B_j$ (Theorem 1.8.20). Define $\gamma((t_1, \dots, t_n)) = \bigcup_{j=1}^{n} \alpha_j(t_j)$. Then $\gamma([0, 1]^n)$ is an n–cell contained in $\mathcal{U}$. Since $n \geq 3$, $\mathcal{U}$ cannot be a 2–cell.

Next, suppose $n = 2$. Suppose that $\mathcal{D}$ is a 2–cell in $\mathcal{C}_2(X)$ with nonempty interior in $\mathcal{C}_2(X)$. Then there exists $D \in Int_{\mathcal{C}_2(X)}(\mathcal{D})$ such that D has two nondegenerate components, D_1 and D_2 (Theorem 6.1.7 and Corollary 1.7.23). Let $\varepsilon > 0$ such that if $E \in \mathcal{C}_2(X)$ and $\mathcal{H}(E, D) < \varepsilon$, then $E \in Int_{\mathcal{C}_2(X)}(\mathcal{D})$.

Let $\mu \colon \mathcal{C}(D_2) \to [0, 1]$ be a Whitney map. Let $\varphi \colon \mathcal{C}(D_2) \to \mathcal{C}_2(X)$ be given by $\varphi(B) = D_1 \cup B$. Then φ is an embedding of $\mathcal{C}(D_2)$ into $\{G \in \mathcal{C}_2(X) \mid G \subset D\}$. Let $t_0 \in (0, 1)$ such that if $\mu(B) \geq t_0$, then $\mathcal{H}(\varphi(B), D) < \varepsilon$. Let $\mathcal{B} = \{\varphi(B) \mid t_0 < \mu(B) < 1\}$. Then $\mathcal{B}$ is a 2–dimensional subset of $Int_{\mathcal{C}_2(X)}(\mathcal{D})$, $\dim(\mathcal{B}) = 2$ is seen using the fact that no zero–dimensional set separates $\mathcal{C}(D_2)$ by ((2.15) of [40]). Hence, $Int_{\mathcal{C}_2(X)}(\mathcal{B}) \neq \emptyset$ (10.2 of [42]). Thus, letting $B_0 \in \mathcal{C}(D_2)$ such that $\varphi(B_0) \in Int_{\mathcal{C}_2(X)}(\mathcal{B})$, we have that $\varphi(B_0)$ is not arcwise accessible from $\mathcal{C}_2(X) \setminus \mathcal{B}$. However, let $\beta \colon [0, 1] \to \mathcal{C}(D_1)$ be an order arc such that $\beta(0) \in \mathcal{F}_1(D_1)$ and $\beta(1) = D_1$ (Theorem 1.8.20). Then $\beta(s) \cup B_0 \notin \mathcal{B}$ for any $s \in [0, 1)$ and $\beta(1) \cup B_0 = D_1 \cup B_0 = \varphi(B_0) \in \mathcal{B}$, a contradiction.

Therefore, every 2–cell in $\mathcal{C}_n(X)$ is nowhere dense.

Q.E.D.

We end this chapter with the following Theorem, which gives conditions on an indecomposable continuum X in order to have its n–fold hyperspaces homeomorphic to a cone over a finite–dimensional continuum.

6.9.10. Theorem. *Let X be a continuum containing a nondegenerate indecomposable subcontinuum A. Let $n \geq 2$ be an integer, and let Z be a finite–dimensional continuum such that $K(Z)$ is homeomorphic to $\mathcal{C}_n(X)$. If $h \colon \mathcal{C}_n(X) \to K(Z)$ is a homeomorphism, then*

(1) $h(A) = \nu_Z$;

(2) Z has uncountably many arc components. In particular, Z is not locally connected;

(3) $\dim(\mathcal{C}_n(X)) \geq 2n$ and $\dim(Z) \geq 2n - 1$;

(4) Each point z of Z is contained in an arc in Z and some points of Z belong to locally connected subcontinua of Z whose dimension is at least $2n - 1$;

(5) No point of $K(Z) \setminus \{\nu_Z\}$ arcwise disconnects $K(Z)$;

(6) If $A = X$, then X does not contain a nondegenerate proper terminal subcontinuum;

(7) Z is not irreducible. In particular, Z is decomposable.

Proof. (1) The proof is similar to the proof of Theorem 6.9.8.

(2) By Lemma 6.9.7, $\mathcal{C}_n(X) \setminus \{A\}$ has uncountably many arc components. Since $h(A) = \nu_Z$ (by (1)), $K(Z) \setminus \{\nu_Z\}$ has uncountably many arc components. Thus, since $K(Z) \setminus \{\nu_Z\}$ is homeomorphic to $Z \times [0, 1)$, we conclude that Z has uncountably many arc components.

(3) By Theorem 6.9.8, each subcontinuum of X, distinct from A, is decomposable. Thus, $\mathcal{C}_n(X)$ contains a $2n$–cell (Corollary 6.1.12). Hence, $\dim(\mathcal{C}_n(X)) \geq 2n$. Since $K(Z)$ is homeomorphic to $\mathcal{C}_n(X)$ and $\dim(K(Z)) = \dim(Z) + 1$ ((8.0) of [40]), we have that $\dim(Z) \geq 2n - 1$.

(4) Let z be any point of Z. Let $\pi \colon K(Z) \setminus \{\nu_Z\} \twoheadrightarrow Z$ be the projection map. We consider two cases.

First, suppose there exists $t_0 \in [0, 1)$ such that $h^{-1}((z, t_0)) \in \mathcal{C}_n(X) \setminus \mathcal{C}(X)$. Let $B = h^{-1}((z, t_0))$ and let $B_1, \ldots, B_k$ be the components of B, where $k \in \{2, \ldots, n\}$. By Corollary 1.7.23, for each $j \in \{1, \ldots, k\}$, there exists a subcontinuum C_j of X containing B_j properly. We assume, without loss of generality, that $C_j \cap C_\ell = \emptyset$ if $j \neq \ell$.

For each $j \in \{1, \ldots, k\}$, let $\alpha_j \colon [0, 1] \to \mathcal{C}(X)$ be an order arc (Theorem 1.8.20) such that $\alpha_j(0) = B_j$ and $\alpha_j(1) = C_j$. Let $\alpha \colon [0, 1]^k \to \mathcal{C}_n(X)$ be given by $\alpha((t_1, \ldots, t_k)) = \bigcup_{j=1}^{k} \alpha_j(t_j)$. Let $\mathcal{D} = \alpha\left([0, 1]^k\right)$. Then $\mathcal{D}$ is a k–cell such that $B \in \mathcal{D}$ and $A \notin \mathcal{D}$.

Thus, $h(\mathcal{D})$ is a k–cell containing the point (z, t_0) and not containing ν_Z. Hence, $\pi(h(\mathcal{D}))$ is a locally connected subcontinuum of Z containing z. Since $k \geq 2$, $\pi(h(\mathcal{D}))$ is nondegenerate. Thus, z is contained in an arc by (8.23 of [41]).

Next, suppose that $h^{-1}((z, t)) \in \mathcal{C}(X)$ for each $t \in [0, 1)$. Since $h(A) = \nu_Z$, there exists $t' \in [0, 1)$ such that $h^{-1}((z, t')) \neq A$ and $h^{-1}((z, t')) \notin \mathcal{F}_1(X)$. Let $E = h^{-1}((z, t'))$. Since $E \neq A$ and E is nondegenerate, E is a decomposable continuum (by Theorem 6.9.8). Hence, there are two proper subcontinua K and H of E such that $E = H \cup K$.

Suppose, first, that A is not contained in E. Take $x_1 \in H \setminus K$ and $x_2 \in K \setminus H$. Let $\beta_j \colon [0, 1] \to \mathcal{C}(X)$ be an order arc such that $\beta_j(0) = \{x_j\}$, $j \in \{1, 2\}$, $\beta_1(1) = H$ and $\beta_2(1) = K$. Let $\beta \colon [0, 1]^2 \to \mathcal{C}_n(X)$ be given by $\beta((t_1, t_2)) = \beta_1(t_1) \cup \beta_2(t_2)$. Let $\mathcal{G} = \beta([0, 1]^2)$. Then $\mathcal{G}$ is a locally connected subcontinuum of $\mathcal{C}_n(X)$ such that $\mathcal{G}$ contains a 2–cell and such that $E \in \mathcal{G}$ and $A \notin \mathcal{G}$. Thus, $h(\mathcal{G})$ is a locally connected subcontinuum of $K(Z)$ containing a 2–cell, such that $(z, t') \in h(\mathcal{G})$ and $\nu_Z \notin h(\mathcal{G})$. Hence, $\pi(h(\mathcal{G}))$ is a nondegenerate locally connected subcontinuum of Z containing z. Thus, z is in an arc by (8.23 of [41]).

Suppose next A is contained in E. Since A is indecomposable, $E \neq A$. Hence, there exists a point $x_1 \in E \setminus A$. Suppose that $x_1 \in H$. Choose a point $x_2 \in K \setminus \{x_1\}$. Then we just repeat the argument in the preceding paragraph to construct a nondegenerate locally connected subcontinuum of Z containing z.

This completes the proof of the first part of (4). We prove the second part of (4) as follows:

By Corollary 6.1.12, there exists a $2n$–cell $\mathcal{E}$ in $\mathcal{C}_n(X)$. We may choose $\mathcal{E}$ such that $A \notin \mathcal{E}$. Let $B \in \mathcal{E}$. Then $h(\mathcal{E})$ is a $2n$–cell such that $h(B) \in \mathcal{E}$. Hence, $\pi(h(\mathcal{E}))$ is a locally connected subcontinuum of Z containing $\pi(h(B))$, and $\dim(\pi(h(\mathcal{E}))) \geq 2n - 1$ (by 20.10 [42] since $\pi(h(\mathcal{E})) \times [0, 1)$ contains $h(\mathcal{E})$ and, thus, has dimension at least $2n$).

(5) By (4), each point of $K(Z)$ lies in the cone over an arc. Hence, (5) follows easily.

(6) This is a consequence of Theorem 6.9.8, Theorem 6.5.4 and part (5) of this theorem.

(7) Suppose there exist two points z_1 and z_2 of Z such that Z is

irreducible between them.

First, we prove that both z_1 and z_2 belong to $\pi(h(\mathcal{F}_n(X)))$. Suppose this is not true. Note that $\mathcal{F}_n(X)$ intersects all the arc components of $\mathcal{C}_n(X) \setminus \{A\}$. Hence, there are arcs α_1 and α_2 in Z such that one end point of α_j is z_j and the other point of α_j is in $\pi(h(\mathcal{F}_n(X)))$, for $j \in \{1, 2\}$. Hence, by irreducibility, $Z = \alpha_1 \cup \alpha_2 \cup \pi(h(\mathcal{F}_n(X)))$. On the other hand, Z cannot contain free arcs (otherwise, $\mathcal{C}_n(X)$ contains 2–cells with nonempty interior, which contradicts Theorem 6.9.9). Thus, we have proved that z_1 and z_2 belong to $\pi(h(\mathcal{F}_n(X)))$.

Let t_1 and t_2 be points of $[0, 1)$ such that (z_1, t_1) and (z_2, t_2) belong to $h(\mathcal{F}_n(X))$. Let B_1 and B_2 be the elements of $\mathcal{F}_n(X)$ such that $h(B_1) = (z_1, t_1)$ and $h(B_2) = (z_2, t_2)$. Take $x_1 \in B_1$ and $x_2 \in B_2$. Let

$$\mathcal{B}_1 = \{\{x_1\} \cup B \mid B \in \mathcal{F}_{n-1}(X)\}$$

and

$$\mathcal{B}_2 = \{\{x_2\} \cup B \mid B \in \mathcal{F}_{n-1}(X)\}.$$

Then $\mathcal{B}_1$ and $\mathcal{B}_2$ are subcontinua of $\mathcal{C}_n(X)$ containing B_1 and B_2, respectively; also, $\mathcal{B}_1 \cap \mathcal{F}_{n-1}(X) \neq \emptyset$ and $\mathcal{B}_2 \cap \mathcal{F}_{n-1}(X) \neq \emptyset$. Hence, $\mathcal{B}_1 \cup \mathcal{B}_2 \cup \mathcal{F}_{n-1}(X)$ is a subcontinuum of $\mathcal{C}_n(X)$ which does not intersect all the arc components of $\mathcal{C}_n(X) \setminus \{A\}$, which we prove as follows: By Lemma 6.9.7, $\mathcal{C}_n(X) \setminus \{A\}$ has uncountably many arc components. Let $a_1, \ldots, a_n$ be n points of A in n distinct composants, $\kappa_1, \ldots, \kappa_n$, of A such that $\{x_1, x_2\} \cap \bigcup_{j=1}^{n} \kappa_j = \emptyset$; if $A \neq X$, by Lemma 6.9.7, we may take n composants that are not accessible from $X \setminus A$. Let $\mathcal{G}$ be the arc component of $\mathcal{C}_n(X) \setminus \{A\}$ containing $\{a_1, \ldots, a_n\}$. Then $\mathcal{G} \cap (\mathcal{B}_1 \cup \mathcal{B}_2 \cup \mathcal{F}_{n-1}(X)) = \emptyset$.

Since $\mathcal{B}_1 \cup \mathcal{B}_2 \cup \mathcal{F}_{n-1}(X)$ does not intersect all the arc components of $\mathcal{C}_n(X) \setminus \{A\}$, we have that $h(\mathcal{B}_1 \cup \mathcal{B}_2 \cup \mathcal{F}_{n-1}(X))$ is a subcontinuum of $K(Z)$ containing both (z_1, t_1) and (z_2, t_2), which does not intersect all the arc components of $K(Z) \setminus \{\nu_Z\}$. Then $\pi(h(\mathcal{B}_1 \cup \mathcal{B}_2 \cup \mathcal{F}_{n-1}(X)))$ is a proper subcontinuum of Z containing z_1 and z_2, a contradiction. Therefore, Z is not irreducible.

Q.E.D.

REFERENCES

[1] G. Acosta, Continua with Unique Hyperspace, in *Continuum Theory: Proceedings of the Special Session in Honor of Professor Sam B. Nadler, Jr.'s 60th Birthday.* Lecture Notes in Pure and Applied Mathematics Series, Vol. 230, Marcel Dekker, Inc., New York, Basel, 2002. pp. 33–49. (eds.: Alejandro Illanes, Ira Wayne Lewis and Sergio Macías.)

[2] D. P. Bellamy, The Cone Over the Cantor Set–continuous Maps From Both Directions, Proc. Topology Conference (Emory University, Atlanta, Ga., 1970), J. W. Rogers, Jr., ed., 8–25.

[3] D. E. Bennett, Aposyndetic Properties of Unicoherent Continua, Pacific J. Math., 37 (1971), 585–589.

[4] R. H. Bing, Partitioning a Set, Bull. Amer. Math. Soc., 55 (1949), 1101–1110.

[5] K. Borsuk, *Theory of Retracts,* Monografie Mat. Vol. 44, PWN (Polish Scientific Publishers), Warszawa, 1967.

[6] E. Castañeda, A Unicoherent Continuum for Which its Second Symmetric Product is not Unicoherent, Topology Proceedings, 23 (1998), 61–67.

[7] E. Castañeda, *Productos Simétricos*, Tesis Doctoral, Facultad de Ciencias, U. N. A. M., 2003.

[8] J. J. Charatonik, On fans, Dissertationes Math. (Rozprawy Mat.), 54 (1967), 1–37.

[9] D. Curtis and N. T. Nhu, Hyperspaces of Finite Subsets Which are Homeomorphic to $\aleph_0$–dimensional Linear Metric Spaces, Topology Appl., 19 (1985), 251–260.

[10] C. H. Dowker, Mapping Theorems for Non–compact Spaces, Amer. J. Math., 69 (1947), 200–242.

[11] C. A. Eberhart, A Note on Smooth Fans, Colloq. Math. 20 (1969), 89–90.

[12] J. T. Goodykoontz, Some Retractions and Deformation Retractions on 2^X and $\mathcal{C}(X)$, Topology Appl., 21 (1985), 121–133.

[13] J. Grispolakis, S. B. Nadler, Jr. and E. D. Tymchatyn, Some Properties of Hyperspaces with Applications to Continua Theory, Can. J. Math., 31 (1979), 197–210.

[14] J. Grispolakis and E. D. Tymchatyn, Weakly Confluent Mappings and the Covering Property of Hyperspaces, Proc. Amer. Math. Soc., 74 (1979), 177–182.

[15] C. L. Hagopian, Indecomposable Homogeneous Plane Continua are Hereditarily Indecomposable, Trans. Amer. Math. Soc., 224 (1976), 339–350.

[16] J. G. Hocking and G. S. Young, *Topology*, Dover Publications, Inc., New York, 1988.

[17] A. Illanes, The Hyperspace $\mathcal{C}_2(X)$ for a Finite Graph X is Unique, Glansnik Mat., 37(57) (2002), 347–363.

[18] A. Illanes, Comparing n–fold and m–fold hyperspaces, Topology Appl., 133 (2003), 179–198.

[19] A. Illanes, Finite Graphs Have Unique Hyperspaces $\mathcal{C}_n(X)$, Topology Proc., 27 (2003), 179–188.

[20] A. Illanes, A Model for the Hyperspace $\mathcal{C}_2(\mathcal{S}^1)$, Q. & A. in General Topology, 22 (2004), 117–130.

[21] A. Illanes and S. B. Nadler, Jr., *Hyperspaces: Fundamentals and Recent Advances,* Monographs and Textbooks in Pure and Applied Math., Vol. 216, Marcel Dekker, New York, Basel, 1999.

[22] F. B. Jones, Certain Homogeneous Unicoherent Indecomposable Continua, Proc. Amer. Math. Soc., 2 (1951), 855–859.

[23] J. L. Kelley, Hyperspaces of a Continuum, Trans. Amer. Math. Soc., 52 (1942), 22–36.

[24] J. Krasinkiewicz, Curves Which Are Continuous Images of Tree–like Continua Are Movable, Fun. Math., 89 (1975), 233–260.

[25] K. Kuratowski, *Topology*, Vol. II, Academic Press, New York, N. Y., 1968.

[26] M. Levin and Y. Sternfeld, The Space of Subcontinua of a 2–dimensional Continuum is Infinitely Dimensional, Proc. Amer. Math. Soc., 125 (1997), 2771–2775.

[27] S. Macías, On Symmetric Products of Continua, Topology Appl., 92 (1999), 173–182.

[28] S. Macías, Aposyndetic Properties of Symmetric Products of Continua, Topology Proc., 22 (1997), 281–296.

[29] S. Macías, On the Hyperspaces $\mathcal{C}_n(X)$ of a Continuum X, Topology Appl., 109 (2001), 237–256.

[30] S. Macías, On the Hyperspaces $\mathcal{C}_n(X)$ of a Continuum X, II, Topology Proc., 25 (2000), 255–276.

[31] S. Macías, On Arcwise Accessibility in Hyperspaces, Topology Proc., 26 (2001–2002), 247–254.

[32] S. Macías, Fans Whose Hyperspaces Are Cones, Topology Proc., 27 (2003), 217–222.

[33] S. Macías and S. B. Nadler, Jr., n–fold Hyperspaces, Cones and Products, Topology Proc., 26 (2001–2002), 255–270.

[34] S. Macías and S. B. Nadler, Jr., Smoothness in n–fold Hyperspaces, Glasnik Mat., 37(57)(2002), 365–373.

[35] S. Macías and S. B. Nadler, Jr., Fans Whose Hyperspace of Subcontinua are Cones, Topology Appl., 126 (2002), 29–36.

[36] S. Mardešić and J. Segal, ε–mappings onto Polyhedra, Trans. Amer. Math. Soc., 109 (1963), 146–164.

[37] J. van Mill, *Infinite–Dimensional Topology*, North Holland, Amsterdam, 1989.

[38] E. E. Moise, Grille Decomposition and Convexification Theorems for Compact Locally Connected Continua, Bull. Amer. Math. Soc., 55 (1949), 1111–1121.

[39] S. B. Nadler, Jr., Arcwise Accessibility in Hyperspaces, Dissertationes Math., 138 (1976), 1–29.

[40] S. B. Nadler, Jr., *Hyperspaces of Sets,* Monographs and Textbooks in Pure and Applied Math., Vol. 49, Marcel Dekker, New York, Basel, 1978.

[41] S. B. Nadler, Jr., *Continuum Theory: An Introduction,* Monographs and Textbooks in Pure and Applied Math., Vol. 158, Marcel Dekker, New York, Basel, Hong Kong, 1992.

[42] S. B. Nadler, Jr., *Dimension Theory: An Introduction with Exercises*, Aportaciones Matemáticas, Textos # 18, Sociedad Matemática Mexicana, 2002.

[43] J. T. Rogers, Jr., Dimension and the Whitney Subcontinua of $\mathcal{C}(X)$, Gen. Top. and its Applications, 6 (1976), 91–100.

[44] A. H. Wallace, *Algebraic Topology, Homology and Cohomology,* W. A. Benjamin Inc., 1970.

[45] G. T. Whyburn, *Analytic Topology,* Amer. Math. Soc. Colloq. Publ., vol. 28, Amer. Math. Soc., Providence, R. I., 1942.

[46] S. Willard, *General Topology*, Addison–Wesley Publishing Co., 1970.

[47] M. Wojdisławski, Rétractes Absolus et Hyperespaces des Continus, Fund. Math., 32 (1939), 184–192.

Chapter 7

QUESTIONS

In this small chapter we include some open questions related to the topics of the rest of the book.

7.1   Inverse Limits

7.1.1. Question. *Is it true that X is the inverse limit of graphs with monotone simplicial retractions as bonding maps if and only if X is locally connected and each cyclic element of X is a graph? (B. B. Epps)*

Comment: If M is a semi–locally connected continuum a *cyclic element* of M is either a cut point of M, an end point of M or a nondegenerate subset of M which is maximal with respect to being a connected subset without cut points. A map $f\colon X \to Y$ between graphs is *simplicial* if it is a map between the set of vertixes of the graph and it is a linear extension on the edges.

7.1.2. Question. *Let $\{X_n, f_n^{n+1}\}$ be an inverse sequence of poly-hedra with bonding maps $f_n^{n+1}\colon X_{n+1} \to X_n$ such that the inverse limit is a hereditarily indecomposable continuum. Let $g_n\colon \mathcal{S}^2 \to X_n$*

347

be a map such that g_n is homotopic to $f_n^{n+1} \circ g_{n+1}$ for each $n \in \mathbb{N}$. Is g_1 homotopic to a constant map? (J. Krasinkiewicz)

7.1.3. Question. *If X and Y are continua and $X \times Y$ is disk–like, then must X be arc–like? (C. L. Hagopian)*

7.1.4. Question. *If a product of two continua is disk–like, must it be embeddable in $\mathbb{R}^3$? (C. L. Hagopian)*

7.1.5. Question. *Does every disk–like continuum have the fixed point property? (S. Mardešić and C. L. Hagopian, independently)*

Comment: An affirmative answer to this question would imply that every nonseparating plane continuum has the fixed point property. (C. L. Hagopian)

7.1.6. Question. *Does each tree–like homogeneous continuum have the fixed point property? (C. L. Hagopian)*

Comment: For the plane, Oversteegen and Tymchatyn have shown that this question has a positive answer. (W. Lewis)

7.1.7. Question. *Does every triod–like continuum have the fixed point property? (D. Bellamy)*

7.1.8. Question. *Is there a map $f \colon [0,1] \to [0,1]$ such that f has a periodic orbit of odd period larger than one, f has no periodic orbit of period three and $\varprojlim\{[0,1],f\}$ is homeomorphic to the pseudo–arc? (L. Block, J. Keesling and V. V. Uspenskij)*

Comment: Let $f \colon X \to X$ be a map. A point $x \in X$ is said to be *periodic* provided that there exists $m \in \mathbb{N}$ such that $f^m(x) = x$. Now, x is said to have *period n* if $n = \min\{m \in \mathbb{N} \mid f^m(x) = x\}$. We say that f has a *periodic orbit of period n* provided that there exists a point $x \in X$ such that x has period n.

7.1.9. Question. *Given* $\lambda \in \left[\dfrac{1}{2}, 1\right]$, *define* $T_\lambda \colon [0,1] \to [0,1]$ *by*

$$
T_\lambda(x) = \begin{cases} 2\lambda x & \text{if } x \in \left[0, \dfrac{1}{2}\right]; \\[2ex] 2\lambda(1-x) & \text{if } x \in \left[\dfrac{1}{2}, 1\right]. \end{cases}
$$

If $\dfrac{1}{2} < \lambda_1 < \lambda_2 \leq 1$, *then is* $\varprojlim\{[0,1], T_{\lambda_1}\}$ *topologically different from* $\varprojlim\{[0,1], T_{\lambda_2}\}$? *(W. T. Ingram)*

7.1.10. Question. *Suppose $n \geq 4$ is a positive integer and H and K are chainable continua having the property that each is an indecomposable continuum with only n end points and every nondegenerate proper subcontinuum is an arc. Are H and K homeomorphic? (W. T. Ingram)*

Comment: This question has been solved by Barge and Diamond for $n = 5$. Kailhoffer and Raines have further partial solutions. (W. T. Ingram)

Comment: Sonja Štimac has announced at the meeting *Geometric Topology II*, celebrated at the Inter–University Centre, Dubrovnik, Croatia (September 29 through October 5, 2002), that in her dissertation (in Croatian) she has a positive answer to this question. No paper has appeared yet. (S. Macías)

7.2   The Set Function $\mathcal{T}$

7.2.1. Question. *If the set function $\mathcal{T}$ is continuous for the continuum X, is it true that X is $\mathcal{T}$–additive? (D. Bellamy)*

Comment: A bushel of *Extra Fancy Stayman Winesap apples* for the solution. (D. Bellamy)

7.2.2. Question. *If $\mathcal{T}$ is continuous for the continuum X, is it true that the collection $\{\mathcal{T}(\{p\}) \mid p \in X\}$ is a continuous decomposition of X such that the quotient space is locally connected? (D. Bellamy)*

Comment: A partial answer to this question is presented in Corollary 5.1.24. (S. Macías)

7.2.3. Question. *If $\mathcal{T}$ is continuous for X and there is a point p in X such that $\mathcal{T}(\{p\})$ has nonempty interior, is X indecomposable? (D. Bellamy)*

7.2.4. Question. *If X and Y are indecomposable continua, is $\mathcal{T}$ idempotent on $X \times Y$? Even for only closed sets in $X \times Y$? (D. Bellamy)*

7.2.5. Question. *If $\mathcal{T}$ is continuous for X and X is a decomposable continuum, is it true that for each $p \in X$, $Int(\mathcal{T}(\{p\})) = \emptyset$? (D. Bellamy)*

7.2.6. Question. *Let X be a continuum. If $X/\mathcal{T}$ denotes the finest decomposition space of X which shrinks each $\mathcal{T}(\{p\})$ to a point, is $X/\mathcal{T}$ locally connected? (D. Bellamy)*

Comment: The answer is affirmative if it is assumed that X is $\mathcal{T}$–additive.

7.2.7. Question. *If X is an indecomposable continuum and W is a subcontinuum of $X \times X$ with nonempty interior, is $\mathcal{T}(W) = X \times X$? (F. B. Jones)*

Comment: C. L. Hagopian has shown that this question has an affirmative answer if X is chainable. (S. Macías)

7.2.8. Question. *If X is an atriodic continuum (or X contains no uncountable collection of pairwise disjoint triods) and X does not have a weak cut point, then is there a continuum $W \subset X$ such that $Int(W) \neq \emptyset$ and $T(W) \neq X$? (H. Cook)*

Comment: Let X be a continuum and let x be a point of X. We say that x is a *weak cut point* of X provided that there exist two points y and z in X such that if W is a subcontinuum of X and $\{y, z\} \subset W$, then $x \in W$.

7.2.9. Question. *If T is continuous for the continuum X and $f \colon X \twoheadrightarrow Z$ is a continuous and monotone surjection, is T continuous for Z also? (D. Bellamy)*

7.2.10. Question. *If X is a strictly point T–asymmetric dendroid, then is X smooth? (D. Bellamy)*

Comment: A continuum X is *strictly point T–asymmetric* if for any two distinct points p and q of X with $p \in T(\{q\})$, we have that $q \notin T(\{p\})$.

7.2.11. Question. *Let X be a continuum. Suppose the restriction of T to the hyperspace of subcontinua of X is continuous. Does this imply that T is continuous for X? (D. Bellamy)*

Comment: A partial answer to this question is presented in Theorem 3.1.32. (S. Macías)

7.2.12. Question. *Do open maps preserve T–additivity? T–symmetry? (D. Bellamy)*

7.3 Homogeneous Continua

7.3.1. Question. *Is every homogeneous, hereditarily indecomposable, nondegenerate continuum a pseudo–arc? (F. B. Jones)*

Comment: The *pseudo–arc* is the only nondegenerate, hereditarily indecomposable chainable continuum.

Comment: J. T. Rogers, Jr. has shown that it must be tree–like.

7.3.2. Question. *Is each nondegenerate, homogeneous, nonseparating plane continuum a pseudo–arc? (F. B. Jones)*

Comment: F. B. Jones and C. L. Hagopian have shown that it must be hereditarily indecomposable. J. T. Rogers, Jr., has shown that it must be tree–like.

7.3.3. Question. *If a homogeneous continuum X contains an arc must it contain a solenoid or a simple closed curve? (C. L. Hagopian)*

7.3.4. Question. *Is the simple closed curve the only nondegenerate, homogeneous, hereditarily decomposable continuum? (P. Minc)*

7.3.5. Question. *Is every decomposable, homogeneous continuum of dimension greater than one aposyndetic? (J. T. Rogers, Jr.)*

7.3.6. Question. *Is every nondegenerate, homogeneous, indecomposable continuum one-dimensional? (F. B. Jones)*

7.3.7. Question. *Is each atriodic, homogeneous continuum circle-like? (C. L. Hagopian)*

7.3.8. Question. *Is each aposyndetic, nonlocally connected, one-dimensional, homogeneous continuum an inverse limit of Menger curves? Menger curves and covering maps? (J. T. Rogers, Jr.)*

7.3.9. Question. *Let X be an arcwise connected homogeneous continuum which is not S^1. Must X contain a simple closed curve of arbitrary small diameter? (L. Lum)*

7.3.10. Question. *Let X be a homogeneous arcwise connected continuum which is not S^1. Let U be an open set in X and let M be an arc component of U. Is M cyclicly connected? (D. P. Bellamy)*

Comment: An arcwise connected space Z is *cyclicly connected* provided that each pair of points of Z lie on a simple closed curve.

7.3.11. Question. *Let X be a homogeneous arcwise connected continuum which is not a simple closed curve. Is every arcwise connected open subset of X also cyclicly connected? (D. P. Bellamy)*

7.3.12. Question. *Can Jones's Aposyndetic Decomposition Theorem be strengthened to give decomposition elements which are hereditarily indecomposable? (J. T. Rogers, Jr.)*

7.3.13. Question. *If X is an arcwise connected homogeneous continuum other than a simple closed curve, must each pair of points of X be the vertices of a θ-curve in X? (D. Bellamy)*

Comment: D. Bellamy and L. Lum have shown that each pair of points of X must lie on a simple closed curve.

7.3.14. Question. *Does each finite subset of a nondegenerate arcwise connected homogeneous continuum lie on a simple closed curve? (D. Bellamy)*

7.3.15. Question. *Does there exist a homogeneous one–dimensional continuum with no nondegenerate chainable subcontinuum? (W. Lewis)*

Comment: If there exists a nondegenerate, homogeneous, hereditarily indecomposable continuum other than the pseudo–arc, the answer is yes.

7.3.16. Question. *Is every continuum a continuous image of a homogeneous continuum? In particular, is the spiral around the triod such an image? (L. Fearnley)*

7.3.17. Question. *Is there a nondegenerate homogeneous plane continuum X not homeomorphic to a simple closed curve, the pseudo–arc or the circle of pseudo–arcs?*

7.4 n–fold Hyperspaces

7.4.1. Question. *Let X be a continuum. If $\mathcal{C}_n(X)$ does not contain $(n+1)$–cells, then is X hereditarily indecomposable? (S. Macías)*

7.4.2. Question. *Is $\mathcal{C}_3([0,1])$ homeomorphic to $[0,1]^6$? (R. Schori)*

7.4.3. Question. *If $n \geq 2$ is an integer and X is a continuum, then is $\mathcal{C}_n(X)$ countable closed aposyndetic? 0–dimensional closed aposyndetic? (S. Macías)*

7.4.4. Question. *Let n be an integer greater than two. Let X be a continuum and let $x \in X$. If $\{x\}$ is arcwise accessible from $\mathcal{C}_n(X) \setminus \mathcal{C}(X)$, then is $\{x\}$ arcwise accessible from $\mathcal{C}_2(X) \setminus \mathcal{C}(X)$? (S. Macías)*

7.4.5. Question. *Does there exist an indecomposable continuum X such that $\mathcal{C}_n(X)$ is homeomorphic to the cone over a finite-dimensional continuum for some integer $n \geq 2$? (S. Macías and S. B. Nadler, Jr.)*

7.4.6. Question. *Does there exist a hereditarily decomposable continuum X that is neither an arc nor a simple m–od such that $\mathcal{C}_n(X)$ is homeomorphic to the cone over a finite–dimensional continuum for some $n \geq 3$? (S. Macías and S. B. Nadler, Jr.)*

7.4.7. Question. *Find a geometric model for $\mathcal{C}_2(X)$, where X is a simple triod. (A. Illanes)*

REFERENCES

[1] D. P. Bellamy, Questions In and Out of Context, talk given in the VI Joint Meeting AMS–SMM, Houston, TX, May 13–15, 2004 (LaTeX edition by Jonathan Hatch).

[2] L. Block, J. Keesling and V. V. Uspenskij, Inverse Limits Which are the Pseudoarc, Houston J. Math., 26 (2000), 629–638.

[3] H. Cook, W. T. Ingram and A. Lelek, *A List of Problems Known as Houston Problem Book*, in *Continua with the Houston Problem Book*, Lectures Notes in Pure and Applied Mathematics, Vol. 170, Marcel Dekker, Inc., New York, Basel, Hong Kong, 1995. (eds. H. Cook, W. T. Ingram, K. T. Kuperberg, A. Lelek, and P. Minc)

[4] C. L. Hagopian, Mutual Aposyndesis, Proc. Amer. Math. Soc., 23 (1969), 615–622.

[5] C. L. Hagopian, Disk–like Products of λ Connected Continua, I, Proc. Amer. Math. Soc., 51 (1975), 448–452.

[6] C. L. Hagopian, Disk–like Products of λ Connected Continua, II, Proc. Amer. Math. Soc., 52 (1975), 479–484.

[7] A. Illanes, A Model for the Hyperspace $\mathcal{C}_2(\mathcal{S}^1)$, Q. & A. in General Topology, 22 (2004), 117–130.

[8] W. T. Ingram, Inverse Limits on $[0,1]$ Using Tent Maps and Certain Other Piecewise Linear Bonding Maps, in *Continua with the Houston Problem Book*, Lectures Notes in Pure and Applied Mathematics, Vol. 170, Marcel Dekker, Inc., New York, Basel, Hong Kong, 1995, pp. 253–258 (eds. H. Cook, W. T. Ingram, K. T. Kuperberg, A. Lelek, and P. Minc)

[9] W. Lewis, Continuum Theory Problems, Topology Proc., 8 (1983), 361–394.

[10] W. Lewis, The Classification of Homogeneous Continua, Soochow J. of Math., 18 (1992), 85–121.

[11] S. Macías, On the Hyperspaces $\mathcal{C}_n(X)$ of a Continuum X, Topology Appl., 109 (2001), 237–256.

[12] S. Macías, Fans Whose Hyperspaces Are Cones, Topology Proc., 27 (2003), 217–222.

[13] S. Macías and S. B. Nadler, Jr., n–fold Hyperspaces, Cones and Products, Topology Proc., 26 (2001–2002), 255–270.

[14] S. Mardešić, Mappings of Inverse Systems, Glasnik Mat. Fiz. Astronom. Ser. II, 18 (1963), 195–205.

Index